New Developments in Unsupervised Outlier Detection

Xiaochun Wang · Xiali Wang · Mitch Wilkes

New Developments in Unsupervised Outlier Detection

Algorithms and Applications

西安交通大学出版社
XI'AN JIAOTONG UNIVERSITY PRESS

Springer

Xiaochun Wang
School of Software Engineering
Xi'an Jiaotong University
Xi'an, Shaanxi, China

Xiali Wang
School of Information Engineering
Chang'an University
Xi'an, Shaanxi, China

Mitch Wilkes
Department of Electrical Engineering and
Computer Science
Vanderbilt University
Nashville, TN, USA

ISBN 978-981-15-9518-9 ISBN 978-981-15-9519-6 (eBook)
https://doi.org/10.1007/978-981-15-9519-6

Jointly published with Xi'an Jiaotong University Press
The print edition is not for sale in China (Mainland). Customers from China (Mainland) please order the print book from: Xi'an Jiaotong University Press.

This Springer imprint is published by the registered company Springer Nature Singapore Pte Ltd.
The registered company address is: 152 Beach Road, #21-01/04 Gateway East, Singapore 189721, Singapore

Foreword

Being an active research topic in data mining, outlier detection aims to discover observations in a dataset that deviate from other observations so much as to arouse suspicions that they are generated by a different mechanism and is of utmost importance in many application domains. Unsupervised outlier detection plays a crucial role in the outlier detection research and sets out enormous theoretical and applied challenges to advanced data mining technology using unsupervised learning techniques. This monograph addresses unsupervised outlier detection in a local setting of k-nearest neighborhood. Unlike traditional distribution-based outlier detection techniques, k-nearest neighbor-based outlier detection approaches, typified by distance-based and density-based outlier detection methods, have become more and more popular. However, the problems with these methods are that they are very sensitive to the value of k, may have different rankings for top outliers, and doubts exist in general whether they would work well for high-dimensional datasets. To partially circumvent these problems, the algorithms of choice proposed for unsupervised outlier detection in the current research combine k-nearest neighbor-based outlier detection methods and genetic clustering algorithms.

Distance-based outliers and density-based outliers denote two different kinds of definitions for outlier detection algorithms. Distance-based outlier detection methods can identify more globally oriented outliers while density-based outlier detection methods can identify more locally distributed outliers. In this book, several new global outlier factors and new local outlier factors have been proposed, and efficient and effective outlier detection algorithms have been developed upon them that are easy to implement and can provide competing performances with existing solutions.

Having been exploited in outlier detection research for years, distance-based and density-based outlier detection methods work theoretically by calculating k-nearest neighbors for each data point, computing outlier scores for them, ranking all the objects according to their scores, and finally returning data points with top largest scores as outliers. However, there is no reason to assume that this must be the case. To take this aspect into account, several outlier indicators are introduced to judge whether distance-based and density-based outliers exist or not. By this way, outliers can be not only detected but also discriminated from boundary points.

It is generally agreed that learning, either supervised or unsupervised, can provide the best possible specification of known classes and offer inference for outlier detection by a dissimilarity threshold from the nominal feature space. Novel object detection can take a step further by investigating whether these outliers form new dense clusters in both the feature space and the image space. By defining a novel object to be a pattern group that has not been seen before in the feature space and the image space, a nonconventional approach is proposed for multiple novel object detection applications.

Time series often contain outliers and structural changes. These unexpected events are of the utmost importance in fraud detection, as they may pinpoint suspicious activities. The presence of such unusual activities can easily mislead conventional time series analysis and yield erroneous conclusions. Traditionally, time series data are first divided into small chunks. k-nearest neighbor-based outlier detection approaches are then applied for monitoring behavior over time in data mining. However, time series data are very large in size so they can not be scanned multiple times. Further, as they are produced continuously, new data are arrived. To cope with the speed they are coming, a simple statistical parameter-based anomaly method is proposed for environmental time series data fraud detection applications.

The chapters cover such topics as distance-based outlier detection, density-based outlier detection, clustering-based outlier detection, and the applications of these techniques toward boundary point detection, novel object detection, and fraud detection in environmental time series data. Overall, the book features a perspective on bridging the gap between k-nearest neighbor-based outlier detection and clustering-based outlier detection, laying the groundwork for future advances in unsupervised outlier detection research. I hope new developments in unsupervised outlier detection algorithms and applications will serve as an invaluable reference for outlier detection researchers for years to come.

Xi'an, China
May 2020

Xubang Shen
Chinese National Academician

Preface

Data mining represents a complex of technologies that are rooted in many disciplines: mathematics, statistics, computer science, physics, engineering, biology, etc., and with diverse applications in a large variety of different domains: business, health care, science and engineering, etc. Basically, data mining can be seen as the science of exploring large datasets for extracting implicit, previously unknown and potentially useful information. Recently, outlier detection as a research area in data mining has advanced dramatically. A multitude of data mining techniques has been developed with impact on unsupervised outlier detection areas. Our aim in writing this book is to provide a friendly and comprehensive guide for those interested in exploring this fascinating domain. In other words, the purpose of this book is to provide easy access to the recent contributions to unsupervised outlier detection theory and to assess its impact on the field and its implications for theory and practice. It is also intended to be used as an introductory text for advanced undergraduate-level or graduate-level courses in computer science, engineering, or other fields. In this regard, the book is intended to be largely self-contained, although it is assumed that the potential reader has a quite good knowledge of mathematics, statistics, and computer science.

The book is organized as follows. The first part of this book aims to review the state-of-the-art unsupervised techniques used in outlier detection. The material presented in the second part of this book is an extended version of several selected conference articles and represents some of the most recent important advancements in the field of unsupervised outlier detection. In the third part of this book, outlier detection techniques are applied to practical applications. More specifically, the first part consists of two chapters. In Chap. 1, an overview of the book chapters and a summary of contributions are presented. First, the research issues on unsupervised outlier detection are explained. The overview of the book is then followed. Finally, contributions are highlighted. In Chap. 2, some well-known unsupervised outlier detection techniques and models are reviewed. This chapter begins with an overview of some of the many facets of outlier analysis. Then, it investigates some standard outlier detection approaches. Finally, the problem of evaluating the performance of different outlier detection models is discussed. The second part consists of five chapters, which provide an ever-growing list of unsupervised outlier detection models. In Chap. 3, a divisive hierarchical clustering algorithm is explored as a solution for fast distance-based outlier detection problems. In Chap. 4, a new *k*-nearest neighbor centroid-based outlier detection method is proposed for both distance-based and density-based outlier detection tasks. In Chap. 5, we present a new fast minimum spanning tree-inspired algorithm for outlier detection tasks. In Chap. 6, an efficient spectral clustering-based outlier detection algorithm is proposed to extract information from data in such a way that distribution-based outlier detection techniques can be employed for multi-dimensional data. In Chap. 7, an outlier indicator is proposed to enhance outlier detection in which the selection of appropriate parameters is less difficult but more meaningful. The performances evaluated on some standard datasets demonstrate the effectiveness and efficiency of these methods. The third part of this book is concerned with the applications of outlier detection techniques in real-life problems. Following the techniques discussed in the second part, we devote an entire chapter, that is Chap. 8, to a boundary point detection problem, another, that is Chap. 9, to a novel object detection problem, and finally, the third one, that is, Chap. 10, to a time series fraud detection problem. An extensive bibliography is included, which is intended to provide the reader with useful information covering all the topics approached in this book.

Last, but certainly not least, it is our hope that graduate students, young and senior researchers, and professionals from both academia and industry will find the book useful for understanding and reviewing current approaches in unsupervised outlier detection research.

Xi'an, China — Xiaochun Wang
Xi'an, China — Xiali Wang
Nashville, USA — Mitch Wilkes
June 2020

Acknowledgements

First and foremost, the authors would like to thank National Natural Science Foundation of China for its valuable support of this work under award 61473220 and Natural Science Foundation of Shaanxi Province, China, for its valuable support of this work under award 2020JM-046. Without the supports, this work would not have been possible.

The authors gratefully acknowledge the contribution of many people. First of all, they would like to take this opportunity to acknowledge the work of the graduate students of School of Software Engineering at Xi'an Jiaotong University, Yiqin Chen, Yongqiang Ma, Yuan Wang, and Jia Li for their diligence and quality work through these projects. More specifically, Y. Chen developed a *k*-nearest neighbor centroid-based outlier detection algorithm and applied it to boundary point detection. Y. Ma developed a miniMST-based outlier detection algorithm. Y. Wang proposed a spectral clustering-based outlier detection algorithm. J. Li accomplished all the outlier detection experiments for spectral clustering-based outlier detection on real multi-dimensional datasets. The authors would also like to thank Yuan Bao of Xi'an Jiaotong University Press for her timely suggestions and encouragement with the preparation of the manuscript.

Finally, the authors wish to express their deep gratitude to their families for their assistance in many ways for the successful completion of this book.

Contents

About the Authors

Xiaochun Wang received her B.S. degree from Beijing University and the Ph.D. degree from the Department of Electrical Engineering and Computer Science, Vanderbilt University. She is currently an associate professor of School of Software Engineering at Xi'an Jiaotong University. Her research interests are in computer vision, signal processing, data mining, machine learning and pattern recognition.

Xiali Wang received the Ph.D. degree from the Department of Computer Science, Northwest University, China, in 2005. He is a faculty member in the School of Information Engineering, Changan University, China. His research interests are in computer vision, signal processing, intelligent traffic system, and pattern recognition.

Mitch Wilkes received the BSEE degree from Florida Atlantic, and the MSEE and Ph.D. degrees from Georgia Institute of Technology. His research interests include digital signal processing, image processing and computer vision, structurally adaptive systems, sonar, as well as signal modeling. He is a member of the IEEE and a faculty member at the Department of Electrical Engineering and Computer Science, Vanderbilt University.

Abbreviations

ABOD	Angle-based outlier detection
ANN	Approximate nearest neighbor
ARIMA	Autoregressive integrated moving average
ARMA	Autoregressive moving average
AS	Anomaly score
AUC	Area under curve
CBLOF	Cluster-based local outlier factor
CD	Core distance
CDM	Compression-based dissimilarity measure
CFAR	Constant false alarm rate
CHB-*K*-means	Constraint-based high-dimensional bisecting *K*-means
CIRC	Consulting industry research center
COF	Connectivity-based outlier factor
CUSUM	CUmulative SUM Statistics
DB	*k*-nearest neighbor average distance-based outlier detection
DB(p,D)-outlier	Distance-based outlier
DB-Max	*k*-th nearest neighbor distance-based outlier detection
DB-outlier	Distance-based outlier
DBSCAN	Density-based spatial clustering of applications with noise
DENCLUE	DENsity-based CLUstEring
DHCA	Divisive hierarchical clustering algorithm
ECG	Electrocardiogram
EDLOS	Efficient density-based local outlier detection for scattered data
EKNN	Exhaustive *k*-nearest neighbor approach
EM	Expectation maximization
EMST	Euclidean minimum spanning tree
ESD	Generalized extreme studentized deviate
E-SVDD	Efficient SVDD
EWMA	Exponentially weighted moving average
FPR	False positive rate
FS	Fast search
FSA	Finite state automata

F-SVDD	Fast support vector data description
GESR	General extreme student deviated residual
GNN	General nearest neighbor
GWR	Grow-When-Required
HMMs	Hidden Markov models
HSV	Hue, saturation, and value
I/O	Input/output
ICA	Independent component analysis
IMM	Interpolated Markov models
INFS-DBSCAN	Infinite feature selection DBSCAN
KDE	Kernel density estimation
KDEOS	Kernel density estimation-based outlier score
kNN	*k*-nearest neighbors
LCS	Longest common subsequence
LDC	Local deviation coefficient
LDF	Local density factor
LDOF	Local distance-based outlier factor
LOCI	Local outlier integral
LOF	Local outlier factor
LSH	Locality-sensitive hashing
MDEF	Multi-granularity deviation factor
MGD	Multivariate Gaussian distribution
ML	Machine learning
MLM-KHNN	Multi-local mean-based *k*-harmonic nearest neighbor classifier
MOGAAL	Multiple-objective generative adversarial active learning
MRI	Magnetic resonance imaging
MST	Minimum spanning tree
NEO-*k*-means	Nonexhaustive overlapping *k*-means
NR	Normalized residual
ODC	Outlier detection and clustering
ODIN	Outlier detection using in-degree number
OEDP	Outlier-eliminated differential privacy
ORC	Outlier removal clustering
PBOD	Projection-based outlier detection
PCA	Principal component analysis
PET	Positron emission tomography
PIDC	Parameter independent density-based clustering
PQ	Product quantization
RBDA	Rank-based outlier detection algorithm
RBRP	Recursive binning re-projection
RCMLQ	Rough clustering based on multi-level queries
RDOS	Relative density-based outlier score
RGB	Red, green, and blue
RNN	Reverse nearest neighbors
ROC	Receiver operating characteristic curve

SI	Sequential initialization
SLOM	Spatial local outlier measure
SNN	Shared nearest neighbors
SOGAAL	Single-objective generative adversarial active learning
SOS	Stochastic outlier selection
SVDD	Support vector data description
TI	Triangle inequality
TPR	True positive rate
UCN	Unique closest neighbor
UNS	Unique neighborhood set
VA-file	Vector approximation file
VARMA	Vector autoregressive moving average
VWC	Various width clustering
WSPD	Well-Separated Pair Decomposition

Part I
Introduction

Chapter 1
Overview and Contributions

Abstract Outliers are observations that behave in an unusual way with respect to the majority of data, and outlier detection techniques have become an extremely important research branch of modern advanced data mining technologies. Many popular outlier detection algorithms have been developed. The purpose of this book is to introduce some new developments in the unsupervised outlier detection research and some corresponding applications from a k-nearest-neighbor-based perspective. In this chapter, an overview of this book is presented. First, the research issues on unsupervised outlier detection are introduced. Then, the content for each chapter is described. Finally, a summary of our contributions is presented.

Keywords Research issues on unsupervised outlier detection · Content for each chapter · Contributions

1.1 Introduction

In general, data obtained by one or more application processes may either reflect the activities in the system or be observations collected about entities. When unusual behaviors happen in the generating process, outliers are created. Correspondingly, an outlier often contains useful information about the abnormal characteristics of the systems and the entities that impact the data-generation process. The detection of outlier can provide useful application-specific insights for the recognition of such unusual characteristics and, therefore, has been found to be directly applicable in large number of domains. For one example, an anomalous traffic pattern in a computer network could mean that a hacked computer is sending out sensitive data to an unauthorized destination, giving rise to intrusion detection. For another example, outlier detection techniques are widely applied to public health data to detect anomalous patterns in patient medical records which could be symptoms of new diseases. Similarly, outliers existing in credit card transaction data could indicate credit card theft or misuse, those in a satellite image of enemy, that is, the presence of unusual region, could indicate enemy troop movement in military surveillance, and the anomalous readings from spacecraft would signify a fault in some of the craft. More specifically, according to the Wiki news, by data mining methods in general and outlier

X. Wang et al., *New Developments in Unsupervised Outlier Detection*,
https://doi.org/10.1007/978-981-15-9519-6_1

detection methods in specific, the 9/11 attack leader and three other hijackers had supposedly been identified as possible members of an Al-Qaeda cell operating in the USA by an US Army intelligence unit more than a year before the attack. However, it appears that this information was not taken into consideration by the authorities. Finally, Editor Ramon C. Barquin narrated in the 'Foreward' of his book, *The Data Warehousing Institute Series* (Prentice Hall), that his telephone provider told him that, although he spent all the time the day before in Cincinnati, telephone calls were made from Kennedy Airport, New York to La Paz, Bolivia, and to Lagos, Nigeria using his calling card and PIN number. Since these facts did not fit his usual calling patterns, the phone company had detected them as the fraudulent actions using their data mining program and informed him that they had reason to believe his calling card had been stolen. To conclude, the outlier detection as an important data mining technique seeks to discover a limited amount of irregular observations that are unusual with respect to the majority of data under the circumstances of certain assumptions.

The oldest root of the outlier detection problem is in *Statistics* from which it borrowed techniques and terminology, and without which outlier detection would not have been existing. In the data mining community, the research on outlier detection algorithms was initiated by the seminal work of Knorr and Ng in 1997. Since then, new models have continually been proposed to characterize and identify outliers. Unlike statistics, machine learning (ML) utilizes the techniques that allow the computer to learn. Supervised and unsupervised learning are two machine learning techniques used extensively in data mining. In supervised learning, a correspondence (function) is established using a training dataset composed of labeled objects and seen as a 'past experience' of the model so as to predict the value (output) of the function for any new object (input) after completion of the training process. With examples of previous anomalies being available, the supervised outlier detection scenario is a special case of the classification problem. Unlike supervised learning, in unsupervised learning, the learner is fed with only unlabeled objects to establish the model. In this book, outlier analysis is conducted as an unsupervised problem in which labels of previous examples of anomalies and normal data points are not available.

In the following, the research issues on unsupervised outlier detection are discussed in Sect. 1.2. The content description for each chapter is next presented in Sect. 1.3. The contributions made to the current research are then summarized in Sect. 1.4. Finally, conclusions are given in Sect. 1.5.

1.2 Research Issues on Unsupervised Outlier Detection

The goal of outlier detection is to discover unusual patterns that exist in a data. To detect outliers, assumptions are made about outliers versus the rest of the data. According to the assumptions made, unsupervised outlier detection methods usually fall into three main categories, that is, statistical methods, proximity-based methods, and clustering-based methods. On the one hand, the classic statistical methods employ well-defined descriptive statistics (distributions, classical statistical parameters such as mean, median, standard deviation, and etc.) as well as the most powerful and attractive methods of data visualization (histograms of all kinds, box plots, scatter plots, contour plots, matrix plots, icon plots, etc.) to define and detect outliers. On the other hand, a distance measure can usually be defined for a given set of objects in the feature space to quantify the dissimilarity between objects. Therefore, based on the intuition that objects that are far away from others can be regarded as outliers and the assumption that the proximity of an outlier object to its nearest neighbors significantly deviates from the proximity of normal objects to most of the other objects in the data set, the proximity-based outlier detection methods are categorized into two types, namely the distance-based methods and the density-based methods. Finally, believing that an outlier is an object that belongs to a small and remote cluster, or does not belong to any cluster, clustering-based approaches detect outliers by examining the relationship between objects and clusters.

More specifically, unsupervised outlier detection methods expect that normal objects happen far more frequently than outliers and make an implicit assumption that they fall into multiple groups, where each group has distinct features. On the contrary, an outlier is expected to occur somewhere in the feature space far away from any of those groups of normal objects, and it is usually assumed that the number of outliers is much fewer than the number of normal objects. Therefore, in clustering-based outlier detection approaches, clusters are first obtained and then outliers are detected based on their relationship to the major clusters. Following this notion, many clustering methods can be adapted for the purposes of unsupervised outlier detection. Unfortunately, having to cluster a large population of the normal objects before one can mine the outliers can often be costly and thus unappealing. As a result, the latest developments in unsupervised outlier detection research try to propose various clever methods to locate outliers directly without explicitly and completely finding clusters.

Having emerged in many applications, from the academic field to the business or medical activities, outlier detection is a research area with not a long history and still disputed by some scientific fields. Attention should be paid to the fact that, although there is a lot of information hidden in data, it is almost impossible to detect outliers by traditional means and using only the human analytic ability. Therefore, several theoretical and applied challenges have been set out to unsupervised outlier detection techniques. Since the applications of outlier analysis are very diverse and extend to a variety of domains such as fault detection, intrusion detection, financial fraud, web log analytics, one challenging question would be **how to define what an outlier is**. Technically speaking, although the choice of a similarity or distance measure as well as a relationship model to describe data objects is critical in outlier detection, such choices are often application-dependent because different applications may have very different requirements. It is outlier detection's high dependency on the application type that makes it impossible to develop a universally applicable outlier detection method. Since it is difficult to opt for a unique definition providing a picture as complete as possible of the phenomenon, outlier detection methods that are dedicated to specific applications should be developed instead.

Another challenging question would be **how to model normal objects and detect outliers effectively**. The effectiveness of an outlier detection method highly depends on the modeling of normal objects and outliers. However, because it is often difficult to know in advance all possible normal behaviors in an application and there is usually not a fine line between data normality and abnormality (outlier), the building of a comprehensive model for both data normality and abnormality is very challenging task. Therefore, a label of either 'normal' or 'outlier' is assigned to each object in the input dataset by some outlier detection methods, while a score measuring the 'outlierness' of the object is assigned to each object by other methods. In other words, the 'outlierness' of a data point is denoted either by a real-valued outlier score or by a binary label.

A third challenging question would be **how to mine outliers efficiently**. An outlier is a data point that is very different from the majority of data. The generation of an outlier score often requires the construction of a model of the normal patterns. Outliers are data points that do not follow this normal model frequently. In modern data mining tasks, a huge amount of data are generated without being systematically unexplored yet and both computing power and computer science have grown exponentially, the demands for revealing information 'hidden' in data using new methods increase. These new methods concern practical aspects of outlier detection corresponding to the cases where the data may be very large, or may have very high dimensionality. The automatic and efficient search for patterns in huge databases, using computationally efficient techniques from statistics, machine learning and pattern recognition, is in strong desire. Metaphorically speaking, 'finding the needle in a haystack' refers to the nontrivial extraction of implicit, previously unknown and potentially useful unusual information from exponentially growing data.

A fourth challenging question would be how to **handle noise in outlier detection**. More often than not, real data sets tend to be of poor quality because noise often unavoidably appears in data collected in many application domains. Outliers are different from noise. However, being present as deviations in attribute values or even as missing values, noise is a 'perfidious enemy' in building an effective model for outlier detection and can bring a huge challenge to the outlier detection since they can distort the data, blur the distinction between normal objects and outliers, and cannot be fully removed (filtered). Low data quality and the presence of noise may 'hide' outliers and reduce the effectiveness of outlier detection. In other words, an outlier may mistakenly be 'disguised' as a noise point, and a noise point may erroneously be identified as an outlier by an outlier detection method.

A fifth challenging question would be the **understandability**. Usually, a user may not only want to perform outlier detection, but also wish for an understanding of why the detected objects are outliers. To satisfy this understandability requirement, it is desired that an outlier detection method can provide some justification of the detection. More specifically, the understanding of the outlier detection model is a desirable ability that makes possible the identification of the factors that lead both to obtaining 'success' as well as 'failure' in the prediction provided by the model.

Finally, **the design of the criteria for evaluating the quality of an outlier detection process** in a data set is a difficult problem because of the challenges associated with the unsupervised nature of outlier detection. Although estimates of the predictive accuracy such as external validation criteria and the receiver operating characteristic curve can be utilized to evaluate the success of an outlier detection algorithm, the comprehensibility of the learned models should be taken into consideration as an important criterion, especially when domain experts have strong expectations on the properties of outlier detection models.

1.3 Overview of the Book

The rest chapters of this book attend to the theory and development of unsupervised outlier detection algorithms based on k-nearest neighbors (kNN) for objects in a data set and deal with key aspects which are essential to the goal of bringing the concept and mechanisms of clustering-based outlier detection methods to a local setting of a data point in terms of its k-nearest neighborhood to meet the requirement of real-time applications. In the individual chapters, the theoretical analysis of the specific technical problems is provided together with numerical analysis, simulation and real experiments on a number of data sets.

To organize high-quality and recent research advances on unsupervised outlier detection that are underway, we divide the chapters of this book into three parts. The first part consists of two chapters and aims to review the state-of-the-art unsupervised techniques used in outlier detection. The second part consists of five chapters, which provide details of the proposed kNN-based unsupervised outlier detection models. The material presented in the second part of the proposed manuscript is an extended version of several selected conference articles and represents some of the most recent important advancements of our work on unsupervised outlier detection. In the third part of this book, there are three chapters and outlier detection techniques are applied to several practical problems. In the following, a summary for all the chapters is provided.

In this chapter, we present an overview of the chapters and a summary of the contributions. First, the research issues on unsupervised outlier detection are introduced. An short introduction to the contents of the book is then followed. Finally, contributions are highlighted. In Chap. 2, some well-known outlier detection techniques and models are briefed. This chapter begins with an overview of some of the many facets of outlier analysis. Then, it explains some standard outlier detection approaches. Finally, the performance metrics of different outlier detection models are discussed.

In Chap. 3, a divisive hierarchical clustering algorithm is explored as a solution for fast distance-based outlier detection problems. In Chap. 4, a new k-nearest neighbors centroid-based outlier detection method is proposed for both distance-based and density-based outlier detection tasks. In Chap. 5, we present a new fast minimum spanning tree-inspired algorithm for k-nearest neighbor-based outlier detection tasks. In Chap. 6, we propose a new outlier detection method inspired by spectral clustering, which combines the concept of k-nearest neighbors of a data point and spectral clustering techniques to obtain statistically detected outliers by using the information of eigenvalues in the feature space. In Chap. 7, an outlier indicator is proposed to enhance outlier detection by filtering out the majority of the normal data objects in a short time. The performances evaluated on some standard datasets demonstrate the effectiveness of these methods.

In Chap. 8, we propose to apply the state-of-the-art k-nearest neighbor centroid-based outlier detection method to a boundary point detection problem, which can provide high precision and can separate boundary points from outliers at the same time. In Chap. 9, a fast approach for a statistical threshold value extraction process is developed which can provide a robust solution to the nearest neighbor classifier-based vision-based novel object detection problem. In Chap. 10, a new fast outlier detection method based on statistically extracted parameters is explored for fraud detection in environmental time series data.

1.4 Contributions

Unsupervised outlier detection methods make an implicit assumption that normal objects gather together in the feature space in the form of dense clusters. Also known as unsupervised learning, clustering is all about discovering groups at which data points gather. Correspondingly, outlier analysis is all about finding data points that are far away from these groups. Since outliers often tend to occur in small clusters of their own, they can therefore be mined as a side-product of clustering methods, giving rise to the clustering-based outlier detection methodology. Clustering and outlier detection, therefore, share a well-known complementary relationship. A simplistic view is that every data point is either a member of a cluster or an outlier. However, clustering-based outlier detection methods are not an appropriate approach because clustering algorithms are not optimized for outlier detection.

This book advances the unsupervised outlier detection research by proposing several new distance-based and density-based outlier scores in a local setting of k-nearest neighbors with principles inspired by clustering-based outlier detection models. The chapters feature the latest developments in k-nearest neighbor-based outlier detection research and cover such topics as the present understanding of unsupervised outlier detection in general, distance-based and density-based outlier detection in specific, and the applications of this present understanding toward boundary point detection, novel object detection, and environmental time series data fraud detection. The volume mainly features a perspective on bridging the gap between k-nearest neighbor-based outlier detection methodology and clustering-based outlier detection methodology, laying the groundwork for future advances in unsupervised outlier detection research. In the following, an in-depth exploration of the above contributions that will be most valuable to the reader is presented in more detail.

Outlier detection aims to discover observations in a dataset that deviate from other observations so much as to arouse suspicions that they are generated by a different mechanism. Unlike traditional distribution-based outlier detection techniques which are good for single-attribute data, novel k-nearest neighbor-based outlier detection approaches have become more and more promising for multi-attribute data and do not require a presumed distribution model to fit the data. However, these methods are associated with such problems that they are very sensitive to the value of k, may have different rankings for top outliers, and doubts exist in general whether they would work well for high dimensional datasets. To partially circumvent these problems, the algorithms of choice for unsupervised outlier detection combine k-nearest neighbor-based outlier detection methods and the principles of genetic clustering-based outlier detection algorithms and bridge the gap between them.

Distance-based outliers and density-based outliers denote two different kinds of definitions for outlier detection algorithms. Distance-based outlier detection methods can identify more globally-oriented outliers, while density-based outlier detection methods can identify more locally distributed outliers. In this book, we propose several new global outlier factors and new local outlier factors, and efficient and effective outlier detection algorithms developed upon them that are easy to implement and can provide competing performances with existing solutions.

Having been exploited in outlier detection research for years, distance-based and density-based outlier detection methods work theoretically by calculating k-nearest neighbors for each data point, computing outlier scores for them, ranking all the objects according to their scores, and finally returning data points with top largest scores as outliers. However, there is no reason to assume that this must be the case. To take this aspect into account, several outlier indicators are introduced to judge whether distance-based and density-based outliers exist or not. By this way, outliers can be not only detected but also discriminated from boundary points.

Outlier detection techniques and clustering techniques are important areas of data mining, and the study of boundary points is sometimes more meaningful than clusters and outliers. Data points on the boundary regions of a cluster may also be considered weak outliers that are useful in a number of application-specific scenarios. By defining a data object to be a boundary point that has a kNN-based outlier score beyond a threshold, a nonconventional approach is proposed for boundary point detection applications. Further, it is generally agreed that learning, either supervised or unsupervised, can provide the best possible specification for known classes and offer inference for outlier detection by a dissimilarity threshold from the nominal feature space. Novel object detection can take a step further by investigating whether these outliers form new dense clusters in both the feature space and the image space. By defining a novel object to be a pattern group that has not been seen before in the feature space and the image space, a nearest neighbor-based approach is proposed for multiple-novel-object detection applications. Time series data are records accumulated over time. Such data can be viewed as objects with an attribute time. Time series often contain outliers and level shifts or structural changes. These unexpected events are of the utmost importance in fraud detection in time series data, as they may pin-point suspicious activities. By dividing such data into small chunks, conventional *k*NN-based outlier detection techniques can be used for environmental time series data fraud detection applications. However, time series data are very large in size and are produced continuously. To cope with the speed they are coming, a new statistical parameter-based outlier score is proposed for such fraud detection applications.

To summarize, original contributions presented in this book focus on developing algorithms for unsupervised outlier detection and the corresponding applications in the area of k-nearest neighbor-based unsupervised learning paradigms. In particular, some key results obtained in this book are given as follows.

Algorithms

- A new algorithm for the fast distance-based outlier detection purpose is proposed.
- A new *k*-nearest neighbor centroid-based outlier detection method is proposed for both distance-based and density-based outlier detection tasks.
- A new minimum spanning tree clustering-inspired *k*NN-based fast outlier detection algorithm is described.
- A new spectral clustering based on *k*-nearest neighbors of a data point is proposed for outlier detection.
- An outlier indicator is proposed to enhance outlier detection by filtering out boundary points from potential outlier candidates.

Applications

- A boundary point detection algorithm via the concept of *k*-nearest neighbor centroid-based outlier detection is proposed.
- A nearest neighbor-based approach for visual novel object detection is proposed which provides a robust solution to the novel object detection problem.
- A new outlier detection method based on statistic parameters is explored for environmental time series data fraud detection.

1.5 Conclusions

In this book, as solutions to some challenges in unsupervised outlier detection research, we present several novel advanced algorithms which exploit clustering-based outlier detection models in a local setting of *k*-nearest neighbors of a data point to more accurately and efficiently mine the outliers. The topics covered by the chapters include statistics-based outlier detection, distance-based outlier detection, and density-based outlier detection in combination with clustering-based outlier detection models, and their applications to real-world tasks. These different types of models are discussed in detail, and representative algorithms are introduced. The chapters contain key references to the existing literature to provide both an objective overview and an in-depth analysis of some state-of-the-art approaches from the perspective of unsupervised outlier detection and cover both theoretical and practical aspects of real-world outlier detection problems from the perspective of a k-nearest neighbor-based local setting. It is our hope that the presented progresses of this work in both theoretical development and practical applications will be of benefit to those who are interested in the area.

Chapter 2
Developments in Unsupervised Outlier Detection Research

Abstract Being an important task in data mining, outlier detection has high practical value in numerous applications. As a result, many unsupervised outlier detection algorithms have been proposed, including distribution-based, distance-based, density-based, and clustering-based approaches. In this chapter, we first give a review on the fundamental aspects about unsupervised outlier detection techniques used throughout the present book. This is followed by a brief introduction to the performance evaluation metrics of outlier detection algorithms.

Keywords Unsupervised outlier detection · Distribution-based outlier detection · Distance-based outlier detection · Density-based outlier detection · Clustering-based outlier detection · Performance evaluation

2.1 Introduction

Being data points that are significantly different in behavior from the remaining data, outliers are also referred to as abnormalities, discordants, deviants, or anomalies in the data mining and statistics literature. It was Hawkins who gave the most popular definition of an outlier as: 'An outlier is an observation that deviates so much from the other observations as to arouse suspicions that it was generated by a different mechanism' [1]. It is generally believed that an outlier often contains useful information about the abnormal characteristics of systems and entities that impact the data-generation process, the recognition of which can provide useful application-specific insights that can benefit the user. Some examples of relevant applications are given in the following.

- Intrusion detection refers to the detection of malicious activities in networked computer systems. This can be done by examining different kinds of data collected about the operating system calls, network traffic, or other activities within the system, which may show unusual behaviors when malicious activities happen.
- Credit card fraud detection refers to the detection of unauthorized use of credit cards by examining different patterns, such as a buying spree from geographically obscure locations. Such patterns may show up as outliers in credit card transaction data.

X. Wang et al., *New Developments in Unsupervised Outlier Detection*,
https://doi.org/10.1007/978-981-15-9519-6_2

- Sensor networks are spatially distributed sensors to monitor conditions at different locations, such as temperature, sound, vibration, pressure, motion, or pollutants, so as to track various environmental and location parameters in many real applications. Event detection is one of the primary motivating applications in the field of sensor networks. Outlier detection techniques can be applied to detect sudden changes in the underlying patterns which may represent events of interest.
- In medical diagnosis, disease conditions are rare in comparison to normal conditions. In many medical applications, unusual patterns in data are collected from a variety of sources such as magnetic resonance imaging (MRI), positron emission tomography (PET) scans, or electrocardiogram (ECG) time series, typically reflect disease conditions and can be discovered by outlier detection techniques.
- In law enforcement, fraud in financial transactions, trading activity, or insurance claims often happens and should be identified. This typically requires the determination of unusual patterns in the data generated by the actions of the criminal entity. As a result, outlier detection techniques can especially be applied in such law enforcement to search for unusual patterns through multiple actions of an entity over time.
- In earth science, a significant amount of spatio-temporal data has been collected through a variety of mechanisms such as satellites or remote sensing. They provide significant knowledge about weather patterns, climate changes, or land-cover patterns. Outlier detection in such data can discover hidden human or environmental trends that may have caused such anomalies.

All these applications have in common mathematically the basic components of an outlier detection problem: 'given some metric space, in which a dissimilarity measure is defined, and a set of known data points, determine a few fixed or variable number of candidate outlying points, referred to as outliers, so as to optimize a function of the dissimilarity between a few outliers and the majority of normal points.'

To detect outliers, most methods create a model of normal patterns. Data points that do not naturally fit within this normal model are then declared as outliers. To quantify the 'outlierness' of a data point, a numeric value, known as the outlier score, is usually defined. Correspondingly, most outlier detection algorithms produce an output that can be either a real-valued outlier score or a binary label. In the first case, the output real-valued outlier score quantifies the tendency or even a probability value of the likelihood for a data point to be considered as an outlier. Extreme values of the score make it more likely that a given data point is an outlier. In the second case, the output binary value indicates whether a data point is an outlier or not. This type of output can be obtained by imposing a threshold on the outlier scores to convert them into binary labels and, therefore, contains less information than the first one because, unfortunately, the reverse is not possible. Although the outlier scores are more general than binary labels, a binary label score is usually required as the end result in most applications to provide a crisp decision. Both types of models are presented in this book.

In this book, we present solutions to outlier detection problems in an unsupervised way. In order to retrieve a small number of outliers in a given dataset accurately, many difficulties exist such as the interference from the noise, the impact of the creation of a model of normal patterns and so on. Therefore, the research on outlier detection techniques in an unsupervised way is still in its developing stage, and further improvements to many theories are still open. In the sections below, a brief overview of the early developments in outlier analysis is presented and some standard unsupervised outlier detection approaches which have been established in the past are introduced.

2.1.1 A Brief Overview of the Early Developments in Outlier Analysis

The problem of outlier detection has been widely studied in the field of statistics. Usually, users use a statistical distribution to model data points and then use a hypothetical model to determine whether a data point is an outlier based on the distribution of points. Many methods of Discordancy Test for different distributions have been developed. They are applicable to different situations: (1) data distribution; (2) whether the parameters of data distribution are known; (3) the number of abnormal data points; (4) the type of abnormal data (higher or lower than the general sampling value). In this respect, the representative ones include Mikey, Dunn, and Clark's single point diagnosis based on 'mean shift' model and proposed in 1967, Gentleman and Wilk's group diagnosis proposed in 1970, Tietjen and Moore's single sample outlier statistics, E_k, proposed in 1972, Marasinghe's improved E_k statistics, F_k, proposed in 1985, Rosner's ESD (Generalized Extreme Studentized Deviate) method proposed in 1989, and Paul and Fung's work which improved the subjectivity of parameter k selection of ESD method and gave rise to the General Extreme Student Deviated Residual (GESR) method proposed in 1991 for regression analysis. Recently, multi-sample outlier detection methods have also been worked out to some extent. The basic idea is to get a clean set of data without outliers as much as possible and then to perform outlier detection from the remaining data points step by step.

In statistical methods, outliers are likely to be detected by different distribution models. However, statistical methods have their limitations due to the facts that the mechanism of generating these outliers may not be unique and ambiguity often occurs when explaining the meaning of outliers. In other words, statistical methods depend to a large extent on whether the dataset to be mined satisfies a certain probability distribution model. The parameters of a distribution model and the number of outliers are very important for statistical methods; however, these parameters are usually difficult to determine. In real life situations, the above shortcomings greatly limit the applications of the statistical methods, making it mainly restricted to scientific research and calculation, and the portability of the algorithm is poor. In order to overcome these problems, Knorr and Ng initiated distance-based outlier detection research in 1998 [2]. Shortly after, two kNN-based variants of the original distance-based outlier detection method were proposed to suit for different theoretical and practical purposes [3, 4].

Standard distance-based outlier detection methods suffer from the quadratic running time complexity. Fortunately, distance-based outlier score estimates in the brute force nested loop algorithm are a monotonic non-increasing function of the portion of the dataset explored so far [5]. Based on this observation, the goal of finding top n distance-based outliers can be fulfilled by first quickly assigning a good estimate of the outlier scores to each data item and then focusing on top $m \geq n$ ones. By removing all the inliers among them, the required top-n outliers show up. Based on this observation, state-of-the-art distance-based outlier detection methods have employed various clever ways to quickly filter out the normal data, including CPU-efficient ones (the ORCA method [5], the RBRP method [6], the HilOut method [7], and the SolvingSet algorithm [8]), and I/O-efficient ones (including the SNIF method [9]), or both (including the DOLPHIN method [10]). In addition, Hautamaki et al. proposed Outlier Detection using In-degree Number (ODIN) in 2004 which defines outlierness as a low number of in-adjacent edges in the kNN graph [11]. Equivalently, in 2014, Radovanoic et al. proposed a low hubness value to be the outlier score which is defined as the cardinality of the reverse kNN set (the 'RkNNs') [12].

Although simple and elegant by making direct use of the kNN set of each point, unfortunately, distance-based outlier detection techniques only work well for datasets that contain clusters with similar densities and belong to a family of 'global' methods. However, it is beyond its ability to detect all meaningful outliers in many complex situations. A classic situation illustrating this deficiency is shown in Fig. 2.1, where o_1 and o_2 are global outliers and can easily be detected by distance-based methods while o_3 is a local one and cannot be detected by distance-based methods [13]. To overcome this limitation, the notion of local outlier has been proposed.

Taking this fact into account, in 2000, Breunig et al. pioneered the density-based outlier detection research by proposing a density-based outlier score for each data

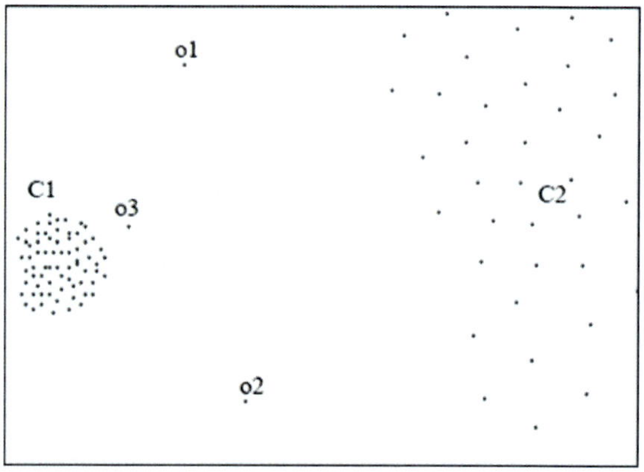

Fig. 2.1 A classic example of a local outlier. Reprinted from Ref. [13], copyright 2015, with permission from Elsevier

item, called the local outlier factor (LOF), which is a ratio between the local density of an object and the average of the local densities of its k-nearest neighbours [14]. The LOF method works by first calculating an LOF value for each object, next ranking data objects by their LOF values, and, finally, returning objects with top-n largest LOF values as outliers. A LOF value of approximately 1 corresponds to objects that are located within a region of homogeneous density. The higher the LOF value of an object O is, the more distinctly O is considered an outlier. Following the notion of local outlier, density-based outlier detection becomes an active research area and many density-based outlier detection methods have been proposed. Connectivity-based outlier factor (COF), proposed by Tang et al. in 2002, is such a score [15]. Compared with LOF, COF uses the cost description of a data point O rather than the distances between O and other objects when calculating the outlier scores. In 2003, Gibbons et al. proposed the local outlier integral (LOCI) which is based on the concept of a multi-granularity deviation factor (MDEF) and is highly effective for detecting outliers and groups of outliers [16]. In 2004, Sun and Chawla proposed a measure, spatial local outlier measure (SLOM), which captures the local behaviour of datum in their spatial neighbourhood [17]. To avoid the problem that comes when outliers are in the location where the density distributions in the neighbourhood are significantly different, Jin et al. presented the INFLO method in 2006 [18]. Unlike COF, INFLO uses the concept of density rather than distance when calculating the scores. In 2007, Latecki et al. proposed a local density factor (LDF) which replaces LOF's density estimate by a variable-width Gaussian kernel density estimation (KDE) modified to use LOF's reachability distance [19]. The resulting estimator is no longer a kernel density in the mathematical sense. In 2009, Kriegel et al. proposed Local Outlier Probabilities (LoOP) which uses a more robust local density estimate based on the quadratic mean distance and normalize the outlier detection score at the same time [20]. As the above methods are ineffective on scattered real-world datasets, Zhang et al. proposed local distance-based outlier factor (LDOF) also in 2009 [21]. To calculate the LDOF, the average of the distances of object O to its k-nearest neighbours and the average of these $k + 1$ objects' pairwise distances are calculated. Then the ratio of these two averages is the LDOF, by which the degree of deviation to O's k-nearest neighbours is obtained. In 2012, Janssens et al. proposed a stochastic outlier selection (SOS) algorithm to quantify the relationship from one point to another by using the concept of affinity, which is proportional to the similarity between two points [22]. Being, an unsupervised outlier selection algorithm, SOS takes as input either a feature matrix or a dissimilarity matrix and outputs for each data point an outlier probability. Intuitively, outliers are those data points with which the other data points have insufficient affinity. In 2013, Huang et al. proposed a new approach for outlier detection named rank-based outlier detection algorithm (RBDA) [23]. It mainly focuses on the question whether an object is the central of its k-nearest neighbours. By RBDA, we need to calculate the rank of object O in its k-nearest neighbours but not the density nearby. In 2014, Schubert et al. proposed a method called KDEOS (kernel density estimation-based outlier score) which uses a z-score with respect to the KDE densities of the kNN set [24, 25]. In 2016, Ru et al. proposed the Normalized residual (NR)-based outlier detection algorithm using the concepts of kernel density estimation (KDE) and constant false-alarm rate (CFAR) [26]. This algorithm can detect not only global outliers but also local outliers through the calculation of weights. Further, a method of calculating a threshold was proposed by randomly dividing the data set into two equal parts. By comparing each data item with a threshold, all outliers can be found directly. In 2017, Tang et al. proposed an outlier detection algorithm named relative density-based outlier score (RDOS) based on the concept of correlation density [27]. This algorithm computes the fractional RDOS that characterizes the degree of outlierness by calculating the local mean of the estimated density associated with the data item O which is then compared to the local density estimate of the data item O itself. The focus of this algorithm is on the determination of relevant points, including not only kNN but also reverse nearest neighbours (RNN) and shared nearest neighbours (SNN). If the kNN of data point P contains data point O, then P is one of the RNNs of O. If both the kNN of data point P and the kNN of data point O contain a certain data point Q, then P is one of the SNNs of O.

To summarize, distance-based as well as density-based outlier scores usually return only top n outliers with two values, the outlier scores and the ranking of the points according to the scores. The problems with these methods are that they are very sensitive to the parameter k, and a small change in k can lead to changes in the scores and, correspondingly, the ranking. To illustrate this problem, take the two-dimensional data set shown in Fig. 2.2 as an example. For distance-based outlier detection techniques, if $k = 6$ nearest neighbors are considered, some of the center points of cluster C_3 will not be detected as outliers, while if $k = 7$, all the data points in cluster C_3 are regarded as outliers. Similar problems exist for density-based outlier detection techniques (such as how to choose k). This situation could be even worse for the detection of outliers in high-dimensional feature space since there is no guarantee that an outlier can be detected visually there.

Fortunately, clustering-based algorithms can sometimes be more intuitive in such cases [13]. Clustering is an important data mining tool to partition a set of data

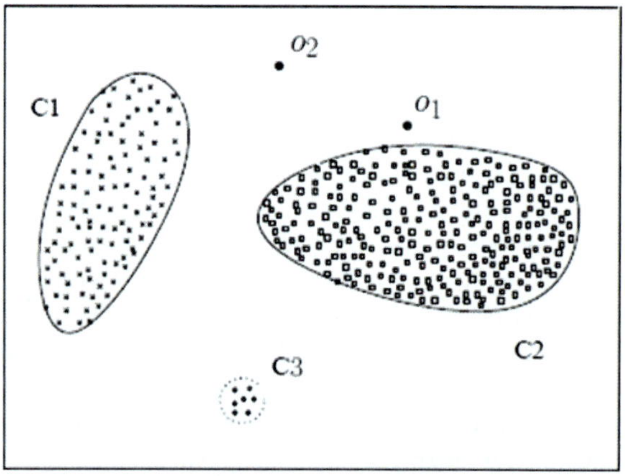

Fig. 2.2 Sample clusters in a 2-D data set. Reprinted from Ref. [13], copyright 2015, with permission from Elsevier

objects into clusters by optimizing some criterion, such as minimizing the intra-cluster distances and maximizing the inter-cluster distances. Outliers are obtained as a by-product of clustering algorithms to be data items in small groups as long as their inter-cluster distances are significantly larger than the intra-cluster distances, making it less sensitive to k, the number of nearest neighbors. Since clustering algorithms consume a significant amount of time on finding the major clusters, such algorithms suitable for outlier detection purpose are limited. Relying on a notion of similarity measure, in addition to K-means [28] and K-medoids [29] algorithms, popular clustering algorithms good for outlier detection include DBSCAN [30], BIRCH [31], DENCLUE [32], CURE [33]. It is well known that the K-means algorithms are very sensitive to the effects of outliers, and hence accurate results may not be obtained. However, K-medoids algorithms are less sensitive to the effects of outliers. In a clustering-based outlier detection method proposed in 2009 by Al-Zoubi [34], K-medoids algorithm is first performed to determine directly small clusters as outlier clusters. The remaining outliers (if any) are then detected based on calculating the absolute distances between the medoid of the current cluster and each one of the points in the same cluster. Being different from the K-means algorithm, minimum spanning tree (MST)-based clustering algorithms, first proposed by Zahn in 1971 [35], are capable of detecting clusters with irregular boundaries. Regarding that data points in smallest clusters formed by cutting the longest edges in an MST may likely be outliers, several MST-based outlier detection techniques have been proposed in 1975, 2001, and 2008, respectively [36–38]. In 2011, John Peter proposed a two-stage MST-based clustering algorithm for detecting outliers [39]. In the algorithm, a new cluster validation criterion is introduced based on the geometric property of data partition of a dataset in order to divide the dataset into an optimal number of clusters. As usual, small clusters are then determined and considered as outliers. The rest of the outliers (if any) are then obtained based on temporarily removing an edge (Euclidean distance between objects) from the dataset and recalculating the weight function. But for modern large and high-dimensional data sets where only a set of N data points is given, MST clustering-based outlier detection algorithms may suffer from the quadratic time complexity required for the construction of an EMST. In the same year, Marghny and Taloba proposed a two-stage algorithm that aims to accomplish outlier detection and data clustering simultaneously [40]. In the first stage, the genetic K-means clustering (IGK) process is enhanced by improving the estimation of centroids of the generative distribution during the process of clustering and outlier discovery, while, in the second stage, the data objects which are far from their cluster centroids are iteratively removed. In 2019, Liu et al. proposed two new kind of outlier detection methods using neural networks, named single-objective generative adversarial active learning (SOGAAL) and its extension multiple-objective generative adversarial active learning (MOGAAL), which is based on generative adversarial learning framework to cope with the problem of lack of prior information of the parametric methods [41].

The outlier detection algorithms presented so far are based on implicit assumptions of relatively low dimensionality of the data and use the distances in full-dimensional space to find outliers. However, data in high-dimensional space are sparse, implying that every point is an equally good outlier from the perspective of distance-based definitions, and the notion of finding meaningful outliers becomes substantially more complex and nonobvious. On the one hand, an angle-based outlier detection (ABOD) method designed especially for high-dimensional data was proposed in 2008 by Kriegel et al. which plays an important role in identifying outliers in high-dimensional spaces [42]. Unfortunately, the ABOD algorithm runs in cubic time (with a quadratic time heuristic). To improve, a novel random projection-based technique was proposed in 2012 that is able to estimate the angle-based outlier factor for all data points in a time near linear to the size of the data [43]. On the other hand, even though the behavior of the data in full dimensionality is usually considered for outlier detection, abnormal deviations may be embedded in some lower-dimensional subspaces [44]. By examining the behavior of the data in subspace, it is possible to detect more meaningful outliers that are specific to the particular subspace in question. In fact, it has been reported in [3] that more interesting outliers on the NBA98 basketball statistics database can be obtained by using fewer features. Based on these observations, in 2005, Aggarwal and Yu proposed new techniques for outlier detection that define a data point to be an outlier if in some lower-dimensional projection it is present in a local region of abnormally low density (referred to as Projection Based Outlier Detection (PBOD) in the following) [45]. To characterize and find such projections, a grid discretization of the data is first performed. Each attribute of the data is divided into φ ranges and, thus, each range contains a fraction $f = 1/\varphi$ of the records. For a k-dimensional cube that is created by picking grid ranges from k different dimensions, the sparsity coefficient $S(C)$ of the cube C is calculated as follows,

$$S(C) = \frac{n(C) - N \cdot f^k}{\sqrt{N \cdot f^k \cdot \left(1 - f^k\right)}} \tag{2.1}$$

where N is the number of data points and $n(C)$ denotes the number of points in the k-dimensional cube. Only sparsity coefficients that are negative indicate cubes in which the presence of the points is significantly lower than expected. Once such patterns have been identified, outliers are defined as those records that have such patterns present in them. An interesting observation is that such lower-dimensional projections can be mined even in datasets that have missing attribute values [46]. The problem with PBOD is the exponentially increasing search space of possible projections with dimensionality. The algorithm is not feasible for a few hundred dimensions. High dimensionality poses significant challenges for outlier detection. Establishing a robust outlier detection model for use in high-dimensional spaces requires the combination of an unsupervised feature extractor and an anomaly detector. Architectures such as deep belief networks (DBNs) are a promising technique for learning robust features. To this end, in 2016, Erfani et al. proposed a hybrid model where an unsupervised DBN is trained to extract generic underlying features and a one-class support vector machine (OCSVM) is trained from the features learned by the DBN

[47, 48]. The hyperplane is used to separate the data sets in high-dimensional space. Since a linear kernel can be substituted for nonlinear ones in the hybrid model without loss of accuracy, the proposed model is scalable and computationally efficient.

2.2 Some Standard Unsupervised Outlier Detection Approaches

The outlier detectors usually begin with the construction of a model of the normal patterns. Some of the most popular models for outlier analysis are introduced in the followings.

1. *Probabilistic models*: A probabilistic model assumes the majority of the data follow a statistical distribution and the degree to which an object may be an outlier can be evaluated based on the likelihood that the object is generated by the same distribution. The smaller the likelihood, the more unlikely the object is from the same statistical distribution, and the more likely the object is an outlier.
2. *Clustering-based models*: Clustering algorithms look for data points that occur together in a group. Being a complementary problem to cluster analysis, clustering models can also be optimized to specifically detect outliers. Clustering-based models look for data points that are isolated from clusters and determine the outliers as a by-product of the clustering algorithm.
3. *Distance-based models*: Distance-based models use the distances of a data point to its k-nearest neighbors to determine whether it is an outlier. In this set of models, a data point is an outlier if its k-th nearest neighbor distance or the average distance to their k nearest neighbors is much larger than those of other data points. Distance-based models work well for datasets that contain clusters with similar densities and belong to a family of 'global' methods.
4. *Density-based models*: Density-based models use the local density of a data point to define its outlier score. More specifically, density-based outlier score is usually a ratio between the local density of an object and the average of the local densities of its k-nearest neighbours. Density-based models belong to a family of 'local' methods.
5. *Time series outlier detection models*: Outlier detection in time series can be defined in two different ways, a point outlier (which is a significant deviation from expected values in a time series at a given timestamp) and shape outliers (which are data points in a contiguous window that may be defined as anomalies when considered together, although no individual point in the series may be considered an anomaly). These models share an interesting relationship with other models.

In the next few subsections, these different types of models are explained in more detail.

2.2.1 Probabilistic Model-Based Outlier Detection Approach

In probabilistic models, outliers happen at the statistical tails of probability distributions. Being very specialized types of outliers, all values in the statistical tails of probability distributions are regarded as outliers. Further, based on Hawkins's definition of generative probabilities, the most isolated point in the dataset should, therefore, also be considered as an outlier from a generative perspective. Although statistical tails are more naturally defined for one-dimensional distributions, a similar argument applies to the case of multivariate data where the outliers lie in the multivariate tail area of the distribution. Even though the basic concept is analogous to that of univariate tails, it is more challenging to formulize the concept of multivariate tails.

After a model distribution is selected, the univariate tails are defined as extreme regions with probability density less than a particular threshold. More specifically, consider the density distribution $f_X(x)$. In general, the tail may be defined as the two extreme regions of the distribution for which $f_X(x) \leq \theta$, for some user defined threshold θ. The concept of density threshold is used to define characteristic of the tail, especially in the case of asymmetric univariate or multivariate distributions since some asymmetric distributions, such as an exponential distribution, may not even have a tail at one end of the distribution. In the following, the methods for univariate and multivariate tail analysis are detailed.

The most commonly used model for univariate data is the normal distribution. As shown in Fig. 2.3, the density function $f_X(x)$ of the normal distribution with mean μ and standard deviation σ is defined as follows:

$$f_X(x) = \frac{1}{\sigma \cdot \sqrt{2 \cdot \pi}} \cdot e^{\frac{-(x-\mu)^2}{2 \cdot \sigma^2}} \tag{2.2}$$

A standard normal distribution is one in which the mean μ is 0 and the standard deviation σ is 1. For other application scenarios, when a large number of data samples are available, the mean and standard deviation may be estimated very accurately. Based on the mean μ and the standard deviation σ, for a random variable, the Z-number z_i of an observed value x_i can be computed as follows:

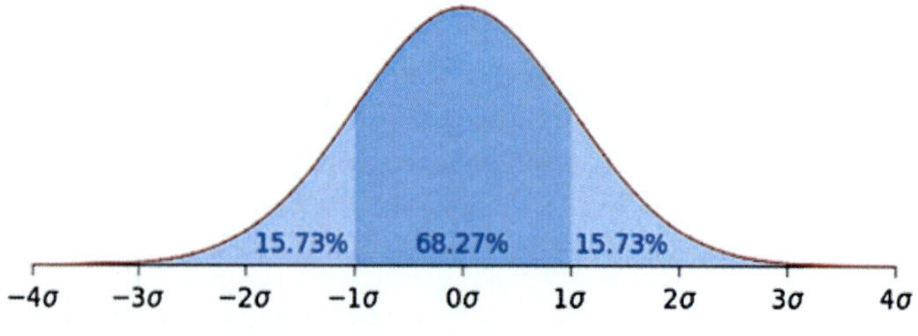

Fig. 2.3 An illustration of a normal distribution

$$z_i = \frac{(x_i - \mu)}{\sigma} \tag{2.3}$$

Based on the normal distribution, large positive values of z_i correspond to the upper tail, whereas large negative values of z_i correspond to the lower tail. Without loss of generality, if the absolute values of the Z-number are greater than 3, the corresponding data points can be considered outliers. The normal distribution of Eq. (2.2) can be expressed directly in terms of the Z-number as follows:

$$f_X(z_i) = \frac{1}{\sigma \cdot \sqrt{2 \cdot \pi}} \cdot e^{\frac{-z_i^2}{2}} \tag{2.4}$$

Let μ be the d-dimensional mean vector of a d-dimensional data set, and Σ be its $d \times d$ covariance matrix, the probability distribution $f\left(\overline{X}\right)$ for a d-dimensional data point $\overline{X}$ can be defined as follows:

$$f\left(\overline{X}\right) = \frac{1}{\sqrt{\left|\sum\right| \cdot (2 \cdot \pi)^{(d/2)}}} \cdot e^{-\frac{1}{2}\cdot\left(\overline{X}-\overline{\mu}\right)\sum^{-1}\left(\overline{X}-\overline{\mu}\right)^T} \tag{2.5}$$

The value of $|\Sigma|$ denotes the determinant of the covariance matrix. If Maha($\overline{X}$, μ, Σ) denotes the Mahalanobis distance between $\overline{X}$ and μ, in terms of the covariance matrix Σ, then the probability density function of the normal distribution can be expressed as,

$$f\left(\overline{X}\right) = \frac{1}{\sqrt{\left|\sum\right| \cdot (2 \cdot \pi)^{(d/2)}}} \cdot e^{-\frac{1}{2}\cdot \mathrm{Maha}\left(\overline{X},\overline{\mu},\sum\right)^2} \tag{2.6}$$

When the probability density drops below a particular threshold, the Mahalanobis distance will be larger than a particular threshold. Thus, the Mahalanobis distance to the mean of the data can be used as an outlier score. Larger values imply more extreme behaviors. Therefore, the utility of the Mahalanobis distance is more effective in using the underlying statistical distribution of the data to infer the outlier behavior of the multivariate data points.

To conclude, statistical tail analysis plays an important role in outlier analysis and can be applied to convert outlier scores to binary labels by identifying those outlier scores that are in the statistical tails. Multivariate statistical tail analysis is often useful in multi-criteria outlier detection algorithms by unifying multiple outlier scores into a single value and also generating a binary label as the output.

2.2.2 Clustering-Based Outlier Detection Approaches

In clustering-based outlier detection models, outliers often tend to occur in small clusters of their own. Let C_1, ..., C_L be a set of clusters of a dataset T discovered by a clustering algorithm and given in the sequence such that $|C_1| \geq |C_2| \geq \cdots \geq |C_L|$. Given parameters α and β, clustering-based outliers are those clusters in C_m through C_L such that: $|C_1|+|C_2|+\cdots+|C_{m-1}| \geq |T|*\alpha$, $|C_1|+|C_2|+\cdots+|C_{m-2}| \leq |T|*\alpha$, and$|C_{m-1}|/|C_m| > \beta$. It is in this sense that there is a well-known complementary relationship between clustering and outlier detection. Therefore, every data point is either a member of a cluster or an outlier. Being one way to assign an outlier score for a data point, in comparison to the case in multivariate tail analysis where the global Mahalanobis distance defines outlier scores, the local Mahalanobis distance can be used with respect to the centroid of the closest cluster. More specifically, the dataset is first partitioned into clusters and then the raw distance of the data point to its closest cluster centroid is computed to be the outlier score.

Consider a data set in which K clusters are obtained by a clustering algorithm. Assume that for the r-th cluster in d-dimensional space, there exist a corresponding d-dimensional mean vector μ_r, and a $d \times d$ covariance matrix Σ_r. The (i, j)-th entry of this matrix is the covariance between the dimensions i and j in that cluster. Then, the Mahalanobis distance Maha($\overline{X}$,μ_r, Σ_r) between a data point $\overline{X}$ and cluster centroid μ_r is defined as follows:

$$\mathrm{Maha}\left(\overline{X}, \overline{\mu_r}, \sum_r\right) = \sqrt{\left(\overline{X} - \overline{\mu_r}\right) \sum_r^{-1} \left(\overline{X} - \overline{\mu_r}\right)^{\mathrm{T}}} \tag{2.7}$$

This distance can be regarded as the outlier score larger values of which indicate a greater outlying tendency. After the outlier scores have been determined, univariate tail analysis may be applied to convert the scores to binary labels.

Unfortunately, it can happen in many applications that the clusters are elongated or have varying densities over the data set. For these cases, the local data distribution often distorts the distances. Therefore, it is not always optimal to use the raw distance. Further, the clustering algorithms in question may not be optimized for outlier detection. Therefore, the detection of outliers as a by-product of clustering methods may not be an appropriate approach. Finally, for cases where the distance to the closest cluster centroid does not often represent their local distribution (or instance-specific distribution), distance-based methods are more effective.

2.2.3 Distance-Based Outlier Detection Approaches

In their original paper [2], given a distance measure on a feature space, the notion of outliers studied by Knorr and Ng is defined as follows: 'An object O in a dataset T

is a distance-based outlier, denoted by DB(p, D)-outlier, if at least a fraction p of the objects in T lies greater than distance D from O, where the term DB(p, D)-outlier is a shorthand notation for a distance-based outlier (detected using parameters p and D).' Following this notion, several variant distance-based outlier definitions have been proposed that are slightly different. Two popular ones are:

1. Given two integers, n and k, outliers are the data items whose distance to their k-th nearest neighbor is among top n largest ones (referred to as the 'DB-Max' method in the following) [3].
2. Given two integers, n and k, outliers are the data items whose average distance to their k-nearest neighbors is among top n largest ones (referred to as the 'DB' method in the following) [4].

Although all the definitions are targeted to mine distance-based outliers, several minor differences exist between them. The first definition only takes into consideration the distance to the k-th nearest neighbor and ignores information about closer points. The second definition makes all the distances to k-nearest neighbors included and, thus, is slower to calculate than the first one. To summarize, the two definitions can be used to locate outliers that are far away from the majority of data objects and, therefore, are targeted to detect global outliers.

The standard distance-based outlier algorithms have a high computational complexity. For a dataset D containing N data points, the determination of the k-th nearest neighbor distance requires $O(N)$ time for each data point and the determination of the outlier scores of all data points may require $O(N^2)$ time. This is clearly not promising for very large datasets. To detect outliers efficiently in modern large multi-dimensional datasets, the index structures can be employed to avoid some unnecessary computations involved and to speed up the computation. However, for the data of a high dimension, the effectiveness of index structures tends to degenerate. Fortunately, in most applications, not all the data points but those with top n outlier scores are required to return and the outlier scores of the rest data points are not interested. In the implementations, based on the property that the distance-based outlier score of each data item is a monotonic nonincreasing function of the portion of the dataset already explored, to reduce the time required for the k-nearest neighbor distance computations by ruling out normal data points quickly that are obviously non-outliers even with an approximate distance computation, it may be possible to terminate a sequential search for an outlier candidate when its current upper bound estimate on the k-th nearest neighbor distance value falls below the n-th best outlier score found so far. This pruning method is referred to as the 'early termination trick'.

2.2.4 Density-Based Outlier Detection Approaches

Distance-based outlier detection models are simple and elegant. However, it is shown in Fig. 2.1 that straightforward measures, such as the Euclidean distance, do not work well for outlying data points when the density and shape of the clusters they are close to vary significantly with data locality. Based on this intuition, Breunig et al. pioneered the density-based outlier detection research by proposing to assign to each object a degree of being an outlier, called the local outlier factor (LOF), for judging the outlierness of every object in the data set based on the local density of an object's neighborhood [14], challenging the traditional binary property of being an outlier. Therefore, different from distance-based outlier detection models being global outlier detectors, density-based outlier detection models are local outlier detectors.

Loosely based on similar principles as density-based clustering, density-based methods try to highlight sparse regions in the underlying data in order to report outliers. Correspondingly, the notion of local outlier factor or kernel density estimate can be used.

2.2.4.1 Local Outlier Factor (LOF)

The *local outlier factor* (LOF) approach adjusts for local variations in cluster density by normalizing the local density of an object with the average of the local densities of its k-nearest neighbours. Before defining LOF, some related concepts are first introduced. For an object q, its k-distance, denoted by k-distance(q), is defined as the distance between q and its kth nearest neighbor; its k-distance neighborhood, denoted by $N_k(q)$, consists of objects whose distance from q is not greater than k-distance(q). It is possible that $|N_k(q)| > k$ if q has more than one neighbor on the boundary of the neighborhood. The reachability distance of an object q with respect to object o, denoted by reach-dist$_k(q, o)$, is defined as $\max\{k\text{-distance}(q), \text{dist}(q, o)\}$. The use of the reachability distance can make the statistical fluctuation of dist(q, o) for all o's close to q significantly reduced. Based on the concept of reachability distance, the local reachability density of an object q is defined as,

$$\text{lrd}_{\text{MinPts}}(q) = 1 / \left[\frac{\sum_{o \in N_{\text{MinPts}}(q)} \text{reach} - \text{dist}_{\text{MinPts}}(q, o)}{|N_{\text{MinPts}}(q)|} \right] \tag{2.8}$$

and finally, the local outlier factor (LOF) is defined as

$$\text{LOF}_{\text{MinPts}}(q) = \left[\frac{\sum_{o \in N_{\text{MinPts}}(q)} \frac{\text{lrd}_{\text{MinPts}}(o)}{\text{lrd}_{\text{MinPts}}(q)}}{|N_{\text{MinPts}}(q)|} \right] \tag{2.9}$$

From the definition formula, it can be observed, being the average ratio of the local reachability density of q and those of q's k-nearest neighbors, the LOF captures the degree to which q is an outlier by looking at the densities of its neighbors: the lower q's local reachability density and the higher the local reachability densities of q's k-nearest neighbors, the higher the LOF value of q's. It has been proved by the authors that the value of LOF is approximately equal to 1 for objects deep inside a cluster.

2.2.4.2 Kernel Density Estimation

In kernel density estimation, a continuous estimate of the density is generated at a given point at which the value of the density is estimated as the sum of the smoothed values of kernel functions $K_h(\cdot)$ associated with each point in the dataset. Each kernel function is associated with a kernel width h that determines the level of smoothing created by the function. The kernel estimation $f(\overline{X})$ based on N data points of dimension d and kernel function $K_h(\cdot)$ is defined as,

$$f(\overline{X}) = \frac{1}{N}\sum_{i=1}^{N} K_h(\overline{X} - \overline{X_i}) \tag{2.10}$$

Thus, each discrete point X_i in the dataset is replaced by a continuous function $K_h(\cdot)$ that peaks at X_i and has a variance determined by the smoothing parameter h. An example of such a distribution is the Gaussian kernel with width h.

$$K_h(\overline{X} - \overline{X_i}) = \left(\frac{1}{\sqrt{2\pi} \cdot h}\right)^d \cdot e^{-\overline{X} - \overline{X_i}^2/(2h^2)} \tag{2.11}$$

The estimation error is defined by the kernel width h, which is chosen in a data-driven manner. The density at each data point is computed without including the point itself in the density computation. The value of the density is reported as the outlier score. Low values of the density indicate greater tendency to be an outlier.

2.2.5 *Outlier Detection for Time Series*

Anomaly detection in time series is a heavily studied area of data science and machine learning. In addition to parametric models for time series outliers [49] which represent the first work on outlier detection for time series data, several other models have subsequently been proposed in the statistics literature, including autoregressive moving average (ARMA), autoregressive integrated moving average (ARIMA), vector autoregressive moving average (VARMA), CUmulative SUM Statistics (CUSUM), exponentially weighted moving average (EWMA), etc. These models can be applied to time series outlier detection by two main types of techniques, to detect outliers over a database of time series and to deal with outliers within a single time series.

2.2.5.1 Outliers in Time Series Databases

Given a database of time series, an outlier score for a time series can either be computed directly to identify a few time series sequences as outliers, or be obtained by first computing scores for overlapping fix-sized windows and then aggregating them to identify a subsequence in a test sequence as an outlier. These techniques are discussed in more detail in the next two subsections.

Direct Detection of Outlier Time Series

To find all anomalies for a given a database of time series using unsupervised parametric approaches, similar to traditional outlier detection methods, usually a model should be first learnt based on the assumption that most of the time series sequences are normal and a few are anomalous in the database, and an outlier score is then computed for each sequence with respect to the model. Popular models for unsupervised parametric approaches include finite state automata (FSA) [50–53], Markov models [54–58] and hidden Markov models (HMMs) [50, 59–62]. A time series sequence is then marked anomalous if the probability of the generation of the sequence from the model is very low.

Unsupervised discriminative approaches employ clustering the time series sequences based on the definition of a similarity function that measures the similarity between two sequences so that within-cluster similarity is maximized while between-cluster similarity is minimized. The anomaly score of a test time series sequence is defined as the distance to the centroid (or medoid) of the closest cluster. The most popular sequence similarity measures are the simple match count-based sequence similarity [62] and the normalized length of the longest common subsequence (LCS) [63, 64]. Popular clustering methods include K-means [65], EM [66], phased K-means [67], dynamic clustering [65], K-medoids [63], single-linkage clustering [68], clustering of multi-variate time series in the principal components space [69], one-class SVM [70, 71], and self-organizing maps [72]. The particular clustering method of choice is application-specific.

Window-Based Detection of Outlier Time Series

The goal of window-based detection of outlier time series is first to break the test sequence into multiple overlapping subsequences (windows), compute anomaly score for each window, and then to find all anomalous time windows and hence anomalous time series based on the anomaly score (AS) for the entire test sequence which is computed in terms of those of the individual windows. In comparison to the approaches in the previous subsection, window-based techniques can perform better for the localization of anomalies. These techniques need the window length as a parameter and usually maintain a normal pattern database as well as a negative pattern or a mixed pattern database by some other approaches.

In approaches based on a normal pattern database, normal sequences are divided into overlapping window subsequences of size w which each is stored in a database with its frequency. For a test sequence, subsequences of size w are obtained, and those subsequences that do not occur in the normal database are considered mismatches. If a test sequence has a large number of mismatches, it is marked as an anomaly [59, 73–75]. For the cases where a subsequence is not in the database, soft mismatch scores can also be computed [62, 76]. In addition to contiguous window subsequences, a look-ahead-based method can also be used for building a normal database [77]. Given a new test sequence, each subsequence is checked with the normal database based on a same look ahead size, and the number of mismatches is computed. In approaches based on negative and mixed pattern databases [72, 78–80], outlier detectors can be generated randomly or by using some domain knowledge of situations that are not expected to occur in the normal sequences. A test sequence is then monitored for the presence of any outlier detector. If any outlier detector matches, the sequence can be considered an outlier.

2.2.5.2 Outliers Within a Given Time Series

Given a single time series, one can find particular elements (or time points) or subsequences within the time series as outliers, referred to as point outliers and shape outliers, respectively. A point outlier is a sudden change in a time series value at a given timestamp. Being closely related to the problem of forecasting in time series data, a point outlier happens if it deviates significantly from its expected (or forecasted) value. On the other hand, shape outliers consist of a consecutive pattern of data points in a contiguous window which be defined as an anomaly when considered together, although no individual point in the series may be considered an anomaly. Such outliers are defined by combining the patterns from multiple data items.

To find shape-based outliers, all nonoverlapping windows of length w are first extracted from the time series by using a sliding-window approach. kNN-based outlier analysis is then performed over these newly created data objects. The windows with the highest kNN-based outlier scores are reported as outliers. Nonoverlapping windows are used to minimize the impact of trivial matches to overlapping windows. To reduce the time complexity of the associated kNN-based outlier detection approaches, some pruning methods like heuristic reordering of candidate subsequences [81], locality sensitive hashing [82], Haar wavelet and augmented tries [83], and SAX with augmented tries [84] can be used.

2.3 Performance Evaluation Metrics of Outlier Detection Approaches

The performance of an outlier mining method is defined by both the quality of the specific detector and the time needed to process the data in order to provide a prediction. Concerning the latter point, the processing speed on using large or very large databases is very important. Concerning the former point, it is desirable to determine the validity of outliers output by a particular algorithm. Due to the nature of being an unsupervised problem, it is hard to validate outlier analysis unless external criteria are synthetically generated or some rare aspects of real data sets are used as proxies. Therefore, without loss of generality, external measures such as precision, recall and the receiver operating characteristic (ROC) curve are often used to evaluate outlier detection methods.

2.3.1 Precision, Recall and Rank Power

Three popular external measures to quantitatively measure the performance of an outlier detection scheme are precision, recall, and rank power [13]. Precision measures the percentage of outliers among top ranked objects returned by a method, while recall measures the percentage of the total outlier set included in top ranked objects. Given a dataset $D = D_o \cup D_n$ where D_o denotes the set of all outliers and D_n denotes the set of all non-outliers and any integer $m \geq 1$, if O_m denotes the set of outliers among the objects in the top m positions returned by an outlier detection scheme, precision and recall are defined as,

$$\text{Precision} = \frac{|O_m|}{m} \tag{2.12}$$

$$\text{Recall} = \frac{|O_m|}{|D_o|} \tag{2.13}$$

Usually, users are not only interested in how many true outliers being returned by a method, but also in where they are placed. Rank power is a metric that considers both the placements and the number of results returned by a method. Suppose that a method returns m objects, n of which are true outliers. For $1 \leq i \leq n$, if L_i denotes the position of the i-th outlier, the rank power of the method with respect to m can be defined as,

$$\text{Rank Power} = \frac{n(n+1)}{2\sum_{i=1}^{n} L_i} \tag{2.14}$$

As can be seen from Eq. (2.14), rank power weighs the placements of the returned outliers heavily. An outlier placed earlier in the returned list adds less to the denominator of the rank power (and thus contributes more to the rank power metric) than those placed later in the list. A value of 1 indicates the best performance and 0 indicates the worst.

2.3.1.1 Receiver Operating Characteristic

In addition to the three external measures introduced in the previous subsection, to evaluate the outlier detection algorithms when the known outlier labels from a synthetic dataset or the rare class labels from a real dataset are used as the ground-truth, the receiver operating characteristic (ROC) curve is often employed. In outlier detection models, a threshold is typically selected on the outlier scores to generate the binary labels. If the threshold is selected too restrictively to minimize the number of declared outliers, true outlier points (false-negatives) can be missed. However, if the threshold is chosen in a more relaxed way, too many false-positives can happen. Therefore, there is a trade-off between the false-positives and false-negatives. For any given threshold t on the outlier score, the declared outlier set is denoted by $S(t)$. The *true-positive rate* is defined as the percentage of *ground-truth* outliers that have been reported as outliers at threshold t,

$$\mathrm{TPR}(t) = \mathrm{Recall}(t) = 100 * \frac{|S(t) \cap G|}{|G|} \tag{2.15}$$

where G represents the true set (ground-truth set) of outliers in the data set. The false positive rate $\mathrm{FPR}(t)$ is defined as the percentage of the falsely reported positives out of the ground-truth negatives,

$$\mathrm{FPR}(t) = 100 * \frac{|S(t) - G|}{|D - G|} \tag{2.16}$$

where G represents the ground-truth positives G for a data set D. The ROC curve is defined by plotting the $\mathrm{FPR}(t)$ on the X-axis, and the $\mathrm{TPR}(t)$ on the Y-axis for varying values of t. Note that the end points of the ROC curve are always at (0, 0) and (100, 100). For a random model, its performance is expected to exhibit along the diagonal line connecting these points. The better the accuracy of the approach, the more the obtained curve is lifted above this diagonal line. The area under the ROC curve can also be utilized as a concrete quantitative evaluation of the effectiveness of a particular method. For cases in which one curve strictly dominates another, it is clear that the algorithm for the former curve is superior. For other cases when algorithms show domination at different parts of the ROC curve, it is hard to say that one algorithm is strictly superior because all parts of the ROC curve may not be equally important for different applications.

2.4 Conclusions

Unsupervised outlier analysis is an important and active research area in outlier detection due to its applicability to a variety of problem domains. Many methods have been developed for unsupervised outlier detection with the most common ones being probabilistic models, distance-based models, density-based models, clustering-based models, and models for time series outlier detection. The probabilistic models assume that the bulk of the data follows some standard distributions and objects that do not fit into the general behaviour are regarded as outliers. However, they often fall short to detect outliers existing in large multi-dimensional data sets. To improve, distance-based models have been proposed. Although simple and elegant, popular distance-based outlier detection methods are computationally more expensive. To speed up, a number of tricks have been proposed to make these models much efficient. Further, distance-based outlier detection methods can only detect global outliers but might not do well for datasets which have complex structures and exhibit very different characteristics in different portions. As a result, density-based models have been proposed so as to be more effective for the detection of local outliers. Models for time series outlier detection are closely related to conventional models, however face the challenge of the trade-off between accuracy and efficiency in the exploration of utilizing conventional models. At the end, due to the challenges associated with the unsupervised nature of outlier detection and the small sample-space problem, the validation of the outlier detection algorithms is a difficult task. On the one hand, external validation criteria such as precision and recall are typically employed. On the other hand, the area under the receiver operating characteristic curve can also provide a quantitative evaluation of the outlier detection algorithm.

Acknowledgements Part of this chapter is from the paper published by our group in Information Systems [13]. The related contents are reused with permission.

References

1. Hawkins, D. M. (1980). *Identification of outliers*. London: Chapman and Hall.
2. Knorr, E. M., & Ng, R. T. (1998). Algorithms for mining distance-based outliers in large datasets. In *Proceedings of the International Conference on Very Large Data Bases (VLDB'98)*, New York, pp. 392–403.
3. Ramaswamy, S., Rastogi, R., & Shim, K. (2000). Efficient algorithms for mining outliers from large data sets. In *Proceedings of the ACM International Conference on Management of Data (SIGMOD'00)*, Dallas, pp. 427–438.
4. Angiulli, F., & Pizzuti, C. (2002). Fast outlier detection in high dimensional spaces. In *Proceedings of the 6th European Conference on Principles of Data Mining and Knowledge Discovery (PKDD'02)*, Helsinki, pp. 15–26.
5. Bay, S. D., & Schwabacher, M. (2003). Mining distance-based outliers in near linear time with randomization and a simple pruning rule. In *Proceedings of the 9th ACM SIGKDD International Conference on Knowledge Discovery and Data Mining (KDD'03)*, Washington, DC, United states, pp. 29–38.
6. Ghoting, A., Parthasarathy, S., & Otey, M. E. (2006). Fast mining of distance-based outliers in high-dimensional datasets. *Data Mining and Knowledge Discovery, 16*(3), 349–364.
7. Angiulli, F., Basta, S., & Pizzuti, C. (2005). Distance-based detection and prediction of outliers. *IEEE Transactions on Knowledge and Data Engineering, 18*(2), 145–160.
8. Angiulli, F., & Pizzuti, C. (2005). Outlier mining in large high-dimensional data sets. *IEEE Transactions on Knowledge and Data Engineering, 17*(2), 203–215.

9. Tao, Y., Xiao, X., & Al, E. (2006). Mining distance-based outliers from large databases in any metric space. In *Proceedings of the 12th ACM SIGKDD International Conference on Knowledge Discovery and Data Mining*, Philadelphia, PA, United states, pp. 394–403.
10. Angiulli, F., & Fassetti, F. (2009). DOLPHIN: An efficient algorithm for mining distance-based outliers in very large datasets. *ACM Transactions on Knowledge Discovery from Data, 3*(1), 1–57.
11. Hautamäki, V., Kärkkäinen, I., & Fränti, P. (2004). Outlier detection using k-nearest neighbor graph. In *Proceedings of the 17th International Conference on Pattern Recognition* (*ICPR'04*), Cambridge, pp. 430–433.
12. Radovanović, M., Nanopoulos, A. and Ivanović, M. (2014). Reverse nearest neighbors in unsupervised distance based outlier detection. *IEEE Transactions on Knowledge and Data Engineering, 27*(5), 1369–1382.
13. Wang, X., Wang, X. L., Ma, Y., & Wilkes, D. M. (2015). A fast MST-inspired kNN-based outlier detection method. *Information Systems, 48,* 89–112.
14. Breuning, M. M., Kriegel, H. P., Ng, R. T., & Sander, J. (2000). LOF: Identifying density-based local outliers. In *Proceedings of the 2000 ACM SIGMOD International Conference on Management of Data (SIGMOD'00),* Dallas, TX, United states, pp. 93–104.
15. Tang, J., Chen, Z., Fu, A. W. C., & Cheung, D. W. (2002). Enhancing effectiveness of outlier detections for low density patterns. In *Proceedings of the 6th Pacific-Asia Conference on Knowledge Discovery and Data Mining* (*PAKDD'02*), Taipei, Taiwan, pp. 535–548.
16. Gibbons, P. B., Papadimitriou, S., Kitagawa, H., & Faloutsos, C. (2003). LOCI: Fast outlier detection using the local correlation integral. In *Proceedings of the IEEE 19th International Conference on Data Engineering* (*ICDE'03*), Bangalore, India, pp. 315–326.
17. Sun, P., & Chawla, S. (2004). On local spatial outliers. In *Proceedings of the 4th IEEE International Conference on Data Mining* (*ICDM'04*), Brighton, UK, pp. 209–216.
18. Jin, W., Tung, A. K. H., Han, J., & Wang, W. (2006). Ranking outliers using symmetric neighborhood relationship. In *Proceedings of the 10th Pacific-Asia Conference on Knowledge Discovery and Data Mining* (*PAKDD'06*), Singapore, pp. 577–593.
19. Latecki, L. J., Lazarevic, A., & Pokrajac, D. (2007). Outlier detection with kernel density functions. In *Proceedings of the 5th International Conference on Machine Learning and Data Mining in Pattern Recognition* (MLDM'07), Leipzig, Germany, pp. 61–75.
20. Kriegel, H. P., Kröger P., Schubert, E., & Zimek, A. (2009). LoOP: Local outlier probabilities. In *Proceedings of the ACM 18th International Conference on Information and Knowledge Management* (*CIKM'09*), Hong Kong, pp. 1649–1652.
21. Zhang, K., Hutter, M., & Jin, H. (2009). A new local distance-based outlier detection approach for scattered real-world data. In *Proceedings of the 13th Pacific-Asia Conference on Knowledge Discovery and Data Mining* (*PAKDD'09*), pp. 813–822.
22. Janssens, J., Huszar, F., Postma, E., & van den Herik, H. (2012). Stochastic outlier selection.
23. Huang, H., Mehrotra, K., & Mohan, C. K. (2013). Rank-based outlier detection. *Journal of Statistical Computation and Simulation, 83*(3), 518–531.
24. Schubert, E., Zimek, A., & Kriegel H. P. (2014). Generalized outlier detection with flexible kernel density estimates. In *Proceedings of the 14th Siam International Conference on Data Mining (SDM'14),* Philadelphia, pp. 542–550.
25. Terrell, G. R., & Scott, D. W. (1992). Variable Kernel Density Estimation. *Annals of Statistics, 20*(3), 1236–1265.
26. Ru, X., Liu, Z., Huang, Z., et al. (2016). Normalized residual-based constant false-alarm rate outlier detection. *Pattern Recognition Letters, 69,* 1–7.
27. Tang, B., & He, H. (2017). A local density-based approach for outlier detection. *Neurocomputing, 241,* 171–180.
28. MacQueen, J. (1965). Some methods for classification and analysis of multivariate observations. In *Proceedings of the 5th Berkeley Symposium on Mathematical Statistics and Probability*, pp. 281–297.
29. Kaufman, L., & Rousseeuw, P. (1990). Finding groups in data: an introduction to cluster analysis. John Wiley & Sons.

30. Ester, M., Kriegel, H.-P., Sander, J., & Xu, X. (1996). A density-based algorithm for discovering clusters in large spatial databases with noise. In *Proceedings of the 2nd ACM SIGKDD International Conference on Knowledge Discovery and Data Mining (KDD'96)*, Portland, Oregon, USA, pp. 226–231.
31. Zhang, T., Ramakrishnan, R., & Livny, M. (1996). BIRCH: An efficient data clustering method for very large databases. In *Proceedings of 1996 ACM SIGMOD International Conference on Management of Data (SIGMOD'96)*, Montreal, Quebec, Canada, pp. 103–114.
32. Hinneburg, A., & Keim, D. A. (1998). An efficient approach to clustering in large multimedia databases with noise. In *Proceedings of 4th International Conference on Knowledge Discovery and Data Mining (KDD'98)*, New York City, New York, USA, pp. 58–65.
33. Guha, S., Rastogi, R., & Shim, K. (1998). CURE: An efficient clustering algorithm for large databases. In *Proceedings of 1998 ACM SIGMOD International Conference on Management of Data (SIGMOD'98)*, Seattle, Washington, USA, pp. 73–84.
34. Al-Zoubi, M. (2009). An effective clustering-based approach for outlier detection. *European Journal of Scientific Research, 28*(2), 310–317.
35. Zahn, C. T. (1971). Graph-theoretical methods for detecting and describing gestalt clusters. *IEEE Transactions on Computers*, C-20(1), 64–82.
36. Rohlf, F. J. (1975). Generalization of the gap test for the detection of multivariate outliers. *Biometrics, 31*, 93–101.
37. Jiang, M. F., Tseng, S. S., & Su, C. M. (2001). Two-phase clustering process for outliers detection. *Pattern Recognition Letters, 22*(6–7), 691–700.
38. Lin, J., Ye, D., Chen, C., & Gao, M. (2008). Minimum spanning tree based spatial outlier mining and its applications. In *Proceedings of the 3rd International Conference on Rough Sets and Knowledge Technology (RSKT'08)*, Chengdu, China, pp. 508–515.
39. John Peter, S. (2011). An efficient algorithm for local outlier detection using minimum spanning tree. *International Journal of Research and Reviews in Computer Science (IJRRCS), 2*(1), 15–23.
40. Marghny, M. H., & Taloba, A. I. (2011). Outlier detection using improved genetic K-means. *International Journal of Computer Applications, 28*(11), 33–42.
41. Liu, Y., Li, Z., Zhou, C., Jiang, Y., Sun, J., Wang, M., & He, X. (2019). Generative adversarial active learning for unsupervised outlier detection. *IEEE Transactions on Knowledge and Data Engineering*. https://arxiv.org/abs/1809.10816.
42. Kriegel, H.-P., Schubert, M., & Zimek, A. (2008). Angle-based outlier detection in high-dimensional data. In *Proceedings of the 14th ACM SIGKDD International Conference on Knowledge Discovery and Data Mining (KDD'08)*, Las Vegas, Nevada, USA, pp. 444–452.
43. Pham, N., & Pagh, R. (2012). A near-linear time approximation algorithm for angle-based outlier detection in high-dimensional data. In *Proceedings of the 18th ACM SIGKDD Conference on Knowledge Discovery and Data Mining (SIGKDD'12)*, Beijing, China, pp.877–885.
44. Beyer, K., Goldstein, J., Ramakrishnan, R., & Shaft, U. (1999). When is nearest neighbors meaningful? In *Proceedings of the 1999 International Conference on Database Theory (ICDT'99)*, pp. 217–235.
45. Aggarwal, C., & Yu, P. (2005). An effective and efficient algorithm for high- dimensional outlier detection. *The VLDB Journal, 14*(2), 211–221.
46. Parthasarathy, S., & Aggarwal, C. C. (2003). On the use of conceptual reconstruction for mining massively incomplete datasets. *IEEE Transactions on Knowledge and Data Engineering, 15*(6), 1512–1531.
47. Schölkopf, B., Platt, J., Shawe-Taylor, J., Smola, A., & Williamson, R. (2001). Estimating support of a high-dimensional distribution. *Neural Computation, 13*(7), 1443–1471.
48. Erfani, S. M., Rajasegarar, S., Karunasekera, S., & Leckie, C. (2016). High-dimensional and large-scale anomaly detection using a linear one-class SVM with deep learning. *Pattern Recognition, 58*, 121–134.
49. Fox, A. J. (1972). Outliers in time series. *Journal of the Royal Statistical Society. Series B (Methodological), 34*(3), 350–363.

50. Chandola, V., Mithal, V., & Kumar, V. (2008). A comparative evaluation of anomaly detection techniques for sequence data. In *Proceedings of the 8th IEEE International Conference on Data Mining (ICDM'08)*, Pisa, Italy, pp. 743–748.
51. Marceau, C. (2000). Characterizing the behavior of a program using multiple-length n-grams. In *Proceedings of the 2000 New Security Paradigm Workshop (NSPW'00)*, pp. 101–110.
52. Michael, C. C., & Ghosh, A. (2000). Two state-based approaches to program-based anomaly detection. In *Proceedings of the 16th Annual Computer Security Applications Conference (ACSAC'00)*, New Orleans, LA, USA, pp. 21–30.
53. Salvador, S., & Chan, P. (2005). Learning states and rules for detecting anomalies in time series. *Applied Intelligence, 23*(3), 241–255.
54. Ye, N. (2000). A Markov chain model of temporal behavior for anomaly detection. In *Proceedings of the 2000 IEEE SMC Information Assurance and Security Workshop* (Vol. 166. pp. 171–174).
55. Yang, J., & Wang, W. (2003). CLUSEQ: Efficient and effective sequence clustering. In *Proceedings of the 19th IEEE International Conference on Data Engineering (ICDE'03)*, Bangalore, India, pp. 101–112.
56. Sun, P., Chawla, S., & Arunasalam, B. (2006). Mining for outliers in sequential databases. In *Proceedings of the 6th SIAM International Conference on Data Mining (SDM'06)*, Bethesda, MD, United states, pp. 94–105.
57. Eskin, E., Lee, W., & Stolfo, S. (2001). Modeling system calls for intrusion detection with dynamic window sizes. In *Proceedings of DARPA Information Survivability Conference and Exposition II* (DISCEX'01), Anaheim, CA, United states, Vol. 1, pp. 165–175.
58. Lee, W., Stolfo, S. J., & Chan, P. K. (1997). Learning patterns from unix process execution traces for intrusion detection. In *Proceedings of the AAAI Workshop on AI Approaches Fraud Detection and Risk Management*, pp. 50–56.
59. Florez-Larrahondo, G., Bridges, S. M., & Vaughn, R. (2005). Efficient modeling of discrete events for anomaly detection using hidden Markov models. In *Proceedings of the 8th International Conference on Information Security (ISC'05)*, Singapore, pp. 506–514.
60. Gao, B., Ma, H.-Y., & Yang, Y.-H. (2002). HMMs (Hidden Markov Models) based on anomaly intrusion detection method. In *Proceedings of the IEEE International Conference on Machine Learning and Cybernetics*, Beijing, China, Vol. 1, pp. 381–385.
61. Qiao, Y., Xin, X., Bin, Y., & Ge, S. (2002). Anomaly intrusion detection method based on HMM. *Electronics Letters, 38*(13), 663–664.
62. Zhang, X., Fan, P., & Zhu, Z. (2003). A new anomaly detection method based on hierarchical HMM. In *Proceedings of the 4th International Conference on Parallel and Distributed Computing, Applications and Technologies* (PDCAT'03), Chengdu, China, pp. 249–252.
63. Lane, T., et al. (1997). Sequence matching and learning in anomaly detection for computer security. In *Proceedings of the AAAI Workshop on AI Approaches Fraud Detection and Risk Management*, pp. 43–49.
64. Budalakoti, S., Srivastava, A. N., & Otey, M. E. (2009). Anomaly detection and diagnosis algorithms for discrete symbol sequences with applications to airline safety. *IEEE Transactions on Systems, Man and Cybernetics Part C: Applications and Reviews, 39*(1), 101–113.
65. Sequeira, K., & Zaki, M. (2002). ADMIT: Anomaly-based data mining for intrusions. In *Proceedings of the 8th ACM SIGKDD International Conference on Knowledge Discovery and Data Mining* (KDD'02), Edmonton, Alta, Canada, pp. 386–395.
66. Nairac, A., et al. (1999). A system for the analysis of jet engine vibration data. *Integrated Computer-Aided Engineering, 6*(1), 53–65.
67. Pan, X., Tan, J., Kavulya, S., Gandhi, R., & Narasimhan, P. (2009). Ganesha: BlackBox diagnosis of mapReduce systems. *Performance Evaluation Review, 37*(3), 8–13.
68. Rebbapragada, U., Protopapas, P., Brodley, C. E., & Alcock, C. (2009). Finding anomalous periodic time series. *Journal of Machine Learninng, 74*(3), 281–313.
69. Portnoy, L., Eskin, E., & Stolfo, S. (2001). Intrusion detection with unlabeled data using clustering. In *Proceedings of the ACM CSS Workshop DMSA*, pp. 5–8.

70. Eskin, E., Arnold, A., Prerau, M., Portnoy, L., & Stolfo, S. (2002). A geometric framework for unsupervised anomaly detection: Detecting intrusions in unlabeled data. In: D. Barbará & S. Jajodia (Eds.), *Applications of Data Mining in Computer Security. Advances in Information Security*, Vol. 6, pp. 77–101.
71. Ma, J., & Perkins, S. (2003). Time-series novelty detection using one-class support vector machines. In *Proceedings of the International Joint Conference on Neural Networks (IJCNN'03)*, Portland, OR, United states, Vol. 3, pp. 1741–1745.
72. Szymanski, B., & Zhang, Y. (2004). Recursive data mining for masquerade detection and author identification. In *Proceedings of the 5th Annual IEEE System, Man and Cybernetics Information Assurance Workshop (SMC'04)*, West Point, NY, United states, pp. 424–431.
73. González, F. A., & Dasgupta, D. (2003). Anomaly detection using real-valued negative selection. *Genetic Programming and Evolvable Machines, 4*(4), 383–403.
74. Cabrera, J. B. D., Lewis, L., & Mehra, R. K. (2001). Detection and classification of intrusions and faults using sequences of system calls. *SIGMOD Record, 30*(4), 25–34.
75. Endler, D. (1998). Intrusion detection applying machine learning to solaris audit data. In *Proceedings of the 14th Annual Computer Security Applications Conference (ACSAC'98)*, Phoenix, AZ, USA, pp. 268–279.
76. Hofmeyr, S. A., Forrest, S., & Somayaji, A. (1998). Intrusion detection using sequences of system calls. *Journal of Computer Security, 6*(3), 151–180.
77. Lane, T., & Brodley, C. E. (1998). Temporal sequence learning and data reduction for anomaly detection. In *Proceedings of the 1998 5th ACM Conference on Computer and Communications Security (CCS-5)*, San Francisco, CA, USA, pp. 150–158.
78. Forrest, S., Hofmeyr, S. A., Somayaji, A., & Longstaff, T. A. (1996). A sense of self for unix processes. In *Proceedings of the 1996 17th IEEE Symposium on Security and Privacy*, Oakland, CA, USA, pp. 120–128.
79. Dasgupta, D., & Nino, F. (2000). A comparison of negative and positive selection algorithms in novel pattern detection. In *Proceedings of the IEEE International Conference on Systems, Man and Cybernetics (SMC)*, Nashville, TN, USA, pp. 125–130.
80. Dasgupta, D., & Majumdar, N. (2002). Anomaly detection in multidimensional data using negative selection algorithm. In *Proceedings of the 2002 Congress on Evolutionary Computation (CEC'02)*, Honolulu, HI, USA, pp. 1039–1044.
81. Keogh, E., Lin, J., & Fu, A. (2005). HOT SAX: Efficiently finding the most unusual time series subsequence. In *Proceedings of the 5th IEEE International Conference on Data Mining (ICDM'05)*, Houston, TX, United states, pp. 226–233.
82. Wei, L., Keogh, E., & Xi, X. (2006). SAXually explicit images: finding unusual shapes. In *Proceedings of the 6th IEEE International Conference on Data Mining (ICDM'06)*, Hong Kong, China, pp. 711–720.
83. Fu, A. W.-C., Leung, O. T.-W., Keogh, E., & Lin, J. (2006). Finding time series discords based on haar transform. In *Proceedings of the 2nd International Conference on Advanced Data Mining and Applications (ADMA'06)*, Xi'an, China, pp. 31–41.
84. Lin, J., Keogh, E., Fu, A., & Van Herle, H. (2005). Approximations to magic: Finding unusual medical time series. In *Proceedings of the 18th IEEE Symposium on Computer-Based Medical Systems (CBMS'05)*, pp. 329–334.

Part II
New Developments in Unsupervised Outlier Detection Research

Chapter 3
A Fast Distance-Based Outlier Detection Technique Using a Divisive Hierarchical Clustering Algorithm

Abstract Today's real-world databases typically have millions of items with many thousands of fields, resulting in data that range in size into terabytes. As a result, traditional distribution-based outlier detection techniques have more and more restricted capabilities and novel approaches that find unusual samples in a data set based on their distances to neighboring samples have become more and more popular. The problem with these *k*-nearest neighbor-based methods is that they are computationally expensive for large datasets. At the same time, today's databases are often too large to fit into the main memory at once. As a result, memory capacity and, correspondingly, I/O cost, become an important issue. In this chapter, we present a simple distance-based outlier detection algorithm that can compete with existing solutions in both CPU and I/O efficiency.

Keywords Distance-based outlier detection · Approximate *k*-nearest neighbor search · Divisive hierarchical clustering · I/O efficiency · Sampling

3.1 Introduction

Outlier detection aims to discover observations that deviate from other observations so much as to arouse suspicions that they are generated by a different mechanism [1]. Due to its important applications in a large number of useful and important tasks such as intrusion detection for cyber-security [2, 3], fraud detection for credit cards, insurance and tax [4], early detection of disease outbreaks in the medical field [5], fault detection in sensor networks for monitoring health, traffic, machine status, weather, pollution and surveillance [6], and so on [7, 8], it has generated enormous interests in recent years. Five popular categories of techniques have been developed for modern outlier detection purposes, namely distribution-based approaches, depth-based approaches, distance-based approaches, density-based approaches and clustering-based approaches.

The notion of distance-based outliers was first introduced by Knorr and Ng in 1998 [9]. Since then, abundant developments in the field have made the distance-based outlier detection a very important branch of modern outlier detection techniques. With no a priori assumptions on the distribution model underlying the data, distance-based

X. Wang et al., *New Developments in Unsupervised Outlier Detection*,
https://doi.org/10.1007/978-981-15-9519-6_3

outlier algorithms use a data point's metric distances to its nearest neighbors as a way to measure its deviation from other data points. They are simple to implement due to a favorable property: the distance-based outlier score of each data item is a monotonic nonincreasing function of the portion of the dataset already explored [10–12]. However, for modern large high-dimensional data sets, the $O(N^2d)$ running time has to be improved significantly while satisfying the constraints imposed by the I/O cost (due to the limited main memory space available). To meet these challenges, distance-based outlier detection techniques have become an active area of research in the past few years and a variety of algorithms have been proposed for this purpose.

Realizing the fact that the majority of the data points are normal and reside in dense regions, while outliers are small in numbers, far away from the normal data points, and should have relatively large distances to some (if not all) of their nearest neighbors, the state-of-the-art distance-based outlier detection algorithms have presented various clever ways to filter out the large amount of normal data efficiently. Aided with smart pruning strategies to make the outliers to show up quickly, these algorithms can be classified into two general categories: the bottom-up approaches (which include the ORCA method [13], the RBRP method [14], and the SNIF method [15]), and the top-down approaches (which include the HilOut method [10], the SolvingSet algorithm [11], and the DOLPHIN method [12]).

Based on our study on these algorithms, in this chapter, we propose an efficient distance-based outlier detection method that embeds a fast approximate k-nearest neighbor search structure into the distance-based outlier detection framework. More specifically, our solution is based on a divisive hierarchical clustering algorithm (DHCA) [16] for the case when the data set fits in memory. We also propose a variant that typically needs to scan the data set twice for the large datasets that cannot be loaded into the main memory at one time. By conducting a feasibility demonstration of such a nearest neighbor search structure, we show that repeated invocations of DHCA can facilitate the mining process by increase the probability of each data item meeting its nearest neighbors in the process. Next, our outlier detection technique follows the basic idea of the SolvingSet algorithm [11] (i.e., to first quickly locate the small number of data items that are potential outlier candidates, followed by subsequent verification and removal of the inliers). Our major contribution in this chapter is the development of a simple outlier detection algorithm that can compete with and, in some situations, outperform the state-of-the-art distance-based outlier detection techniques in running time while consuming much less resources.

In the following, Sect. 3.2 gives an account of related work in terms of the state-of-the-art distance-based outlier detection approaches. We then present the proposed efficient outlier detection algorithm in Sect. 3.3. When enough main memory is available, this algorithm performs only one dataset scan without generating any intermediate files. When the available memory is smaller than the space required by our algorithm, a modified version of our algorithm for I/O efficiency is described in Sect. 3.4. In Sect. 3.5, a performance evaluation is presented to demonstrate the technical soundness of our algorithms based on several large high-dimensional data sets. Finally, conclusions are made and contributions are summarized in Sect. 3.6.

3.2 Related Work

3.2.1 Distance-Based Outlier Detection Research

Given a set of data and a distance measure, a distance-based outlier was defined by Knorr and Ng in their original paper to be: "*An object O in a dataset T is a DB(p,D)-outlier if at least fraction p of the objects in T lies greater than distance D from O*", where the term *DB*(*p*,*D*)-outlier is a shorthand notation for a distance-based outlier (DB-Outlier) detected using parameters *p* and *D* [9]. Following the notion of this definition, several variants with slight differences have been proposed to suit for different theoretical and practical purposes. Three popular ones are:

1. Given a real number *r* and an integer *p*, a data item is an outlier if there are fewer than *p* other data items within distance *r* [17, 18].
2. Given two integers, *n* and *k*, outliers are those data items whose distances to their *k*-th nearest neighbor are among top *n* largest ones [19].
3. Given two integers, *n* and *k*, outliers are those data items whose average distances to their *k* nearest neighbors are among top *n* largest ones [2, 20].

Although all these definitions are targeted to mine distance-based outliers, some minor differences exist among them. The first definition provides no ranking but requires the specification of a distance parameter *r*, which could be difficult to determine and may involve trial and error to guess an appropriate value. To circumvent this inconvenience, the second definition considers only the distance to the *k*-th nearest neighbor. However, the information about closer points is ignored. To improve, the last definition accounts for all the distances to *k*-nearest neighbors (*k*NN) but is slower to compute than the first two. Apparently, all these definitions are based on a nearest neighbor density estimate with those data items in low-probability regions being regarded as outliers.

Physically, the resource consumed by the DB-Outlier calculation includes the space for the partial or whole data set to reside in memory, the space to store the nearest neighbors for each or some data items, and/or the space for some indexing structure if needed. There is a trade-off between the available memory space and the performance of the algorithms: The more the in-memory space is available, the more the useful information (e.g., the affordability of the sophisticated data structures) can be retained, and the better the run time performance can be, constrained by a balance between space and speed.

To quickly remove the large amount of normal data, Bay and Schwabacher's method, named ORCA, maintains a top-*n*-outlier candidate list from samples processed so far and utilized the fact that, in the traditional nest-loop approach, most of the inliers are involved in distance computations with a number much smaller than *N* (the number of data objects in the dataset) for them to be dropped off from the top-*n*-outlier list [13]. To further enhance this possibility, the data were preprocessed to be in random order. The third outlying score definition was used in the paper. The advantage of ORCA is that it is simple to implement and there are not any intermediate files to write to during the whole outlier detection process. The ORCA method is regarded as a bottom-up approach and is summarized in Table 3.1. If the main

Table 3.1 Basic kNN-based bottom-up DB-Outlier search algorithm

Input:			
	The dataset		
	An input number of nearest neighbors		
	An input number of top n outliers		
	A distance measure		
	A DB outlying score definition		
Output:			
	Top *n* outliers		
Procedure:			
1:	Start with the first data point of the dataset sequentially read in		
2:	Compute its distances with the rest of the dataset in a sequential order, after each distance computation, check if it its current DB outlying score is smaller than the smallest score of the current top n outlier candidates		
3:	if yes		
		if all the data points have been searched	
			Return top n DB-outliers and go to end (i.e., exit)
		else	
			Drop this data item, move to the next data item and go to 2
		end	
4:	else		
		if all distances between it and the rest of the data set have been calculated	
			Update the top n outlier list by replacing the data point having the smallest score with this data point
		else	
			Continue its distance computing process with the rest of the data set in the sequential order (i.e., move on to the next data point and go to 2)
		end	
5:	end		

memory can afford to store the whole database, near linear CPU processing time performance was reported by ORCA.

Claimed to be a further improvement, with the aid of a K-means-based clustering (recursive binning) followed by projection (i.e., principal component analysis) in each cluster (re-projection), the RBRP method found a much better way to enhance the probability of each data point being nearer to its nearest neighbors and showed that its performance can outperform the ORCA method by an order of magnitude on average. Unfortunately, there is a penalty paid for its superiority in running time performance in its current implementation. That is, intermediate files of a total size even larger than that of the original database file are generated by RBRP. In other words, for fast processing of nearest neighbors' search, the RBRP method used K-means-based clustering (recursive binning) followed by projection to reorganize the database in the form of bins and wrote these bins out. For outlier detection, the RBRP method operates on these bin files in exactly the same way as the ORCA method operates on pre-randomized database. Both ORCA and RBRP have only an outlying score assigned to each data item. However, neither employ any other facility to remember who their nearest neighbors are. Finding the cutoff threshold from a bottom-up approach, i.e., the ORCA and RBRP methods, could be affected by the slow decay of the cutoff threshold.

To overcome this problem, Angiulli and Pizzuti proposed to rank the data items on the basis of the sum of the distances to the k-nearest neighbors and gave more attention to the top data items on the list which were called the potential outlier candidates [10]. More specifically, the algorithm (referred to as the HilOut method) exploits Hilbert space-filling curves to project the data multiple times onto the interval [0, 1]. Each successive projection improves the estimate of an object's outlier score and progressively reduces the set of candidate outliers. Taking a step further, the SolvingSet algorithm proposed the concept of outlier detection solving set which includes a sufficient number of points and permits the detection of the top outliers by considering only a subset of all the pairwise distances from the data set [11]. These methods are basically top-down approaches and are summarized in Table 3.2. Apparently, the slow decay of the cutoff threshold does not affect these top-down approaches as much as they do on the bottom-up approaches if the initialized outlying scores are tight enough.

Although the ORCA and RBRP methods are claimed to be solutions of a sublinear time complexity, the issue with regard to the I/O efficiency has not been addressed and both methods can incur the quadratic I/O cost [15]. Based on the first definition of the distance-based outliers, the SNIF method uses some pruning rules to avoid the unnecessary storage of some data points and can report all outliers by scanning the database at most twice (in some cases, even once), which the authors claimed significantly outperformed the existing solutions by a factor up to an order of magnitude [15]. In the algorithm, priorities are assigned to objects with relatively few neighbors indicating the likelihood that the object would be an outlier. More specifically, to identify distance-based outliers from the datasets that cannot fit into the main memory at once, the SNIF method randomly selects s samples as centroids to partition the data set. Based on the distances of each data point to the centroids, nonoutliers can be

Table 3.2 Basic kNN-based top-down DB-Outlier search algorithm

Input:	
	The dataset
	An input number of nearest neighbors
	An input number of top n outliers
	A distance measure
	A DB outlying score definition
Output:	
	Top n outliers
Procedure:	
1:	Initialize by finding, for each data point, k neighbors (i.e., k distances) and assigning a score to each data point based on these distances
2:	Start with potential outlier candidates (i.e., those data points whose outlying scores are among top n ones) and search its kNN through the whole database
3:	Check if top potential outlier candidates are verified to be true outliers
4:	If yes, return all top n DB-outliers and go to end (i.e., exit), else, go to 2 and start another round of verification
5:	end

identified and removed easily using their carefully designed heuristics. Only when a data point is an outlier is a complete scan of the database necessary. Although some intermediate files could be generated in the detecting process (as the authors claimed), no such files happened in their experiments. However, the SNIF method is not very CPU efficient because the value of s can not be too small. For $s = 1000$ used in the paper, at least $1000N$ distance computations are needed, where N is the number of data items in the database.

More recently, a new distance-based outlier detection algorithm, called DOLPHIN, has been developed specifically for working with disk-resident datasets. Simultaneously achieving linear CPU time performance and linear I/O cost by scanning very large multidimensional disk-resident datasets only twice while with a small usage of main memory, DOLPHIN gains its efficiency through naturally merging three strategies (which are adopted one way or the other by other up-to-date distance-based outlier detection algorithms, namely, (1) selection policy of objects to be maintained in the main memory, (2) usage of pruning rules, (3) similarity search techniques) [12]. Although DOLPHIN has demonstrated its ability to simultaneously satisfy constraints put on by the CPU cost and the I/O cost, it is based on the first definition of distance-based outlier (i.e., range-based).

3.2.2 *A Divisive Hierarchical Clustering Algorithm for Approximate* **k***NN Search*

Given a data set and an input value for K, the divisive hierarchical clustering algorithm (DHCA) starts with K randomly selected centers, and, by assigning each point to its closest center, create a K-partition. At each subsequent stage in the iteration, for each of these K partitions, DHCA recursively selects K random centers and continues the clustering process, forming at most K^n partitions at the *nth* stage. This procedure can be used for approximate k-nearest neighbor search. To suit for our purpose, the procedure proceeds until the number of elements in a partition is below $K + 2$, at which time, a pure nearest neighbor search is done among all the data items in that partition. Such a strategy ensures that points that are close to each other in space are likely to be collocated in the same partition.

However, since any data point in a partition is closer to its cluster center than to the center of any other partition (in case the data point is equidistant to two or more centers, the partition to which the data point belongs is a random one among partitions represented by these centers), data points in cluster boundaries may not necessarily be classified to their nearest neighborhood. Fortunately, misclassification possibilities can be greatly reduced by multiple runs of DHCA [21]. To remember k neighbors, in the implementation, there exist two arrays: a distance array, which is actually an array of priority queues of size $k + 1$ used to record the distances of each data point to some other data points in the sequentially stored data set, and an index array, which is actually an array of integer arrays of size k used to record the indices of k neighboring data items corresponding to the data entries in the distance array.

Further, we are more interested in those data points that are potential outlier candidates rather than those whose outlying scores are too low for further consideration. Therefore, before each subsequent DHCA, we compute the mean and the standard deviation of all the outlying scores and use their sum as a threshold. To save some computations and give the potential outlier candidates more attention, only when the current outlying score of a data item is larger than the threshold do we carry out a distance computation to classify it into its closest center. Finally, since we are interested in top n outliers, at the first-stage of each DHCA construction, the K cluster centers are chosen among the top K un-verified potential outlier candidates and the process terminates when the true top n outliers are found. Here and hereafter, by "true" top n outliers, we mean the actual top n outliers that would be found in an exhaustive distance-based comparison across the entire dataset in high-dimensional space.

3.2.3 An Efficiency Analysis of DHCA for Distance-Based Outlier Detection

In this subsection, we present an efficiency analysis of DHCA (i.e., why repeated invocations of the DHCA algorithm can make the distance-based outlier mining process more efficient by decreasing the probability of misclassification). A theoretical explanation together with empirical evidence has been given in [21]. The proof is briefly described here.

Repeated invocations of DHCA can be thought of as a set of independent Bernoulli trials where one keeps running DHCA and classifying each data point to its closest cluster center (randomly selected at each stage of the process), until it succeeds (i.e., it hits its nearest neighbor, or at least, its approximate nearest neighbor). If the probability that a random data point hits its nearest neighbor is denoted by p, and the random variable representing the number of trials needed for a random data point to hit its nearest neighbor is denoted by Y, the probability of obtaining a success on trial y is given by

$$P(Y = y) = q^{y-1}p \tag{3.1}$$

where $q = 1 - p$ denotes the probability that a failure occurs. The relationship between p and $P(Y = y)$ is plotted in the left part of Fig. 3.1. From the figure, we can see that, for a randomized process (i.e., $p = 0.5$), at most 50 DHCAs are needed for most of the data points to meet their nearest neighbor. The percentages of correct hits of DHCA trials on some synthetic and real data are presented in the right part of Fig. 3.1, which confirms our expectation.

In the following, evidence will be given to the fact that the size of clusters obtained at each level of DHCA will decrease exponentially as we descend through the hierarchical tree.

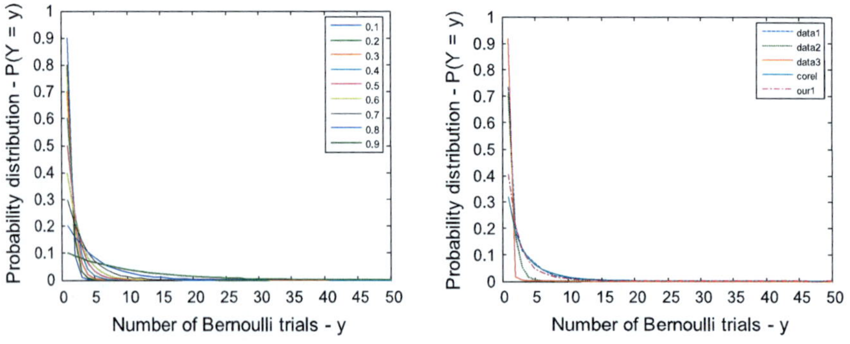

Fig. 3.1 Probability distribution of (left) Bernoulli trials (right) DHCA trials [21]

Theorem 1 Both the geometric size and the number of data points associated with each cluster (or node) generated at each level of the partitioning process will decrease exponentially as we descend through the hierarchical tree produced by the DHCA until the number of associated points is less than a predefined cluster size, upon which the node is made a leaf of the tree.

The proof given next follows a loose notion of induction. First, we indicate that Theorem 1 is true for a two-dimensional quadtree decomposition. Next, we extend it to d-dimensional quadtree decomposition. Finally, we conclude that it is true for the general hierarchical decomposition. More specifically, we first conduct an accurate analysis of a simple but good case performance that can be achieved by the algorithm following a quadtree-based decompositions of data points (uniformly distributed on the regular grids) into subquadrant parts.

Named by Raphael Finkel and J. L. Bentley in 1974, a quadtree is a tree data structure in which each internal node has up to four children and most often used to partition a two-dimensional space by recursively subdividing it into four regions. The regions may be square or rectangular, or may have arbitrary shapes.

Proof Case I: Two-Dimensional QuadTree Decomposition. Suppose there are N two-dimensional data points distributed uniformly on space grid of span $S = s \times s$, where s denotes the length of each of the four sides. At level 0 of the tree, there is one cluster, the root cluster, with totally N data points and space span of size S (i.e., the partition space is of area S). At level 1, there are four clusters, each with $N/4$ data points and space span of $S/4$. At level 2, there are 16 clusters, each with $N/4^2$ data points and space span of $S/4^2$. At level n, there are 4^n clusters, each with $N/4^n$ data points and space span of $S/4^n$.

Case II: d-Dimensional Quadtree Decomposition. For N d-dimensional data points distributed uniformly on space grid of hypercube volume $V = s^d$, where s denotes the length of each side. At level 0 of the tree, there is one cluster, the root cluster, with totally N data points and space span of size V (i.e., the partition space is of volume V). At level 1, there are 2^d clusters, each with $N/2^d$ data points and space span of $\frac{V}{2^d} = \left(\frac{s}{2}\right)^d$. At level 2, there are 2^{2d} clusters, each with $\frac{N}{(2^d)^2}$ data points and space span of $\frac{V}{(2^d)^2} = \left(\frac{s}{2^2}\right)^d$. At level n, there are 2nd clusters, each with $\frac{N}{(2^d)^n}$ data points and space span of $\frac{V}{(2^d)^n} = \left(\frac{s}{2^n}\right)^d$.

Case III: d-Dimensional m-ary Tree Decomposition. A more detailed analysis of the d-Dimensional quadtree decomposition for arbitrary hyperrectangles (given in [22]) demonstrates that, for m-ary tree for d-dimensional data points, there are $\frac{N}{(m)^n}$ data points in each cluster at level n.

Case VI: Nonuniform Distribution with Outliers. Uniform distribution corresponds to a complete tree (leaves always happen at the lowest level). For more real situations where data points are distributed with different densities in different regions (i.e., boundary effects happen) and outliers exist, the partitioning results in an unbalanced tree as illustrated by Fig. 3.2. In other words, each node has a maximum

Fig. 3.2 A quadtree decomposition when outliers exist

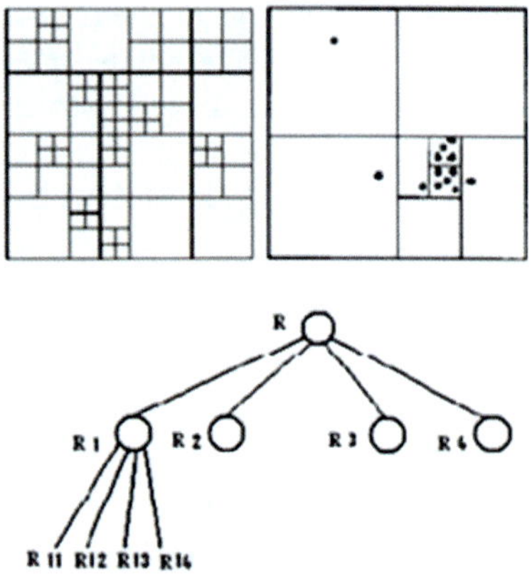

capacity (i.e., the maximum number of data points in each node). When the maximum capacity is reached, the node splits. Put in another way, since an outlier is a data item for which there are fewer than m other data items within node grid of size r (i.e., the original definition of distance-based outliers), leaf nodes can show up at much early stages of the partitioning trees (i.e., before the majority of the leaf nodes happen at the bottom level). It is in this way that the tightness of the upper bounds can be restricted by the grid sizes at the early stages of the partitioning tree. In other words, when outliers exist in the data set, in DHCA partitioning, leaf nodes appear at much early stages of the partitioning process.

In reality, when several groups exist in the data set, boundary effects (separation between densely populated regions) can misclassify some boundary points into wrong clusters (since each data point is closest to its center, its nearest neighbors may be assigned to a different cluster). As a result, choosing different cluster centers in each consecutive DHCA trial will decrease the probability of wrong classifications, particularly when we are only interested in the small number of outliers.

3.3 The Proposed Fast Distance-Based Outlier Detection Algorithm

In this section, we propose an alternative procedure based on the third definition of distance-based outliers. As is described in the following, we can attribute the advantages of our top-down outlier-finding approach to the power of multiple invocations of a modified version of the divisive hierarchical clustering algorithm (DHCA) for a CPU-efficient k-nearest neighbor search procedure.

3.3.1 A Simple Idea

Given a set of d-dimensional data, in general, the user is interested in top n outliers, where n can be expected to be very small and is relatively independent of the majority of the data set, making them easier for the user to locate. In d-dimensional space, there exists a distance between each pair of the data items. For example, since the data are always stored sequentially, a priority queue of size $k + 1$ can be used for each item to remember the distances to its k immediate predecessors or successors. These initial distances, whatever they are, provide an upper bound for the distances of each data item to its k-nearest neighbors.

Since the distance-based outlying scores are a monotonic nonincreasing function of the portion of the dataset already explored [10–12], the goal of finding top n DB-Outliers can be achieved by first quickly finding outlying scores tight enough for all the data items and then repeatedly reducing the outlying scores of the top n ones to their true values until there are no changes, at which time, all DB-Outliers show up.

From the above explanation, it is easy to see that the quality of our strategy (as well as top-down DB-Outlier detection techniques) heavily depends on how to provide tight distance upper bounds efficiently so that a minimal number of distance computations would be enough to finally discover top n outliers. In order to quickly provide the tight distance upper bounds, we propose to run the divisive hierarchical clustering algorithm (DHCA) multiple times [21].

3.3.2 The Proposed CPU-Efficient DB-Outlier Detection Method

Based on the findings presented in the previous sections, our DB-Outlier detection strategy can be developed,

1. Sequentially read the data set in, run DHCA once, then compute the one-dimensional array of the DB-Outlier scores.
2. Sort the one-dimensional score array in a nonincreasing order, compute the threshold (i.e., mean + std), select top K unverified data items as the first-level cluster centers and run DHCA to update the k neighbors for those unverified data items whose outlying score is larger than the threshold.
3. If all true top n outliers show up, i.e., top n candidates in the sorted one-dimensional outlying score array are all verified outliers (i.e., exhausted search has been carried out for these data items), stop. Otherwise, go to Step 2.

3.3.3 Time Complexity Analysis

From the description presented in the previous subsections, our algorithm includes multiple runs of DHCA. Since DHCA is basically a hierarchical partitioning algorithm (i.e., a $O(N\log N)$ algorithm), we expect the repeated invocation of DHCA (i.e., no thresholding involved) to scale as $O(fN\log N)$, where f denotes the number of DHCA constructed before the terminating condition is satisfied. Since in our implementation, at each step of the updating using DHCA, before we assign a data item to a cluster, if its current outlying score is smaller than the threshold, we ignore it, the time complexity is actually $O(f(xN)\log(xN))$, where x is between 0 and 1 and denotes the proportion of the data set that are involved in the calculations of DHCA process.

Further, because the number of desired outliers is much smaller than the dataset size N, for sufficiently small values of x, the time complexity could be near linear on average. Therefore, we expect the algorithm to scale as $O(Ne)$, where e denotes the number of data points checked before top n outliers are discovered, although the worst case time complexity still could be $O(N^2)$.

3.3.4 Data Structure for Implementing DHCA

To organize the nodes generated during the process of the DHCA, a C++ data structure called Node is used [21]. The member variables of the Node class include an array of integers remembering the indices of the subset being clustered into it from its parent level, an integer array of size K remembering the indices of the cluster centers chosen from its own set for its descendants. The member function is one that operates on the member variables to generate at most K new nodes as the result of the partitioning.

The clustering process creates a Node array which has only one element (the top Node) at the beginning. This top Node has every data item in the data set as its samples. Subsequently, with the new Nodes being generated on the fly and pushed to the back of the Node array, they will be processed in order until no new Nodes are generated and the end of the Node array is reached. Totally, the DHCA distance update procedure as shown in Table 3.4 can be embedded into the C++ Node class which is shown in Table 3.3.

3.4 Scale to Very Large Databases with I/O Efficiency

Our method proposed in the previous section is efficient when the dataset can wholly reside in the main memory. More often than not, the datasets are too large to fit in, and the minimization of I/O cost becomes a major concern in any algorithm design. Our objective in this section is to develop a strategy that can adapt our proposed

Table 3.3 Node class

Name	Explanation
Public data members:	
sampleNumbers:	An array holding the indices of all samples in the cluster
centroidSeeds:	An array holding the randomly chosen cluster centers
Public methods:	
void DHCA	Our modified DHCA

approach for the identification of top n outliers with I/O overhead being near linear to the number of database scans. As a straightforward adaptation, our strategy is to apply our method to each manageable part of the whole database that is loaded sequentially into the main memory.

As a direct solution to reduce the I/O cost, our adapted method uses the first scan through a very large database to find for each data point its k approximate nearest neighbors, and to locate top $m \geq n$ potential outlier candidates. During the second scan through the whole database, verification can be performed for each of the m candidates so as to finally mine top n verified ones.

To describe our methods, suppose, in addition to having enough space for holding 2H data items, the memory can also accommodate the necessary distance and index arrays for the kNN of the 2H data items. Here 2H refers to the maximum in-memory page size left for holding data items. If the data set has N data items, there need $M = N/\mathrm{H} - 1$ times of data loading during the first scan. More specifically, at the beginning, we load 2H data items from the database into the main memory, perform the DHCA and identify the regional top $m \geq n$ outliers from the data items in the memory so far. Next, we sort these 2H outlying scores and retain in the main memory those data items whose outlying scores are among the top H ones (and delete the rest H data items and their corresponding neighbors). After that, we load another H data items into the memory, perform multiple runs of DHCA on the 2H data items currently in the memory, identify the top m potential outliers from these data items, sort the 2H outlying scores and finally retain top H data items. These steps will be repeated until the end of the database is reached.

From the procedure just described, it can be observed that top H data items that are retrained in the memory at the end of the first scan through the database will meet the rest of the database during the second scan. Therefore, as long as these retained top H data items include all the true top n outliers, the second scan will fulfill the outlier detection task and output the desired outliers.

Then, the question is how large H should be for the true top n global outliers to be included in the retained H data items at the end of first scan. To estimate H, the first step is to quantify the number of data points which do not have their k-nearest neighbors loaded with them at the same time (i.e., regional potential outlier candidates). To do so, we follow a similar procedure presented in [15]. To formulize it, n_{oi} is used to denote the number of data points from all subsets of 2H data items,

Table 3.4 DHCA member function

Procedure Name:		DHCA			
Input:					
dist_knn, edge_knn		the arrays to remember kNN for each data item			
kNN		the number of NNs of a data item			
nodeArray		an array of the Node structures			
currentNode		the current *Node* of the *Node* array			
k		the number of clusters at each step			
data		the input data set			
maxclustersize		the maximum size of each cluster			
threshold		the value used to filter			
Output:					
updated dist_knn, edge_knn					
newly generated $\leqq k$ Nodes pushed back to the back of *nodeArray*					
Begin					
	randomly selected k centers from *sampleNumbers* of *currentNode*				
	generate k new Nodes				
	for each sample i in sampleNumbers of currentNode that is not a center {				
	{				
		find its nearest center j out of k;}			
		if (dist_knn[i].max) > distance(i, j) {			
		{			
			update dist_knn, edge_knn;}		
		}			
		if (dist_knn[i].average_distance) > threshold) {			
		{			
			assign *sampleNumbers*[i] to groups of center j;}}		
		}			
	}				
	for each new Node j = 1 to k {				
	{				
		if (newNode[j].*sampleNumbers*.size() > maxclustersize) {			
			push newNode[j] to the end of nodeArray;}		
	}				
End					

when loaded into the main memory, whose average distance to their k approximate nearest neighbors met so far do not exceed the cutoff threshold (i.e., the minimum outlying score among the true top n outliers). Further, since any 2H data points residing in the main memory are obtained from the database following the sampling-with-replacement scheme, there are totally N^{2H} possible patterns to reside in the main memory. Let us denote them as $T_1, T_2, \cdots, T_{N^{2H}}$, respectively.

Then, we construct a two-dimensional array with N rows and N^{2H} columns, where the i-th row concerns data point o_i and the j-th column corresponds to T_j. In each cell c_{ij} at the i-th row and the j-th column, we fill in '0' if the average distance of o_i to its k neighbors met so far is less than the cut-off threshold (i.e., the minimum outlying score among the true top n outliers); otherwise, we fill in '1'.

If we add up the cell values at the j-th column, the sum denoted by col_j equals the number of potential outlier candidates that should be retained in memory for the second scan through the database to verify. Hence, the expected number of such outlier candidates (given an arbitrary 2H data subset) is the average sum of all columns:

$$\frac{1}{N^{2H}} \sum_{j=1}^{N^{2H}} \mathrm{col}_j \tag{3.2}$$

Note that $\sum_{j=1}^{N^{2H}} \mathrm{col}_j$ in the above formula is exactly the number of $1'$s in the array. To find out how much it is, we count the $1'$s in an alternative "row-oriented" manner. Let row_i be the number of the $1'$s at the i-th row. Clearly, row_i is the number of subsets of size 2H in which the average distance of data point o_i to its regional k-nearest neighbors met so far (which is either the same as or larger than the average distance of o_i to its true k-nearest neighbors among the whole set) is larger than the cutoff threshold (i.e., the minimum outlying score among the true top n outliers). On the other hand, since no more than (N-n_{oi}) data points in the whole dataset are further away from o_i than the cut-off threshold (i.e., the minimum outlying score among the true top n outliers), there are totally no more than $\left(N - n_{o_i}\right)^{2H}$ different number of regional potential outlier candidates. Therefore, $\sum_{i=1}^{N} \mathrm{row}_i = \sum_{i=1}^{N} \left(N - n_{o_i}\right)^{2H}$ is also the number of $1'$s in the array. As a result, the expected number of outlier candidates produced using our strategy is,

$$\frac{1}{N^{2H}} \sum_{j=1}^{N^{2H}} \mathrm{col}_j = \sum_{i=1}^{N} \frac{\left(N - n_{o_i}\right)^{2H}}{N^{2H}} = \sum_{i=1}^{N} \left(1 - \frac{n_{o_i}}{N}\right)^{2H} \tag{3.3}$$

Clearly, the chance that an inlier is mistakenly regarded as an outlier due to memory space restriction decreases exponentially with 2H (i.e., with the increase of memory space). Put in another way, for any non-outlier object o_i, as long as $\left(1 - n_{o_i}/N\right)$ is a non-trivial selectivity, $\left(1 - n_{o_i}/N\right)^{2H}$ evaluates to a negligible value. For example, if 2H is of the order of 1000 and $\left(1 - n_{o_i}/N\right) = 0.99$,

$(1 - n_{o_i}/N)^{2H}$ becomes less than 5 × 1.0^{-5}. While if 2H is of the order of 10,000 and $(1 - n_{o_i}/N) = 0.99$, $(1 - n_{o_i}/N)^{2H}$ becomes less than 3 × 1.0^{-44}. The use of these facts can be formalized by the following two assumptions: (1) the number of outliers, n, is much smaller than the size of memory, (2) the number of dense clusters in the database is much smaller than the size of memory.

With H's selection being solved, the next question is how to set a value for top m outlier candidates so that there is no need for a third scan. Our strategy is to sacrifice a little more distance computations for I/O efficiency. In other words, we propose to retain $m = Mn$ for each 2H-data block's in memory process. To summarize, at the beginning, we load 2H data items from the database into the main memory and use our CPU efficient outlier detection method to identify the top Mn regional outliers from them. Next, we sort the 2H outlying scores and retain in the main memory those data items whose outlying scores are among top m. Then, we load another H data items from the database and perform the same set of operations on the 2H data items currently in the main memory. This process proceeds until the end of the database is reached and top H potential outlier candidates are kept for the second scan of the database to verify and to finally locate the true top n outliers.

3.5 Performance Evaluation

In this section, we present the results of experiments performed to evaluate the efficiency of our fast DB-Outlier detection algorithms. First of all, we describe the characteristics of the data sets employed in our experiments. Next, we analyze the sensitivity of our CPU-efficient method to the input parameter K to the DHCA. Our method is then compared with state-of-the-art DB-Outlier detection algorithms. After that, we study the effectiveness of multiple runs of DHCA on the nearest neighbor search, which is followed by a discussion on how the curse of dimensionality affects the performance of the algorithms. Finally, the sensitivity to the buffer size of our method (adapted for I/O efficiency) is studied.

All the algorithms are implemented in C++. All the experiments were performed on a computer with Intel Core 2 Duo Processor E6550 2.33 GHz CPU and 2 GB RAM. The operating system running on this computer is Ubuntu Linux. In all the experiments, we use the total execution time in seconds as the performance metric (the timer utilities defined in the C standard library are employed to report the CPU time), and the total execution time accounts for all the phases of our DB-Outlier detection algorithm, including those spent on the DHCAs and the rest. Each result we show was obtained as the average over 10 runs of the program for each data set. The results show the superiority of our DHCA-based algorithm over ORCA, RBRP and SNIF. We would like to indicate here that the implementations of the comparing algorithms (i.e., RBRP and SNIF) are all obtained from the original authors and that all the DB-Outliers mined using our algorithm are exact for all the in-memory algorithms as well as the I/O-efficient versions.

Table 3.5 Sets of data

Data name	Data size	# of dimension
outData	67,798	10,041
Corel	68,040	32
IPUMS	88,443	61
Covertype	581,012	55
Poker	1,000,000	11
USCensus	2,458,285	68
KDDCup	4,898,430	42

3.5.1 Data Characteristics

Table 3.5 summarizes all the datasets used in the experiments. All of them but the first one are downloaded from the UCI KDD Archive [23].

Corel Histogram consists of examples that encode the color histogram of images from a Corel image collection. Covertype represents the type of forest coverings in the Rocky Mountain region, and each example contains attributes such as the dominant tree species, the elevation, the slope and the soil type of the region, and so on. KDDCUP 1999 consists of network connections to a military computer network where there have been intrusions by unauthorized users. The IPUMS data set contains unweighted PUMS census data from the Los Angeles and Long Beach areas for the years 1970, 1980, and 1990, respectively. Each record of the "Poker Hand" (or simply, the "Poker") data is an example of a hand consisting of five playing cards drawn from a standard deck of 52. The USCensus1990 data set was obtained from the USCensus1990 raw data set and has 68 attributes. Our working data (denoted by ourData) contains 65,798 highly sparsed small-valued feature vectors, which are extracted from 20 color images captured along a hallway and each of which contains a 10,000-dimensional color histogram and 41 more texture measures (the value in each dimension is in the range 0–1) [21]. All the categorical features in these data sets, if existing, have been casted to integer values.

3.5.2 The Impact of Input K on Running Time

In this subsection, the sensitivity of the proposed outlier detection algorithm to the parameter K (the number of partition centers at each stage of the DHCA) is studied. Figure 3.3 shows the total running time used to mine top 50 outliers with k (the number of nearest neighbors) being set to 30 on all the data sets as we vary K from 2 to 30.

From the figure, it can be seen that, overall, for large K, our algorithm incurs larger number of distance computations and the running time increases with K. However, as K increases, the running time actually decreases at the beginning and

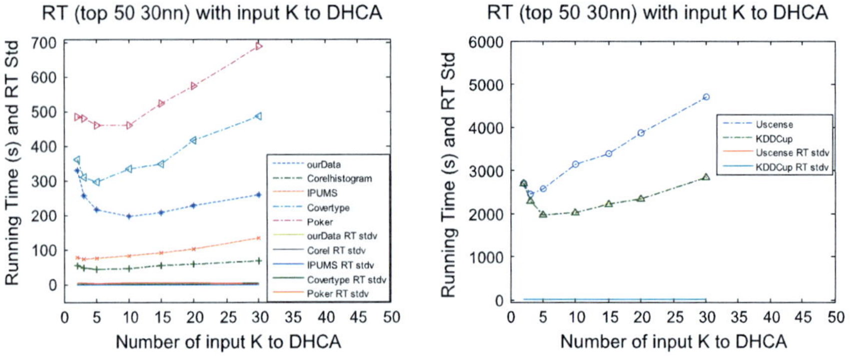

Fig. 3.3 Run time (mean and the corresponding standard deviation) variations with input K to DHCA (left) for the first five data sets, (right) for the rest two data sets

then increases with larger K's on the whole. This is because, when K is small, the overhead of constructing the DHCA exceeds the small increases in the number of distance computations. In other words, for small K's, the total number of Nodes in the whole execution of DHCA may decrease a little with small increases in K, which makes our algorithm have even better performance. This phenomenon can be seen in the data when K increases from 2 to 10. When K gets larger and larger, more and more distance computations to the partitioning centers need to be done and the increases in distance computations eventually dominate. Similar behavior has been observed for other values of kNNs (e.g., $k = 2, 10, 20, \ldots$).

3.5.3 Comparison with Other Methods

In this subsection, we show the comparison of our CPU-efficient algorithm with two state-of-the-art in-memory DB-Outlier detection algorithms, namely ORCA and RBRP. The comparisons are done through the scalability analysis of two aspects: the scalability with k (the number of nearest neighbors used in the calculation of the outlying scores) and the scalability with the sizes of the data sets. The parameters of the three algorithms are set to achieve their best performances. Figure 3.4 through Fig. 3.11 show the total running time used to mine top 50 outliers with k (the number of nearest neighbors) set to 10, 20, and 30 on all the data sets (with the K set to 5 in our algorithm).

The ourData dataset has 65,798 data items all with dimension of 10,041 which is equivalent to $65{,}798 \times 10{,}041 = 6.6 \times 10^8$ float numbers and cannot be stored in memory in our computer. Since it is highly sparse and therefore, sparse coding is used to only remember the nonzero features. As a result, the experiments are feasible only with our method and ORCA (This is also true for Figs. 3.11, 3.12, and 3.20 in the following). In Fig. 3.4, the solid blue line denotes the run time performance of the

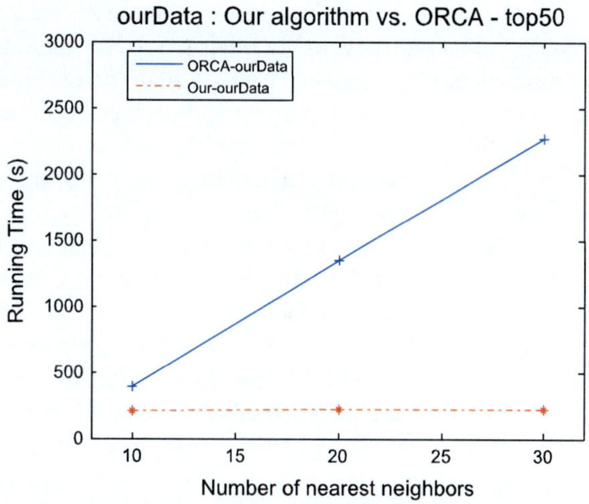

Fig. 3.4 Run time performance of our algorithm in comparison with ORCA on our Data

ORCA method while the dashdot red line represents the run time performance of the proposes method. It can be seen that the proposed method significantly outperforms the ORCA method.

In Fig. 3.5 through Fig. 3.10, the run time performances on Corel Histogram, Covertype, IPUMS, Poker, USCensus and KDDCup data of the proposed method, denoted by the dashdot red line, are shown in comparison with those of the ORCA and RBRP methods, denoted by the solid blue line and dash magenta line, respectively. From the figures, it can be seen that our algorithm outperforms ORCA on all the datasets. As to the comparison with RBRP, we can see that our algorithm obviously does a better job than RBRP for Corel Histogram, Covertype, and KDDCup but has quite a similar performance for IPUMS, Poker and USCensus. This phenomenon is due to the fact that data points in IPUMS, Poker and USCensus are distributed in very

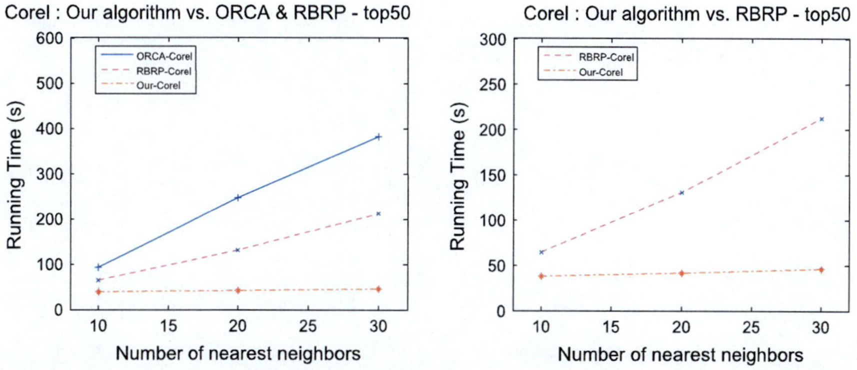

Fig. 3.5 Run time performances of our algorithm on Corel Histogram data in comparison with (left) ORCA and RBRP (right) RBRP

dense clusters and, therefore, more and more distance computations need to be done for the outliers to show up. In other words, for these kinds of datasets, the distance computations consumed in the most CPU-efficient bottom-up approach, RBRP, meet those consumed in our top-down approach and eventually dominate (Figs. 6, 7, 8, 9 and 10).

In Fig. 3.11, we summarize the relative performances of the ORCA method and our method by showing their run time ratios, that is, the run times of ORCA over those of our algorithm. From the figure, we can see that the proposed method can outperform the ORCA by a factor of 100 on the Covertype data and by a factor of 5 to 10 on some of the rest datasets.

Figure 3.12 through Fig. 3.18 report the CPU execution times of ORCA, RBRP and our algorithm (with the parameter K set to 5) on all the dataset to mine top 50 outliers with the parameter k set to 30, while the size of the dataset is varied between 10% and 100% of the whole data size. In the figures, the performances

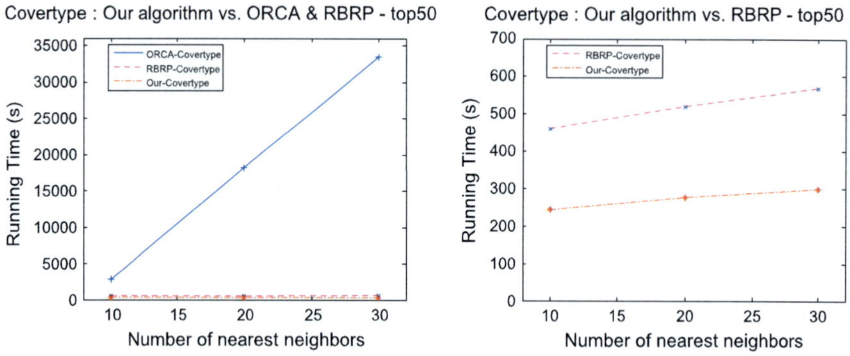

Fig. 3.6 Run time performance of our algorithm on Covertype data in comparison with (left) ORCA and RBRP (right) RBRP

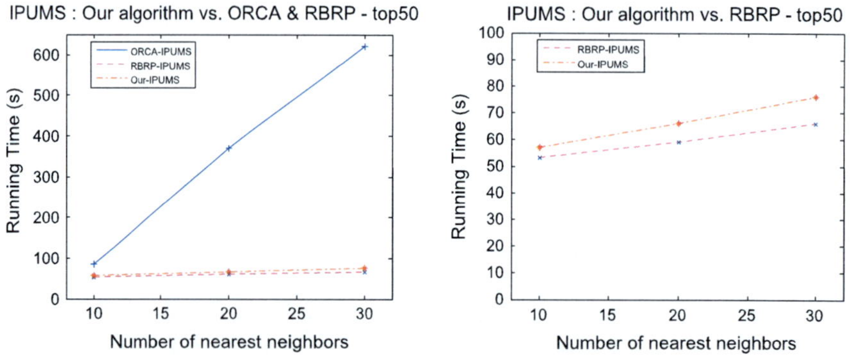

Fig. 3.7 Run time performance of our algorithm on IPUMS data in comparison with (left) ORCA and RBRP (right) RBRP

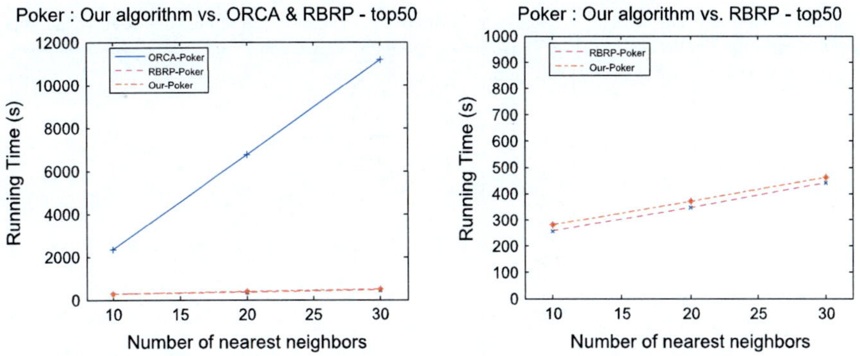

Fig. 3.8 Run time performance of our algorithm on Poker data in comparison with (left) ORCA and RBRP (right) RBRP

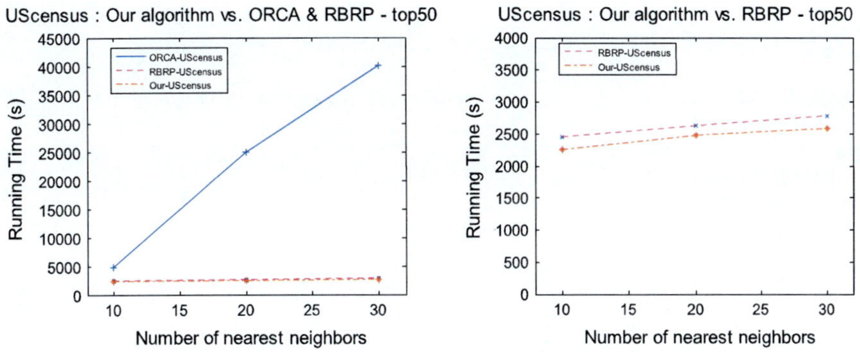

Fig. 3.9 Run time performance of our algorithm on USCensus data in comparison with (left) ORCA and RBRP (right) RBRP

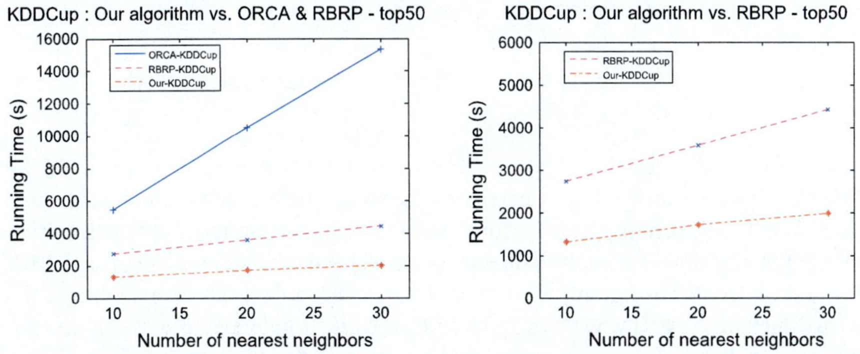

Fig. 3.10 . Run time performance of our algorithm on KDDCup data in comparison with (left) ORCA and RBRP, (right) RBRP

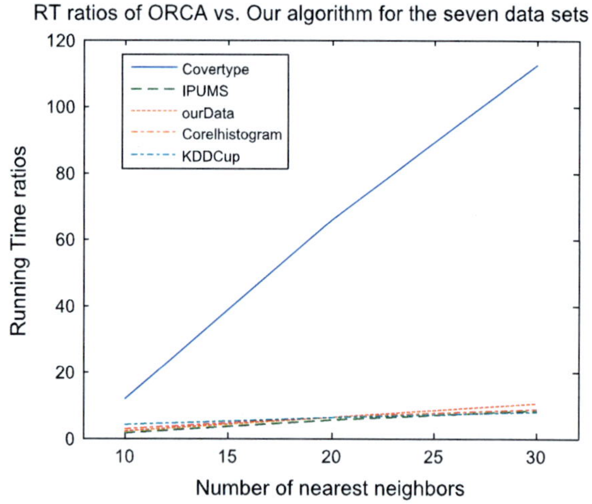

Fig. 3.11 Run time ratios of our algorithm over ORCA for five datasets

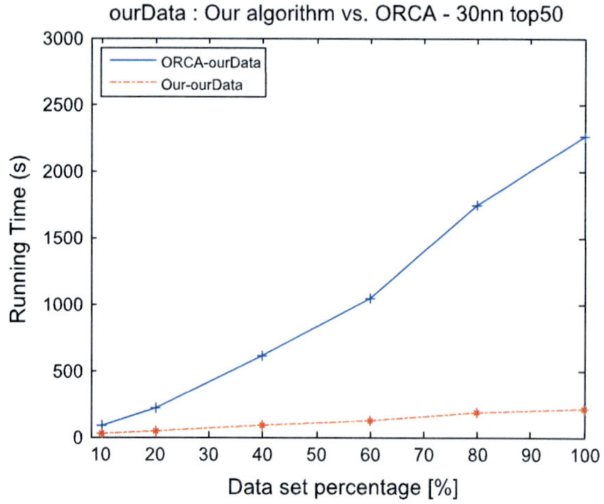

Fig. 3.12 Run time performance of our algorithm on ourData data with varied data sizes in comparison with ORCA

of the proposed method are denoted by the dashdot red line, while those of the ORCA and RBRP methods are denoted by the solid blue line and dash magenta line, respectively. The running time performance of our algorithm remains superior over the ORCA algorithm for all the data sets we have tested while is better than the RBRP method on Corel Histogram, Covertype and KDDCup data but comparable to the RBRP method on IPUMS, Poker, and USCensus data. Interestingly, it can be seen that our algorithm scales more linearly with respect of the dataset sizes (Figs. 13, 14, 15, 16, 17, and 18).

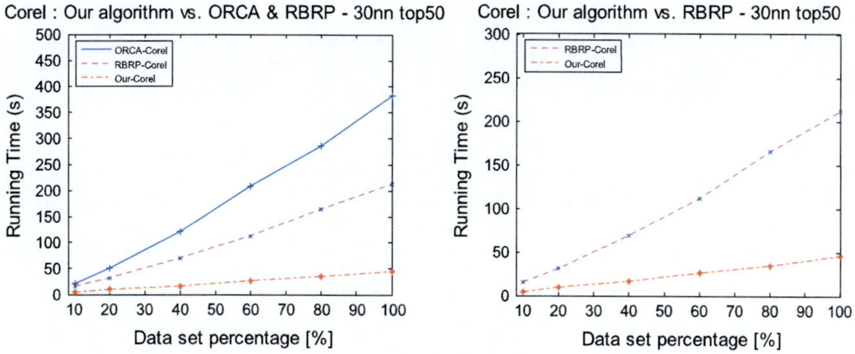

Fig. 3.13 Run time performance of our algorithm on Corel Histogram data with varied data sizes in comparison with (left) ORCA and RBRP (right) RBRP

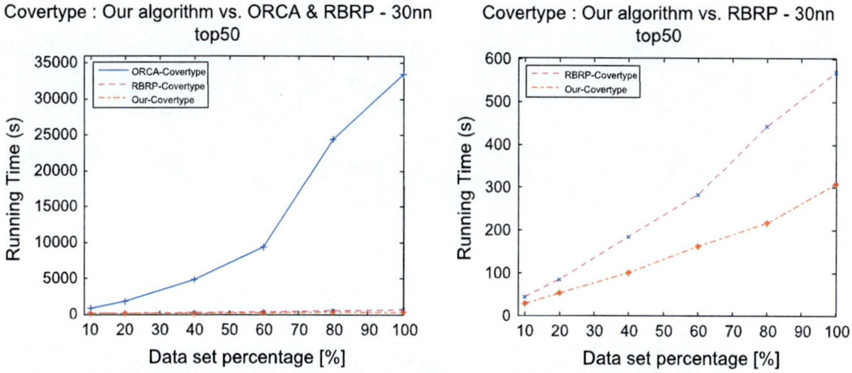

Fig. 3.14 Run time performance of our algorithm on Covertype data with varied data sizes in comparison with (left) ORCA and RBRP (right) RBRP

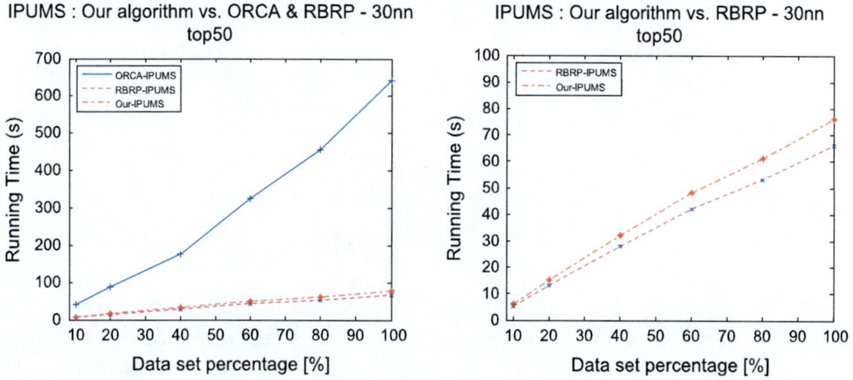

Fig. 3.15 Run time performance of our algorithm on IPUMS data with varied data sizes in comparison with (left) ORCA and RBRP (right) RBRP

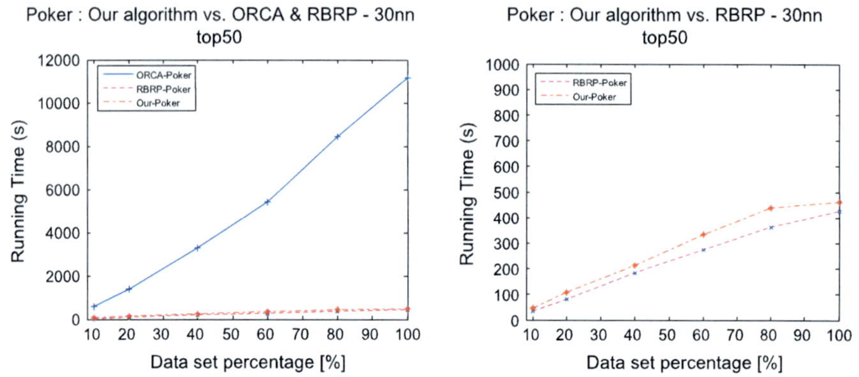

Fig. 3.16 Run time performance of our algorithm on Poker data with varied data sizes in comparison with (left) ORCA and RBRP, (right) RBRP

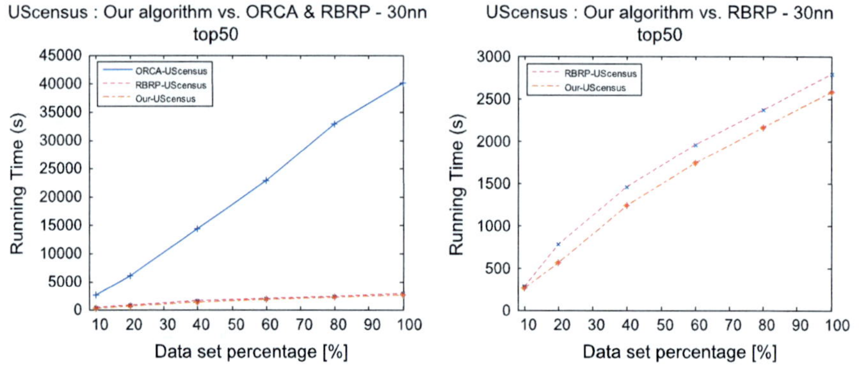

Fig. 3.17 Run time performance of our algorithm on UScensus data with varied data sizes in comparison with (left) ORCA and RBRP, (right) RBRP

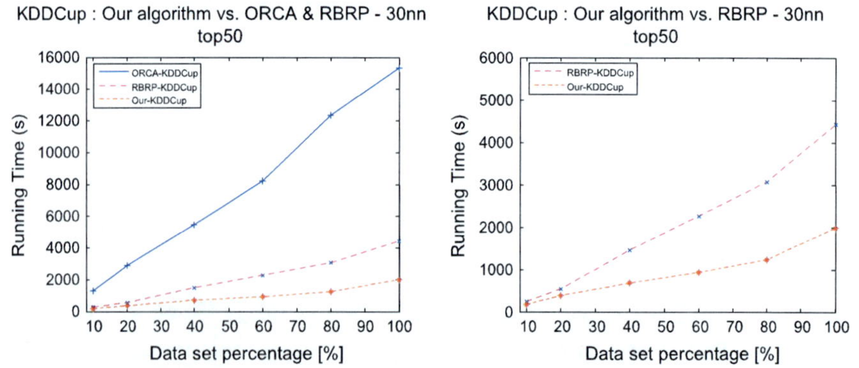

Fig. 3.18 Run time performance of our algorithm on KDDCup data with varied data sizes in comparison with (left) ORCA and RBRP (right) RBRP

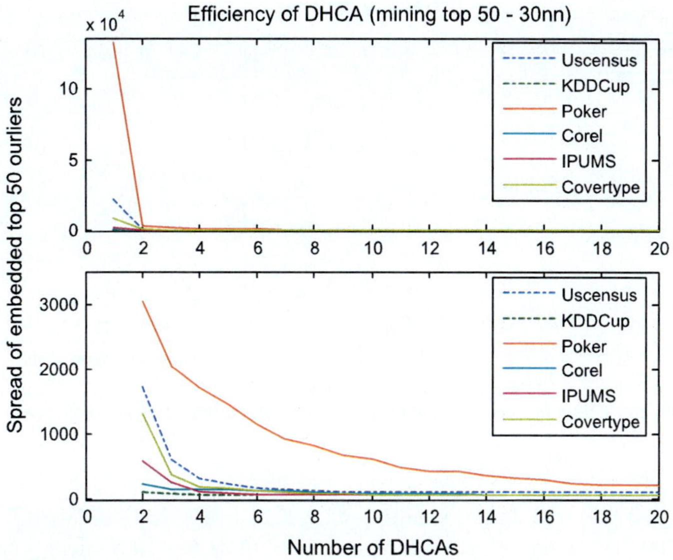

Fig. 3.19 Reduction to the core set with increasing number of DHCA trials

3.5.4 Effectiveness of DHCA for kNN *Search*

From the previous sections, we can see that our algorithm has good performances. To further test the efficiency of multiple runs of DHCA, we run the DHCA ($K = 5$) 20 times consecutively and count the minimum number of data points in which the true top 50 outliers ($k = 30$) are embedded. Of course, this is a number larger than 50 since it also includes false-positive data points. The results are shown in Fig. 3.19 for the six data sets (i.e., Corel Histogram, IPUMS, Covertype, Poker, USCensus and KDDCup). From the figure, we can see that the number of data points which embed the true top outliers (i.e., the core set) is decreasing with the increasing number of DHCA trials and converges to 50 eventually. The results agree with our intuition and successfully provide an empirical evidence for the effectiveness of DHCA.

3.5.5 The Impact of Curse of Dimensionality

It is generally believed that the performances of similarity search methods degrade as the dimensionality of the data increases. In this section, we analyze how the performance of our method is affected by the curse of dimensionality. In order to test the impact of the increasing dimensionality on the overall execution time of ORCA and our algorithm, six data have been made, each consisting 65,798 highly sparsed small-valued feature vectors (i.e., color histograms, which are extracted from 20

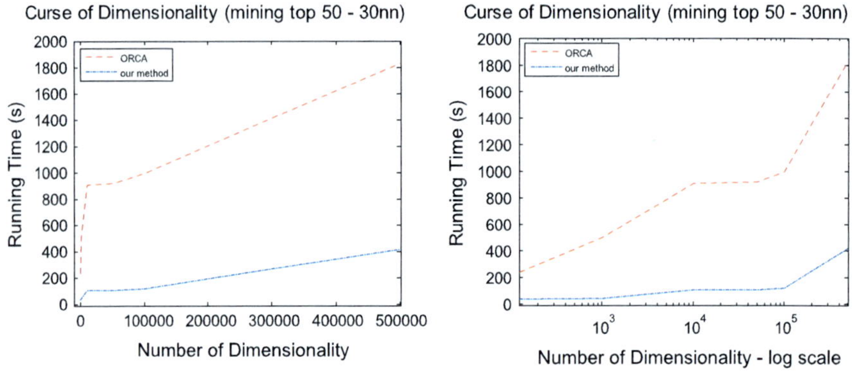

Fig. 3.20 Running time performances to mine top 50 30nn outliers in (left) linear scale, (right) log scale

color images) with the dimensionality being 125, 1,000, 10,000, 50,000, 100,000, and 500,000. The running time performances to mine top 50 outliers with k set to 30 are presented in Fig. 3.20. There are two graphs in the figure. These two graphs have two lines. The upper line denoted by the dash red line represents the execution time to find top 50 outliers with k set to 30 by ORCA. The lower line denoted by the dashdot blue line represents the running time of our algorithm. Clearly, our method has a better scalability with dimensionality for these data sets than ORCA.

3.5.6 Scale to Very Large Databases with I/O Efficiency

To test the I/O scaling performance of our methods on four data sets with size larger than 10^5 (since today's main memory can afford Corel Histogram, ourData and IPUMS data easily and our CPU efficient algorithm can handle them quite well, they are not used in this comparison). Without loss of generality, the default memory size is set to equal 10% of the space occupied by the underlying database (i.e., $M = N/(2H) = 10$), assuming that our memory space is constrained and pretending that the algorithm is only aware of 2H data points at each time during loading the databases.

First of all, for each 2H data block of data sets, we try to mine regional top $m = n, 2n, \ldots, Mn$ to see what kind of final top n outliers we can get. We find that, for USCensus and KDDCup, mining regional top n outliers is enough to get all correct top n outliers at the end of second scan. However, for Covertype and Poker, we have to compute more. Summarized in Table 3.6 are the number of regional top outliers (in term of the number of n's) that have to be mined to get the final global top n outliers correctly for these four data sets.

These results are the best we can get using our I/O efficient algorithm. It can be seen that different data sets have different disk-storing behavior. For USCensus

Table 3.6 Number of regional top Mn for each 2H data block

	$k = 10$ nn	$k = 20$ nn	$k = 30$ nn
USCensus	$m = 1$ n	$m = 1$ n	$m = 1$ n
KDDCup	$m = 1$ n	$m = 1$ n	$m = 1$ n
Poker	$m = 1$ n	$m = 1$ n	$m = 2$ n
Covertype	$m = 2$ n	$m = 3$ n	$m = 8$ n

and KDDCup, most nonoutlier data in these data sets are stored with their nearest neighbors on the same part of disk, while, for Covertype and Poker, the situation is a little more complicated.

Finally, the running time for mining n outliers ($n = 50$, $k = 10$ and $K = 5$ for this set of the experiments) in different data sizes (and $M_{\%} = N_{\%}/(2\text{H})$) and its comparison with SNIF method (corresponding parameters for SNIF are extracted from the results of our CPU efficient method with $n = 50$, $k = 10$ and $K = 5$ for this set of the experiments) are shown in Figs. 3.21 and 3.22, in which the black

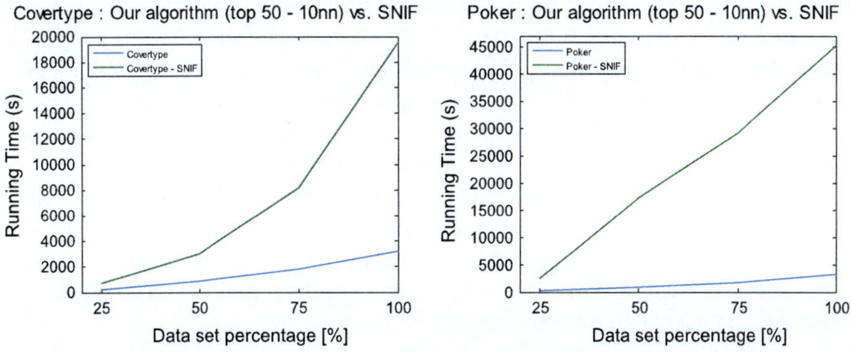

Fig. 3.21 Scaling analysis of SNIF and our method for (left) Covertype data and (right) Poker data

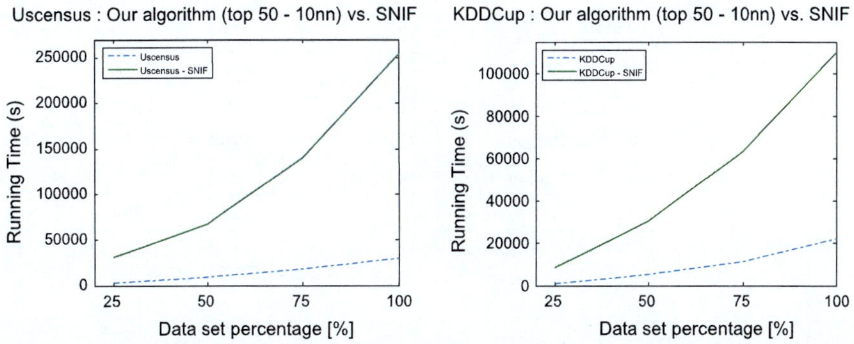

Fig. 3.22 Scaling analysis of SNIF and our method for (left) USCensus data and (right) KDDCup data

lines represent the results for the SNIF method, while the blue lines in Fig. 3.21 and Fig. 3.22 represent the result of the proposed method. From the figures, we can see that our method can outperform SNIF for all the four data sets.

3.5.7 Discussion

From the experimental results related to the comparison with RBRP, we can see that our algorithm obviously does not do much of a better job than RBRP. However, as mentioned in Sect. 3.2 that, for the RBRP method, intermediate files are generated during the first phase of the outlier mining process, which are then used to facilitate outlier detection in the second phase. In Table 3.7, the typical sizes of these intermediate bin files for these data sets are listed. It can be observed that they are even larger than the original data set sizes. Therefore, it is fair to say our method performs better than RBRP if resources are constrained. Further, compared to RBRP method, our method does not involve complex calculation (such as principal component projection) except distance computations. By using an index array in our implementation to remember the indices of each data item's kNN, our method can be easily extended to density-based outlier detection techniques. Although not considered in this chapter, some indexing structures [24, 25] can be employed to save some computation in the verification process of each outlier candidate. And we use sparse coding to efficiently code the highly sparsed ourData so as for it to reside in memory.

Finally, we would like to point out that, in this chapter, we do not compare our algorithm with the DOLPHIN algorithm presented in [12]. However, we have compared ours with the RBRP method, which was not done in [12]. Further, the largest size of the data sets used in [12] is 10^6 while larger-sized datasets (i.e., USCensus and KDDCup) are used in our experiments.

Table 3.7 Sizes of original datasets versus intermediate bin files from RBRP

Data name	Data size (MB)	# of bin	Total size of bins(MB)
Corel	19.5	2	24.5
IPUMS	13.5	402	36.9
Covertype	71.7	4398	121.9
Poker	23.4	5298	65.4
USCensus	344.6	4698	647.1
KDDCup	655.5	411	784.8

3.6 Conclusions

In this chapter, an efficient distance-based outlier detection method together with its adaptation for disk-resident datasets has been presented. The I/O cost of the algorithm is very low, corresponding to the cost of sequentially reading the input dataset file twice. Both theoretical justification and empirical validation of the efficiency of the methods have been conducted. An efficiency illustration of the DHCA algorithm for k approximate nearest neighbor search has been presented. It has been shown that a small number of consecutive runs of DHCA can detect the top outliers, which are only a small fraction of the overall dataset. As far as the temporal cost of the algorithm is concerned, it has been shown that our method has near linear-time performances with the dataset sizes, which is better than those of the state-of-the-art algorithms and is more efficient than competitors, thus confirming the qualitative analysis. In order to deal with the scenario in which the available memory is smaller than the standard requirements of the storage of large high-dimensional data sets, a modification of the basic schema of our algorithms has been introduced. The experimental results have shown that the increase in execution time is reasonable and our method behaves better than competing methods.

To summarize, our contributions include:

- Our algorithm has been shown to be very fast and able to efficiently handle enormous disk-resident collections of data.
- Our method does not generate intermediate files in the outlier detection process.
- Our algorithm does not require any complex calculation except the distance computation.
- Our method can be easily extended to density-based outlier detection techniques by remembering kNN for each data item in the implementation.
- Our method uses sparse coding (by only remembering those nonzero features) to efficiently code some highly sparsed datasets so as for them to reside in memory, which could not happen using normal coding.

In our future work, a lot of aspects concerning the algorithm are worth further exploration. For example, in this work, we simply return top 50 outliers, but there is no reason to assume that this must be the case. In the future, we will further study the properties of the existing outlier definitions and come up with reasonable termination conditions for finding distance-based outliers.

References

1. Hawkins, D. M. (1980). *Identification of Outliers*. London: Chapman and Hall.
2. Eskin, E., Arnold, A., Prerau, M., Portnoy, L. & Stolfo, S. (2002). A geometric framework for unsupervised anomaly detection: Detecting intrusions in unlabeled data. In: Barbará, D., Jajodia, S. (eds) *Applications of Data Mining in Computer Security. Advances in Information Security*, vol. 6, pp. 77–101.

3. Lane, T. & Brodley, C.E. (1998). Temporal sequence learning and data reduction for anomaly detection. In *Proceedings of the 1998 5th ACM Conference on Computer and Communications Security* (CCS-5), San Francisco, CA, USA, pp. 150–158.
4. Bolton, R. J., & David, J. H. (2002). Unsupervised profiling methods for fraud detection. *Statistical Science, 17*(3), 235–255.
5. Wong, W., Moore, A., Cooper, G. & Wagner, M. (2002). Rule-based anomaly pattern detection for detecting disease outbreaks. In *Proceedings of the 18th National Conference on Artificial Intelligence*, Edmonton, Alta., Canada, pp. 217–223.
6. Sheng, B., Li, Q., Mao, W. & Jin, W. (2007). Outlier detection in sensor networks. In *Proceedings of ACM International Symposium on Mobile Ad Hoc Networking and Computing*, pp. 219–228.
7. Hodge, V. J., & Austin, J. (2004). A survey of outlier detection methodologies. *Artificial Intelligence Review, 22*(2), 85–126.
8. Chandola,V., Banerjee, A. & Kumar, V. (2009). Anomaly detection: A survey. *ACM Computing Surveys*, 41(3): 15.1–15.58.
9. Knorr, E. M. & Ng, R. T. (1998). Algorithms for mining distance-based outliers in large datasets. In *Proceedings of the International Conference on Very Large Data Bases* (VLDB'98), New York, pp. 392–403.
10. Angiulli, F., & Pizzuti, C. (2005). Outlier mining in large high-dimensional data sets. *IEEE Transactions on Knowledge and Data Engineering, 17*(2), 203–215.
11. Angiulli, F., Basta, S., & Pizzuti, C. (2006). Distance-based detection and prediction of outliers. *IEEE Transactions on Knowledge and Data Engineering, 18*(2), 145–160.
12. Angiulli, F., & Fassetti, F. (2009). DOLPHIN: An efficient algorithm for mining distance-based outliers in very large datasets. *ACM Transactions on Knowledge Discovery from Data, 3*(1), 1–57.
13. Bay, S. D. & Schwabacher, M. (2003). Mining distance-based outliers in near linear time with randomization and a simple pruning rule. In *Proceedings of the 9th ACM SIGKDD International Conference on Knowledge Discovery and Data Mining* (KDD '03), Washington, DC, United states, pp. 29–38.
14. Ghoting, A., Parthasarathy, S., & Otey, M. E. (2006). Fast mining of distance-based outliers in high-dimensional datasets. *Data Mining and Knowledge Discovery, 16*(3), 349–364.
15. Tao, Y., Xiao, X. & Al, E. (2006). Mining distance-based outliers from large databases in any metric space. In *Proceedings of the 12th ACM SIGKDD International Conference on Knowledge Discovery and Data Mining* (KDD '06). Philadelphia, PA, United states, pp. 394–403.
16. Wang, X., Wang, X. L. & Wilkes, D. M. (2008). A fast distance-based outlier detection technique. In *Poster Proceedings of the 8th Industrial Conference on Data Mining*, Leipzig, Germany, pp. 25–44.
17. Knorr, E. M. & Ng, R. T. (1999). Finding intensional knowledge of distance-based outliers. In *Proceedings of the 25th International Conference on Very Large Data Bases* (VLDB'99), Edinburgh, Scotland, pp. 211–222.
18. Knorr, E. M., Ng, R. T., & Tucakov, V. (2000). Distance-based outliers: Algorithms and applications. *The VLDB Journal, 8*(3–4), 237–253.
19. Ramaswamy, S., Rastogi, R. & Shim, K. (2000). Efficient algorithms for mining outliers from large data sets. In *Proceedings of the 2000 ACM International Conference on Management of Data* (SIGMOD'00), Dallas, pp. 427–438.
20. Angiulli, F. and Pizzuti, C. (2002). Fast outlier detection in high dimensional spaces. In *Proceedings of the 6th European Conference on Principles of Data Mining and Knowledge Discovery* (PKDD'02), Helsinki, pp. 15–26.
21. Wang, X., Wang, X. L., & Wilkes, D. M. (2009). A divide-and-conquer approach for minimum spanning tree-based clustering. *IEEE Transactions on Knowledge and Data Engineering, 21*(7), 945–958.
22. Palmer, C. R., Gibbons, P. B. & Faloutsos, C. (2002). Fast approximation of the "neighborhood" function for massive graphs. In *Proceedings of the 8th ACM SIGKDD International Conference on Knowledge Discovery and Data Mining* (KDD'02), Edmonton, Alta, Canada, pp. 81–90.

23. UCI: The UCI KDD Archive. (https://kdd.ics.uci.edu/). Irvine, CA: University of California, Department of Information and Computer Science.
24. Yu, C., Ooi, B. C., Tan, K. L. & Jagadish, H. V. (2001). Indexing the distance: An efficient method to KNN processing. *In Proceedings of the 27th International Conference on Very Large Data Bases* (VLDB'01), Roma, Italy, pp. 421–430.
25. Yu, C., Ooi, B. C., Tan, K. L., & Jagadish, H. V. (2005). iDistance: An adaptive B+-tree based indexing method for nearest neighbor search. *ACM Transactions on Database Systems, 30*(2), 364–397.

Chapter 4
A k-Nearest Neighbor Centroid-Based Outlier Detection Method

Abstract Detecting outliers in multi-dimensional datasets is a challenging data mining task. To distinguish exceptional outliers from regular objects based on measuring the degree of deviation for outlier ranking, there have been well-established methods among which k-nearest neighbor-based approaches have become more and more popular. However, most k-nearest neighbor-based methods have a shortcoming in parameter detection. That is, they are very sensitive to the value of k and may have different rankings for top outliers with varying k's. Further, for modern large high-dimensional datasets, not only k-nearest neighbor search becomes very computationally expensive, but also concepts like proximity, distance, or nearest neighbor become less meaningful with increasing dimensionality. To partially circumvent these problems, motivated by the centroid concept of K-means clustering algorithms, in this chapter, we introduce a simple and effective method for automatically determining the input parameter k and propose a k-nearest neighbor centroid-based outlier detection method that is easy to implement and can provide competing performances with existing solutions. Experiments performed on real datasets demonstrate the efficacy of our method.

Keywords Distance-based outlier detection · Density-based outlier detection · K-means clustering-based outlier detection · kNN-centroid-based outlier detection

4.1 Introduction

By indicating irregular patterns that deserve special attentions, outlier detection techniques have found immense use in many application domains [1–7]. The task of finding outliers in a data set has long been an area of active research, and many different outlier detection techniques, such as distribution-based, distance-based, density-based, and clustering-based approaches, have been developed.

Most of distance-based methods [8] and density-based methods [9] are based on k-nearest neighbors (kNN) for the assessment of differences in deviation among objects in the full-dimensional Euclidean data space. State-of-the-art k-nearest neighbors (kNN)-based outlier detection algorithms are simple to implement and can locate the small number of outliers quickly. Unfortunately, they face several challenges

X. Wang et al., *New Developments in Unsupervised Outlier Detection*,
https://doi.org/10.1007/978-981-15-9519-6_4

for which solutions are still open. First of all, these methods usually compute a kNN-based outlier score or factor for each data point, rank the points according to their scores, and finally return data points with top n scores as outliers. As a result, different methods (i.e., distance-based methods and density-based methods) may have different rankings for top n outliers. Secondly, it is generally agreed that kNN-based outlier definitions are sensitive to the parameter k and a small change in k can lead to changes in the scores and, correspondingly, the ranking. Finally, in high-dimensional data space, these methods are bound to deteriorate due to the notorious 'curse of dimensionality' where concepts like proximity, distance, or nearest neighbor become less meaningful with increasing dimensionality. The issue exists whether the notion of these outliers is still meaningful for high-dimensional data.

To provide partial solutions to some of these challenging problems, motivated by the centroid concept of K-means clustering algorithms, we propose in this chapter a new kNN-based centroid concept for outlier detection approach that overcomes the above limitations to some extent. Being classic clustering algorithms, K-means [10] and K-medoids [11] methods group data points around some 'centers' and take into consideration the fact that data points deep inside a cluster are well surrounded by other points and can be the center of these surrounding points when clustering. More specifically, if a data point is positioned inside a cluster, it will be surrounded by other points in all possible directions. On the contrary, outliers are positioned outside of some sets of points that are grouped together and positioned only in certain directions. The distances of normal data points to their cluster centroid or center are small, while those of outliers to their cluster centroid or center can be rather large (usually in terms of several standard distance deviations away from a cluster center). Unfortunately, K-means clustering algorithms do not work well when clustering datasets with irregular boundaries. To provide an effective solution that can detect outliers in a large multi-dimensional dataset arising in real applications where a completely unsupervised method is desirable, we propose to hybrid K-mean clustering-based outlier detection and kNN-based outlier detection so as to increase the robustness and consistency of the detecting results.

Basically, our method starts by finding k-nearest neighbors for each data point. Next, a centroid is computed upon a data point's k-nearest neighbors. The distance between the data point and its kNN-based centroid is then computed and utilized in the determination of the outlying scores. Finally, a small number of outliers are identified relatively within each point's kNN using our proposed outlier scores. We expect that the proposed method can give a new angle to view kNN-based outlier detection techniques. Our first contribution in this chapter is a proposal of an outlier detection method which is developed upon two new kNN-centroid-based outlier scores, a global one and a local one. Our second contribution in the current research is a study of a set of current outlier detection algorithms when applied to some multi-dimensional datasets. Thirdly, to be as general as possible, our algorithm has no specific requirements on the dimensionality of data sets and can be applied to outlier detection in large high-dimensional datasets. Finally, a number of experiments on both synthetic and real data sets demonstrate the robustness and effectiveness of the proposed approach in comparison with state-of-the-art outlier detection algorithms.

The rest of this chapter is organized as follows. In Sect. 4.2, we review some existing work on K-means clustering algorithms and K-means clustering-based outlier detection. We next present our proposed approach in Sect. 4.3. Evaluation of our algorithm is then given in Sect. 4.4. Finally, conclusions are made in Sect. 4.5.

4.2 *K*-means Clustering and Its Application to Outlier Detection

There are two parts of the unsupervised learning literature that are related to our study: K-means clustering and K-means clustering-based outlier detection.

4.2.1 K-*means Clustering*

Within the context of machine learning, clusters correspond to hidden patterns in the data and searching for clusters is typically an unsupervised learning activity. Cluster analysis can be achieved by various algorithms that differ significantly in their understanding of what constitutes a cluster and how to efficiently find them. A wide range of models have been developed although no singe model is appropriate for all data sets. Typical cluster models include partitioning-based models, hierarchical models, density-based models, graph-based models, and neural network-based models. Partitioning-based models, for example, the K-means clustering algorithm, represent each cluster by a single center vector. Hierarchical clustering builds models based on distance connectivity. Density-based models such as DBSCAN and OPTICS define clusters as connected dense regions in the data space. Graph-based models, that is, a subset of nodes in a graph such that every two nodes in the subset are connected by an edge, can be considered as a prototypical form of cluster. Neural network models, with self-organizing map being the most well-known unsupervised neural network, can usually be characterized as being similar to one or more of the above models and include subspace models when neural networks implement a form of principal component analysis or independent component analysis.

In partitioning-based clustering methods, clusters each are represented by a central vector, which may not necessarily be a member of the data set. When the number of clusters is fixed, K-means clustering gives a formal definition as an optimization problem: Find the K cluster centroids and assign the objects to the nearest cluster center such that the sum of the distances from the data points to their cluster center is minimized.

Partitioning-based clustering starts with K-means, an effective, widely used, and all-around clustering algorithm. After setting the number of clusters, K, the algorithm begins by selecting K points as starting centroids ('centers' of clusters). Then, the following two steps are iteratively repeated:

1. Assignment step: For each point from the dataset, we calculate its distances to the K centroids, and each of the N points is then assigned to a cluster that is represented by the closest of the K centroids.
2. Update step: From the previous step, we have a set of points which are assigned to a cluster. For each cluster, a new centroid is calculated as the mean of all points in the cluster and is declared as a new centroid of the cluster.

After each iteration, the centroids are slowly moving, and the total sum of the distances from each point to its assigned centroid gets lower and lower. The two steps are alternated until convergence, that is, until there are no more changes in cluster assignment. After a number of iterations, the same set of points will be assigned to each centroid, therefore leading to the same set of centroids again. K-means is guaranteed to converge to a local optimum. However, that does not necessarily have to be the best overall solution (global optimum). K-means has a number of interesting theoretical properties. Firstly, it partitions the data space into a structure known as a Voronoi diagram. Secondly, it is conceptually close to the nearest neighbor classification, and as such is popular in machine learning. Thirdly, it can be seen as a variation of model-based clustering, and Lloyd's algorithm as a variation of the expectation–maximization (EM) algorithm.

However, K-means algorithms have a number of theoretical and practical application challenges. On the one hand, the final clustering result can depend on the selection of initial centroids. One simple solution would be just to run K-means a couple of times with random initial assignments and then select the best result by taking the one with the minimal sum of distances from each point to its cluster centroid, that is, the error value that we are trying to minimize in the first place. Other approaches to selecting initial points can rely on selecting distant points. This can lead to better results, but may cause a problem due to the existence of outliers. Since outliers are far away from any meaningful cluster, each such point may end up with being its own 'cluster.' A good balance would be to still pick random points as an initialization, but with probability proportional to square distance from the previously assigned centroids. Points that are further away will have higher probability to be selected as starting centroids. Consequently, if there is a group of points, the probability that a point from the group will be selected also gets higher as their probabilities add up, resolving the outlier problem. On the other hand, most K-means-type algorithms require the number of clusters, K, be specified in advance, which is considered to be one of the biggest drawbacks of these algorithms. Furthermore, the algorithms prefer clusters of approximately similar size, as they will always assign an object to the nearest centroid. This often leads to incorrectly cut borders of clusters (which is not surprising since the algorithm optimizes cluster centers, not cluster borders).

Variations of K-means clustering algorithms often include such optimizations as choosing the best of multiple runs, but also restricting the centroids to members of the data set (i.e., K-medoids), choosing medians (i.e., K-medians), choosing the initial centers less randomly (i.e., K-means++), or allowing a fuzzy cluster assignment (i.e., fuzzy c-means).

4.2.2 K-means Clustering-Based Outlier Detection

In 2001, Jiang et al. proposed a two-phase clustering algorithm for outlier detection [12]. In the first phase, the dataset is preprocessed by the K-means algorithm to form clusters. In the second phase, a minimum spanning tree is constructed upon the cluster centers obtained from the first phase. Finally, clusters in small subtrees are considered as outliers. In 2003, He et al. introduced the concept of cluster-based local outlier and designed a measure, called cluster-based local outlier factor (CBLOF), to identify such outliers [13]. In 2005, Hautamäki et al. proposed a two-stage outlier removal clustering (ORC) algorithm to identify clusters and outliers from a dataset simultaneously [14]. The first stage is a purely K-means algorithm; the second stage iteratively removes the data points that are far away from their cluster centroids. In 2007, Rehm et al. defined outliers in terms of noise distance. The data points that are above the noise distance or further away from any other cluster centers get high membership degrees to be the outlier cluster [15]. In 2009, Zhou et al. proposed a three-stage K-means algorithm to cluster data and detect outliers [16]. In the first stage, the dataset is clustered by the fuzzy c-means algorithm. In the second stage, local outliers are identified and the cluster centers are recalculated. In the third stage, certain clusters are merged and global outliers are identified. In 2011, Pamula et al. used the K-means algorithm to prune some points around the cluster centers and the LDOF measure [17] to identify outliers from the remaining points [18]. In 2013, Ahmed and Naser proposed a modified version of the K-means algorithm, the outlier detection and clustering (ODC) algorithm, to detect outliers [19]. In the ODC algorithm, a data point that is at least p times the average distance away from its centroid is considered as an outlier. In the same year, Chawla and Gionis proposed a generalization of the K-means algorithm to provide data clustering and outlier detection simultaneously, which requires two parameters: the desired number of clusters and the desired number of top outliers [20]. In 2015, Whang et al. proposed nonexhaustive overlapping K-means (NEO-K-means) algorithm, which is also able to identify outliers during the clustering process [21]. In 2016, Yu et al. proposed an outlier-eliminated differential privacy (OEDP) K-means algorithm which removes outliers from the dataset before applying the K-means algorithm [22]. In the same year, Aparna and Nair proposed a constraint-based high-dimensional bisecting K-means (CHB-K-means) algorithm by using a weighted attribute matrix to detect outliers [23]. In 2017, Barai and Dey proposed a general method of detecting outliers in K-means which is based on calculating the maximum and minimum value of pairwise distances for all observations and using the sum of these two values as a threshold to detect outliers [24]. In the same year, Diez-Olivan et al. proposed to combine the constrained K-means clustering for outlier detection and behavior characterization and the fuzzy modeling of distances to normal patterns found. Finally, a LOF-based score is calculated over time, considering the membership degree to resulting in fuzzy sets by a local outlier factor, to provide a fully comprehensive yet accurate prognostics approach [25]. Also in 2017, Gan and Ng proposed to extend the K-means algorithm to provide data clustering and outlier detection simultaneously by introducing an additional 'cluster' to the K-means algorithm to hold all outliers [26]. In 2018, Zhao et al. proposed a user-adaptive algorithm based on K-means clustering, local outlier factor (LOF), and multivariate Gaussian distribution (MGD) for activity recognition [27]. To automatically cluster and annotate a specific user's activity data, an improved K-means algorithm with a novel initialization method is designed. Then, a method based on LOF is proposed to select high confidence samples for personalizing three MGD models. In addition to kNN-based outlier detection algorithms, isolation forest (IF) is another branch of well-known outlier detection algorithms. In 2020, Karczmarek et al. proposed a K-means-based isolation forest method that allows to build a search tree based on many branches in contrast to the only two considered in the original method and enables a user to intuitively determine the more intuitively appealing anomaly score for an individual record of the analyzed dataset [28].

Some of the aforementioned algorithms perform clustering and outlier detection in stages. In these algorithms, a clustering algorithm is used to divide the dataset into clusters and some measure is calculated for the data points based on the clusters obtained to identify outliers. The ODC algorithm, the K-means algorithm, and the NEO-K-means algorithm integrate outlier detection into the clustering process.

4.3 A kNN-Centroid-Based Outlier Detection Algorithm

4.3.1 General Idea

From our study of state-of-the-art distance-based and density-based outlier detection methods [29–34], we have obtained the following observations.

First of all, distance-based outlier detection methods are good at identifying global outliers. They directly use a point's distance to its kth nearest neighbor as a measure of its outlying degree where k is usually set to be a fixed constant for every data point. For an ideal situation of a sample dataset as shown in the left part of Fig. 4.1, there

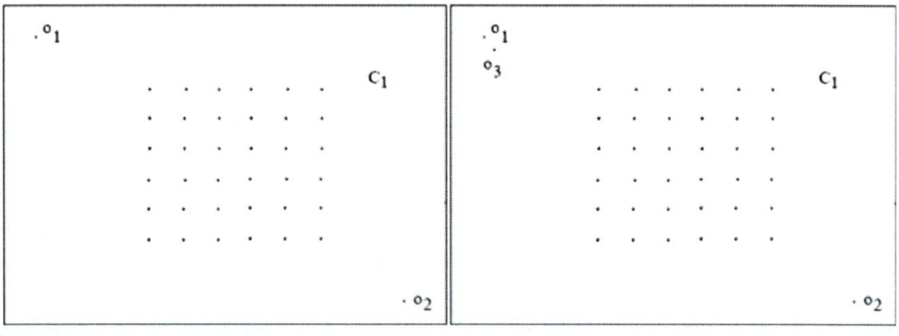

Fig. 4.1 An outlier detection task

are two outliers, o_1 and o_2, which can be easily detected even if k is set to 1. Without loss of generality, suppose outlier o_1 has a larger distance to its first nearest neighbor (i.e., a higher outlier rank) than outlier o_2 to its first nearest neighbor. However, when some random noise, say outlier o_3, appears near outlier o_1 (as shown in the right part of Fig. 4.1), k has to be set to 2 for o_1 to be detected as an outlier. The problem that comes up is that, for $k = 2$, o_2 may have a larger distance to its second nearest neighbor than o_1 (which will have the same distance to its current second nearest neighbor as it is to its first nearest neighbor if o_3 is not there), and therefore is going to be ranked as the top most distance-based outlier. To summarize, since noise could exist, it may be meaningful if different k's are used to detect different distance-based outliers. In other words, we propose to denoise in some way in the first place before outlier detection can proceed so that, after denoising, we are left with only the distance to its first nearest neighbor (i.e., the distance to some boundary point) for each global outlier. This observation is based on the fact that outliers and their noisy companions are mutual nearest neighbors with shorter pairwise distance than their distances to the boundary points.

Secondly, another problem with distance-based outlier detection is that these methods do not take the outlying degrees of a data point's k-nearest neighbors into consideration in the detection process. As a result, false positives can happen. For example, for the sample dataset in Fig. 4.2, although DB or DB-Max outlier scores can be calculated for each data point and four boundary data points of cluster C_2 can be mistakenly detected as DB or DB-Max outliers, there are no outstanding outliers. To prevent the false positives from happening, there must be some ways to differentiate between boundary points and outliers existing in a dataset in the first place. To do so, as a first degree approximation, the distances of each data point and its *k*NN to their first nearest neighbor within a cluster can be assumed to follow a uniform distribution and the corresponding mean and standard deviation can thus be calculated. The ratio of the standard deviation over the mean can be used to judge to some degree whether outliers exist or not. For the sample dataset shown in Fig. 4.2, if k is set to be 2 (i.e., 2NN), the distance of data point o_1 to its first nearest neighbor is the same as that of data point o_2 to its first nearest neighbor and that of data point

Fig. 4.2 An illustration of the false positives

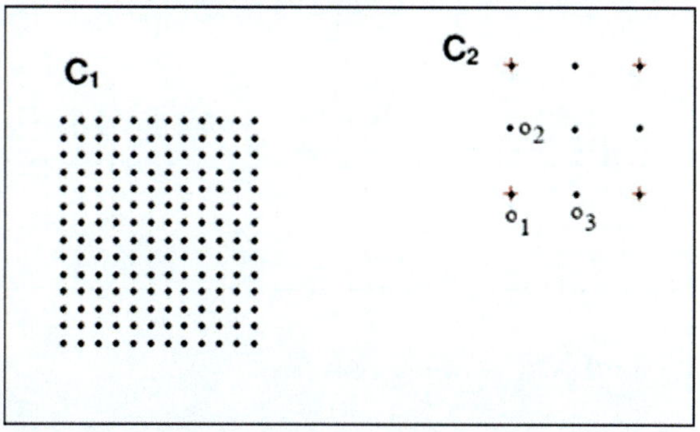

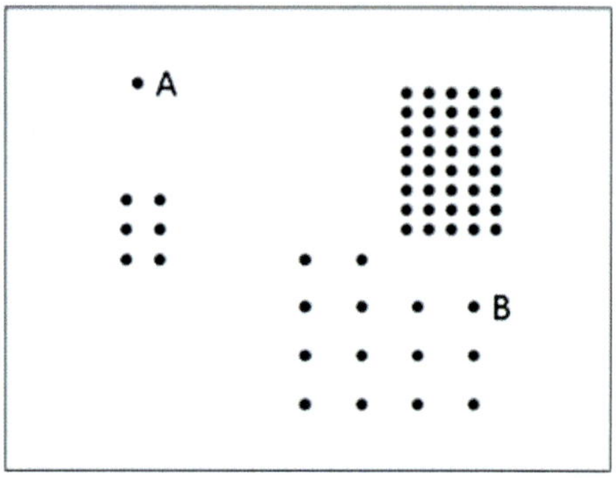

Fig. 4.3 An illustration of a difference between global outlier and local outlier definitions

o_3 to its first nearest neighbor. The corresponding ratio of the standard deviation over the mean is 0, indicating there are not outstanding global outliers.

Thirdly, global outliers and local outliers are based on two different kinds of outlying scores. Density-based outlier scores are computed as a ratio, and algorithms designed to mine such local outliers may not do well in global outlier detection. Taking the dataset in Fig. 4.3 as an example, data point A is farthest away from its six closest neighbors and therefore should be identified as a global outlier. However, for $k = 6$, the LOF-based outlier detection method assigns a higher outlier score to data point B than to A and, therefore, fails to identify A as the most significant global outlier. If global outlier detection is separated from local outlier detection, this will not happen.

Finally, both distance-based outlier definitions and density-based outlier definitions regard outliers to stay in sparse regions. From our observation as illustrated in Fig. 4.4, another kind of outliers may come from a denser region, i.e., cluster C_3. As shown in the left plot of Fig. 4.4, cluster C_3 is a denser one than cluster C_2. If cluster C_3 is under-sampled in reality, as illustrated in the right plot of Fig. 4.4, the four data points in it become outliers. Density-based outlier definitions should take such situations into consideration.

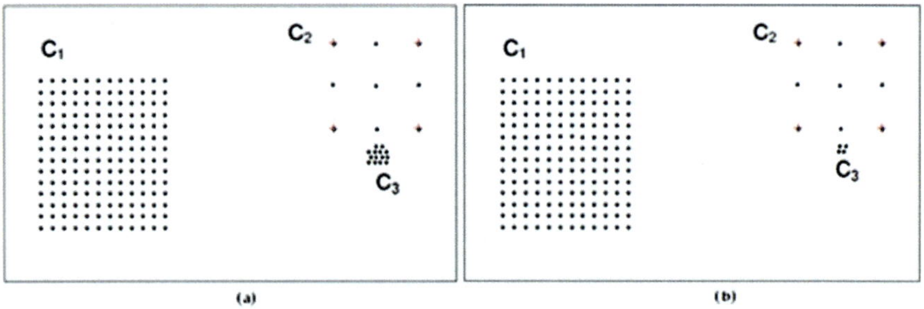

Fig. 4.4 Sample clusters in a 2-D data set

4.3.2 Definition for an Outlier Indicator

Definition 1 (*k-Distance of an object p*) For any positive integer *k*, the *k*-distance of object *p*, denoted as *k*-distance(*p*), is defined as the *distance*(*p*,*o*), or simply, *d*(*p*,*o*), between *p* and an object $o \in D$ such that:

(1) For at least *k* objects $o' \in D \setminus \{p\}$, $d(p,o') \leq d(p,o)$.
(2) For at most $k - 1$ objects $o' \in D \setminus \{p\}$, $d(p,o') < d(p,o)$.

Definition 2 (*k-Nearest Neighbors of an object p*) For any positive integer *k*, given *k*-distance(*p*), *k*-nearest neighbors of *p* contain the first *k* closest objects whose distance from *p* is not greater than *k*-distance(*p*), denoted as $k\mathrm{NN}_{k\text{-Distance}(p)}(p)$, for which *k*NN(*p*) is used as shorthand.

For outlier detection, we are more interested in those data points whose distance to its first nearest neighbor is significantly larger than the average value of the distances of the point's *k*NN to their first nearest neighbors. To quantify the significance of a data point's positioning outside some cluster, the uniform distribution is used as a first-degree approximation for the distances associated with a data point and its *k*NNs to their first nearest neighbors. To distinguish between boundary points and outliers, we therefore formulate an outlier indicator using the distances associated with the first nearest neighbors of a data point and its *k*NNs as in the following to focus our attention on the small number of outstanding global and local outliers.

$$\mathrm{Mean}_{\mathrm{nn-dist}} = \frac{1}{k+1}\sum_{i=0}^{k} \mathrm{dist}[i] \tag{4.1}$$

$$\mathrm{Std}_{\mathrm{nn-dist}} = \sqrt{\frac{1}{k+1}\sum_{i=0}^{k} (\mathrm{dist}[i] - \mathrm{Mean}_{\mathrm{nn-dist}})^2} \tag{4.2}$$

$$\mathrm{SOM}_{\mathrm{nn-dist}} = \frac{\mathrm{Std}_{\mathrm{nn-dist}}}{\mathrm{Mean}_{\mathrm{nn-dist}}} \tag{4.3}$$

where dist[0] denotes the distance of a data point to its first nearest neighbor and dist[*i*] denotes the distance of a data point's *i*-th nearest neighbor to its first nearest neighbor.

$\mathrm{SOM}_{\mathrm{nn-dist}}$ can be used as a quantitative measure of deviation from normality and as a threshold to rule out the large portion of normal data. Based on these ideas, our *k*NN-centroid-based global outlier score is given in the following subsection.

4.3.3 Formal Definition of kNN-Based Centroid

Classic K-means clustering algorithms aim to partition data objects into K clusters in which each object belongs to the cluster with the nearest mean that, in Euclidean distance, can be calculated to be the centroid of the objects in the cluster and serves as a prototype for the cluster. K-means clustering-based outlier detection methods define outliers to be those with distances several standard deviations away from the cluster mean (or, prototype). However, it is well known that traditional K-means clustering algorithms are good for identifying clusters in spherical datasets, but do not work well when the boundaries of the clusters are irregular and/or outliers exist. However, if a centroid is calculated from a small local region, this adverse effect will be diminished. Therefore, we propose to use the distance of each point not to its cluster centroid (i.e., cluster mean) but to the centroid calculated from a point's local neighborhood, say from k-nearest neighbors, to discern between normal data points (well surrounded by other points) and outliers (surrounded by other points only in certain directions).

To illustrate this observation, consider a simple data set as shown in Fig. 4.5. For a point deep inside a cluster, say point O_3 in Fig. 4.5, the data point and its 4NN-based centroid differ slightly (actually zero in this case). The difference will become larger for points at the border of a cluster, say points O_4 and O_5. However, even here the difference is still relatively small compared to the deviation of those for real outliers, say O_1 and O_2, to their first neighbors. Since most points are embedded into some clusters, their deviations to their kNN-based centroids are expected to be small, making outlier detection more effective than using the measures based on the distances to the cluster mean (or, prototype) and the corresponding standard deviation. To summarize, the deviation of a data point to its kNN-based centroid remains rather large for an outlier whereas the deviations are small for the boundary points (i.e., those close to the border of a cluster) and very small for the inner points of a cluster (those deep inside a cluster and well surrounded by others). Therefore, the distance between a data point and its kNN-based centroid can be used to describe how well a point is surrounded by other points in all possible directions. If the

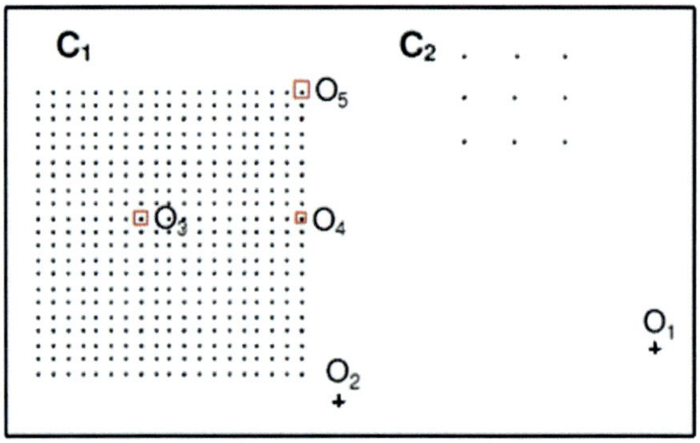

Fig. 4.5 A classic example of centroids

deviation for a point from its *k*NN-based centroid is rather large, other points will be positioned only in certain directions. Thus, rather a large difference for a point from its centroid implies that the point is probably an outlier. The hybridization of *K*-means clustering-based outlier detection and *k*NN-based outlier detection can be utilized to increase the robustness and consistency of the detecting results.

As a result of these observations, to represent the degree a data point is inside a cluster in a setting of *k*-nearest neighbors, the formal definition for *k*NN-based centroid is introduced in the following.

Definition 3 (*Centroid of k-Nearest Neighbors of an object p*) Given the *k*-nearest neighbors of *p*, the centroid of *k*-nearest neighbors of an object *p* is defined as

$$\text{Centroid}_{k\text{NN}(p)} = \frac{\sum_{i=1}^{k} \vec{q}_i}{k} \tag{4.4}$$

where $q_i \in k\text{NN}(p)$ and $1 \leq i \leq k$.

4.3.4 Two New Formulations of Outlier Factors

Based on Definition 3, the corresponding distribution of the distances between each data point and its 4NN-based centroid (i.e., 4 nearest neighbors are considered) for three types of points is illustrated in Fig. 4.6 for the sample data in Fig. 4.5. From Fig. 4.6, it can be seen that global outlier O_1, denoted by the red star, has the largest deviation distance. The next four largest deviation distances correspond to the four corner points (denoted by the red cross) of cluster C_2, that is, data points O_6, O_7, O_8, and O_9. Data point O_2, denoted by the red triangle, has the fifth largest deviation distance. The fifth largest deviation value corresponds to the second outlier O_2. Then, the next fourth largest deviations are associated with the four corner points of cluster C_1.

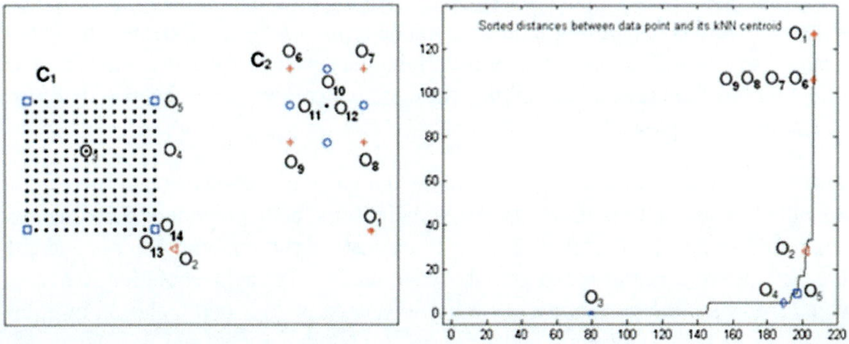

Fig. 4.6 Distribution of deviation distances

To identify the relatively small number of global outliers, kNN for each data point is first computed. Next, the corresponding centroid and the distance between the centroid and the data point are calculated for every data point. Then, those data points with the largest deviation distances should be located and returned. However, for global outlier detection, there is no reason to assume that this must be the case due to the existence of boundary data points. To distinguish global outliers from global boundary points (both of which have large deviation distances), outlier indicators can be used. This is because, different from global outliers, global boundary points have similar (actually the same in this case) distances to their first nearest neighbor as those of its kNNs to their first nearest neighbors. For example, although it has a larger deviation than data point O_2, data point O_6 has the same distance to its first nearest neighbor (e.g., O_{10}) as those of its 2NN, that is, O_{10} and O_{11}, have to their first nearest neighbor (e.g., O_{12} and O_{12}, respectively), while data point O_2 has a larger distance to its first nearest neighbor (e.g., O_{14}) than those of its 2NN, that is, O_{13} and O_{14}, have to their first nearest neighbor (e.g., O_{14} and O_{13}, respectively). In other words, we are actually more interested in those data points as global outliers which not only have the largest deviation distances but also have dissimilar distances to their first nearest neighbor as those of their kNNs have to the corresponding first nearest neighbor. It can be seen from Fig. 4.6 that, although the corner points of cluster C_2 (i.e., O_6, O_7, O_8, and O_9) have large deviation distances, they all have similar distances (actually the same distance in this case) to their first nearest neighbor.

To find global outliers, following the notion of kNN-based distance outlier definition, a suitable value for k must be determined for each data point to describe its nearest neighborhood. To do so, we first find for each point its kNN and then calculate the kNN-based centroid and the deviation between the data point and its kNN-based centroid. With varying k values, the deviations may have different values. We then search for the k value which corresponds to the minimum deviation value (which should be very close to zero for most normal data points). The k thus determined may be different for different data points. With a k being set for each data point, we then calculate the distance-based outlier factors (i.e., DB-Max) for all the data points, sort them in a nonincreasing order, and search for the largest outlier factor values. If their corresponding outlier indicators are significantly larger than a threshold value, cut-thred, top n data points can be regarded as the global outliers. Finally, these global outliers are ranked according to their distance to their $(k + 1)$-th nearest neighbor, where k may be different for different outlying data points. By using different k for different outliers in the outlier ranking, the denoising process mentioned in the above is actually accomplished.

Definition 4 (*Centroid-based global outlier factor of an object p*) Let a centroid be computed upon a data point p's k-nearest neighbors, dist[i] denote the first nearest neighbor distance associated with the i-th nearest neighbor of the point p, and cut-thred (i.e., kNN-centroid-based global outlier indicator) be a threshold for $\text{SOM}_{\text{nn-dist}}$ which measures the possibility of outlier existence, with k being chosen by minimizing the distance between a data point and its kNN-based centroid and obtained by Eq. (4.7), and a *kNN*-centroid-based global outlier factor is defined to be the distance between the data point and its $(k+1)$-th nearest neighbor as

$$\mathrm{CGOF}(p) = (k+1) - \mathrm{Distance}(p) \tag{4.5}$$

$$\mathrm{SOM} \geq \mathrm{cut} - \mathrm{thred} \tag{4.6}$$

where k is obtained by the following minimization,

$$\min\left\{d\left(\vec{p}, \mathrm{Centroid}_{k\mathrm{NN}(p)}\right)\right\} \tag{4.7}$$

In addition to using outlier indicators to identify boundary points, for data points deep inside local clusters, the ratio of the distance between a data point's $(k+1)$-th nearest neighbor and itself over the distance to its first nearest neighbor can be used as a measure to judge its inside-a-cluster degree.

For local outliers, such as the ideal one, i.e., O_2, as shown in Fig. 4.5, since they are significantly far away only from their nearest neighboring cluster, the local outlier score can be defined as the ratio of the distance to its first nearest neighbor over the distance of the nearest neighbor of the data point (say, O_2's first nearest neighbor) to its nearest neighbor. This definition is simple and elegant. However, for the situation illustrated in Fig. 4.4, this definition will not do. For this case, the local outlier score can be defined as the ratio of the distance of the data point to its $(k+1)$-th nearest neighbor over the distance to its first nearest neighbor. Therefore, for local outliers, we can calculate the above two ratios and use the larger one as their local outlier score. The main challenge resides in the determination of the k-nearest neighbors when noise exists.

Definition 5 (*Centroid-based local outlier factor of an object p*) Let a centroid be computed upon a data point p's k-nearest neighbors, and dist[i] denote the first nearest neighbor distance associated with the i-th nearest neighbor of the point p, and kNN-centroid-based local outlier factor is defineds as

$$\mathrm{CLOF}(p) = \max\left\{\frac{(k+1) - \mathrm{Distance}(p)}{\mathrm{dist}[k+1]}, \frac{(k+1) - \mathrm{Distance}(p)}{\mathrm{dist}[0]}\right\} \tag{4.8}$$

where k is determined by the following minimization,

$$\min\left\{d\left(\vec{p}, \mathrm{Centroid}_{k\mathrm{NN}(p)}\right)\right\} \tag{4.9}$$

The local outlier detection can be assumed to be over when the ratio drops below a threshold, or only top n ones are chosen and returned. Finally, the outlier scores are used to assign the returned data points a degree of being outlier. Centroid-based local outlier factor of an object p captures the degree p is deviated from its nearest neighboring cluster. It is not difficult to see that the larger the CLOF of p is, the further the point p is away from the boundaries of its nearest cluster.

To summarize, given a dataset, our outliers can be found with a two-step approach. In the first step, data points from the database satisfying the outlier conditions are mined. In the second step, the outliers are checked by calculating their outlier indicators so as to remove all the inliers to obtain the final top outliers.

4.3.5 Determination of k

One of the problems with current kNN-based methods is that they are very sensitive to k that is used in the algorithms. From our observation, using a fixed k for all the data points in the outlier score computation is neither effective nor necessary. For the inliers, k can be set to a value which results in the smallest distance between a data point and its kNN-based centroid. For example, in the sample set depicted in Fig. 4.7, we can easily recognize that A, C, and E are boundary points surrounded by two, four, and six different nearest neighbors (in green color), respectively, and unambiguously detect that B and D are within cluster points which are well surrounded by five and

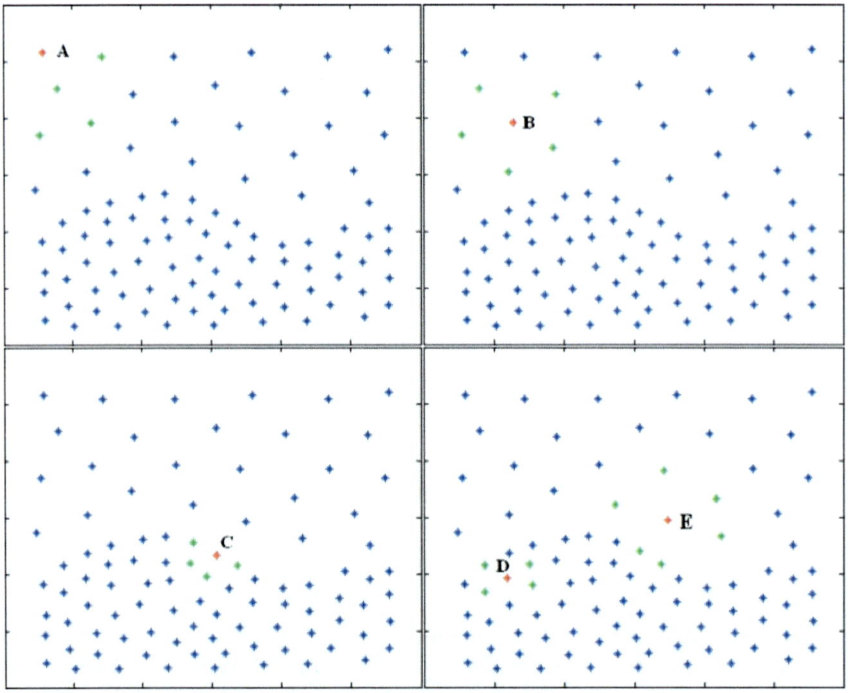

Fig. 4.7 An illustration of five different kinds of data points

four other data points in the sense that five and four neighbors are chosen since they result in the smallest distance between the data points and their kNN-based centroids.

As a result, for the calculation of two outlier factors for all the data points, k's are determined which result in the smallest distance between the data points and their centroids for outlier detection and are denoted by minPoints. For the calculation of the outlier indicators, k can be defined in Eq. (4.10), in which N is the size of the data set [35].

$$k = 5\log_{10} N \tag{4.10}$$

Definition 6 (*Threshold of outlier indicators in a data set*) Let the outlier indicators of all points in the dataset be computed, based on which the average value, *mean*, and the corresponding standard deviation, *std*, are calculated. The threshold of outlier indicators, cut-thred, for finding potential outliers can be defined as

$$\text{cut - thred} = \text{mean} + f \times \text{std} \tag{4.11}$$

where f can be 1, 1.5, 2, 2.5, 3, and more.

4.3.6 The Complexity Analysis

From the description in the previous subsections, it can be seen that our algorithm mainly consists of three steps, an initial k-nearest neighbor search for each of N data points, the calculation of minPoints and three scores, and finally top outliers' mining. Therefore, most of the calculation time is spent on the first two steps, k-nearest neighbor-based minPoints search. As a result, the total time complexity for the average case can be O(N^2) or O(Nlog$N$) if some index structure can be used. Since the number of outliers is small, the third step takes a nearly linear time on average. Physically, the resource consumed by our algorithm includes the space to hold the whole data set in memory, space to store their k-nearest neighbors, and some space for three scores and the index structure if used.

4.3.7 The Proposed Outlier Detection Algorithm

To improve the readability, the algorithms for determining a value of minPoints, kNN-based centroid, CGOF, and CLOF for each data point are presented in a pseudo-code format in Tables 4.1, 4.2, 4.3 and 4.4.

We combine the proposed three factors, global outlier factors, local outlier factors, and outlier indicators to create our kNN-centroid-based outlier detection algorithm.

Table 4.1 Function to find minPoints

Input:				
p, *q*	Data points			
d	Dimension of data			
Output:				
minPoints	minPoints of every point p			
Function distance(q, p) returns distance calculated between data point p and data point q.				
Begin:				
	For each data point p			
	{	*k*?=?3d; find *k*NN of p;		
		//for point q to be the *i*th NN of p		
		for each i?=?1:*k*		
		{	minDistC?=?distance(q, p);	
			//initialize p's centroid C.values	
			for each j?=?1:*d*	
			{	C.values[j]?+??=?q.values[j]/i;}
			DistTemp?=?distance(C, p);	
			if(DistTemp?<?=?minDistC)	
			{	minDistC?=?DistTemp;
				minPoints?=?i;
			}	
		}		
	}			
	return minPoints;			
End				

1. Read in all the data points in a data set, and determine a k for each data point which minimizes the distance between the data point and its kNN-based centroid.
2. Compute two 1-dimensional arrays of the estimated outlier scores, sort them in a nonincreasing order, and compute a 1-dimensional array of the outlier indictors with k being set according to Eq. (4.10).
3. For each data item with top global outlying scores, return it as a global outlier if its outlier indicator is larger than the threshold, i.e., cut-thred.
4. For each data item with top local outlying scores, return it as a local outlier if its outlier score is larger than a given threshold.
5. Repeat steps 3 and 4 until all the required top n outliers are mined.

The outlier detection algorithm is presented in a pseudo-code format in Table 4.5.

Table 4.2 Function to find centroid

Input:			
p	The object we are concerned about this time		
k	The number of nearest neighbors		
points	The data set		
Output:			
centroid			
Begin:			
	kNN = points.getkNN(p,k);		
	For(point q:kNN){		
	{	for(int i = 0;i < dimension;i++){	
			c.values[i] += q.values[i]/k;
	}		
	Return c;		
End;			

Table 4.3 Function to calculate CGOF for each object

Input:			
p	The object we are concerned about this time		
k	The number of nearest neighbors		
points	The data set		
Output:			
CGOF			
Begin:			
	kNN = points.getkNN(p,k + 1);		
	CGOF(p) = dist(p, kNN.get(k));		
	Return CGOF;		
End;			

4.4 A Performance Study

In this section, we present the results of three sets of experiments performed to evaluate our *k*NN-centroid-based outlier detection algorithm. In the first set of experiments, two 2-dimensional synthetic data sets are used to show that our proposed outlier detection method can outperform classic outlier detection algorithms in classification accuracy. Further, it is important for an outlier detection method to work effectively on real-world data sets. Therefore, in the second set of experiments, twelve real multi-dimensional data sets obtained from UCI Machine Learning Repository

Table 4.4 Function to calculate CLOF for each object

Input:		
points, p	The data set and the object we are concerned about this time	
k	The number of nearest neighbors	
Output:		
CLOF		
Begin:		
	kNN = points.getkNN(p,k + 1);	
	p1 = kNN.get(0);	
	kNN1 = points.getkNN(p1,1);	
	dist1 = dist(p1,kNN1.get(0));	
	p(k + 1) = kNN.get(k + 1);	
	kNN(k + 1) = points.getkNN(p(k + 1),1);	
	dist(k + 1) = dist(p(k + 1),kNN(k + 1).get(0));	
	if(dist1 < dist(k + 1)){	
		disti = dist1
	else	
	{	disti = dist(k + 1);
	}	
	CLOF(p) = CGOF(p)/disti;	
	if CLOF(p) < 1	
		CLOF(p) = 1/CLOF(p)
	Return CLOF;	
End;		

[36] are used to check the effectiveness of this study and to illustrate the effectiveness of our method in real-world situation. Finally, in the third set of experiments, three real high-dimensional data sets obtained from UCI Machine Learning Repository [36] are used to check the technical robustness of our method in high-dimensional data space by comparing it with the state-of-the-art outlier detection methods, that is, DB-Max [29], DB [30], LOF [9], COF [31], INFLO [32], LDOF [17], RBDA [33], and RDOS [34]. All the data sets are briefly summarized in Table 4.6.

All the algorithms are implemented in Java and run on a computer with Intel i5-6300HQ CPU and 2 GB RAM. The operating system running on this computer is Windows 10. The results show that, overall, our proposed kNN-centroid-based outlier detection algorithm is competent with state-of-the-art outlier detection algorithms in classification accuracy. For consistency, we only use the parameters k to represent the neighborhood size in the investigation of the state-of-the-art methods.

Table 4.5 Outlier detection algorithm

Input:			
data	A set of N data points		
kNNMap	The set of kNN for all the data points		
minPointsMap	The set of minPoints for all the data points		
m	The number of top outliers		
list	The set of outlier indicators for all the data points		
Output:			
set	A set of top m outliers		
Function distance(q, p) returns distance calculated between data point p and data point q			
Begin			
	for each point p in data		
	{	minPoints = minPointsMap.get(p);	
		kNNlist = kNNMap.get(p).getTop(minPoints);	
		point pk = kNNlist.get(minPoints + 1);	
		CGOF = distance(p, pk);	
		p1 = kNNlist.get(0);	
		dist1 = distance(p1, kNNMap.get(p1).get(0));	
		distk = distance(pk, kNNMap.get(pk).get(0));	
		if(dist1 < distk)	
			disti = dist1;
		else	
			disti = distk;

(continued)

Table 4.5 (continued)

Input:			
		CLOF = CGOF/disti;	
		GMap.put(p, CGOF);	
		LMap.put(p, CLOF);	
	}		
	threshold = average(list) + std(list);		
	while(set.size < m)		
	{		
		if(GMap.valueset.next > threshold)	
			set.put(GMap.keyset.next);
		if(LMap.valueset.next > threshold)	
			set.put(LMap.keyset.next);
	}		
	return set;		
End			

Table 4.6 Sets of data

Dataset name	Size (N)	Dimensionality
Dataset1	197	2
Dataset2	155	2
Heart2	270	13
Horse colic	368	27
Tae	151	5
Sonar	208	60
Glass identification	214	10
Australian	690	14
Ecoli	336	8
Hepatitis	155	19
Spect heart	267	22
Dermatology	366	33
Breast Cancer Wisconsin (wdbc)	569	32
Abalone	4177	8
Arrhythmia	452	279
Internet Advertisements(IAD)	3279	1558
Musk (version 2)	6598	168

4.4.1 Performance on Synthetic Datasets

In this subsection, as shown in Fig. 4.8, we use two 2-dimensional synthetic data sets to show that the proposed method can efficiently identify local and global outliers in various scenarios. In each dataset, there are multiple clusters and six outliers (*A*, *B*, *C*, *D*, *E*, and *F*) are planted in the vicinities of the clusters. A particular challenging feature of these data sets is that clusters are of different sizes and have different densities. We exclusively use 2-dimensional data sets due to their convenience for

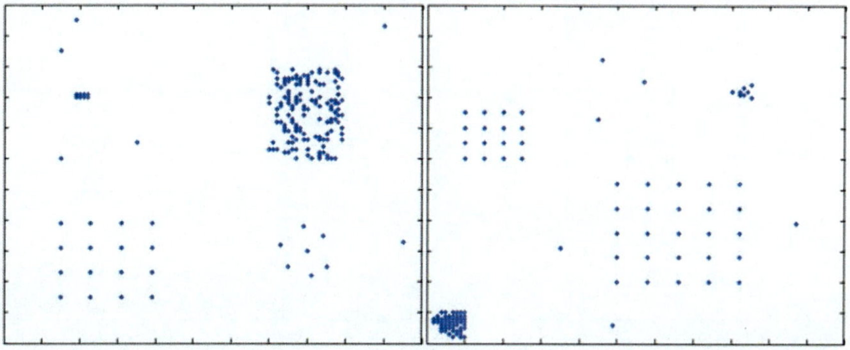

Fig. 4.8 (Left) Synthetic Dataset 1 and (right) Synthetic Dataset 2

visual inspection. The relative effects of our proposed outlier detection algorithm are studied in comparison with DB, DB-Max, LOF, COF, INFLO, LDOF, RBDA, and RDOS methods.

Synthetic Dataset 1 consists of 197 instances, including six planted global outliers, two small outlying clusters (containing 7 data points and 8 data points, respectively), and two large normal clusters. This is a global outlier detection task.

When k is set to 3 for all the other algorithms in this case, the experimental results are depicted in Fig. 4.9 in which 6 top outliers returned by each algorithm are marked with red plus, red star, red cross, red triangle, red circle, and red square in order, respectively. From Fig. 4.9, it can be seen that DB, DB-Max, LOF, and our method (i.e., CENTROID) correctly detect all the outliers, although the rankings are slightly different for different methods. As for other competing algorithms, INFLO successfully detected *B*, *C*, *D*, *E*, *F* but missed *A*. The other comparison algorithms COF, LDOF, RBDA, and RDOS do not perform as well. More specifically, COF misses *A*, *C*, *D*, *E*. Since the average of k-nearest distances and that of the $k(k-1)$ pairwise inner distances are both large for *A*, *B*, *C*, *D*, *E*, their ratios are near 1 and they are missed by LDOF. RBDA misses four global outliers (i.e., *A*, *C*, *D*, *E*). RDOS detects 2 right for top 6 outliers. Finally, in the algorithms that successfully detect all outliers, the orders of the returned top 6 outliers of DB-Max and centroid methods

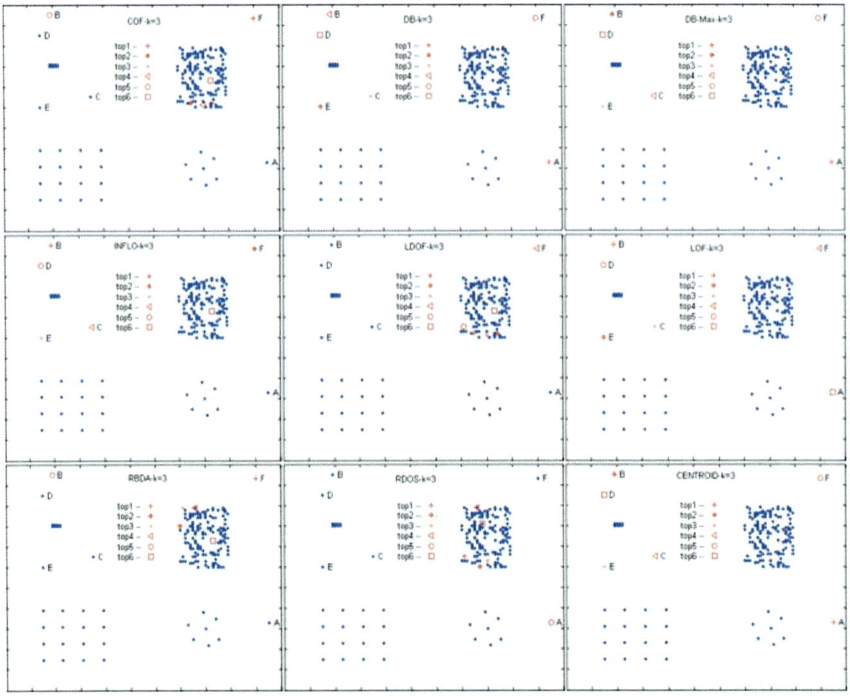

Fig. 4.9 Outlier detecting results on Synthetic Dataset 1 for $k = 3$

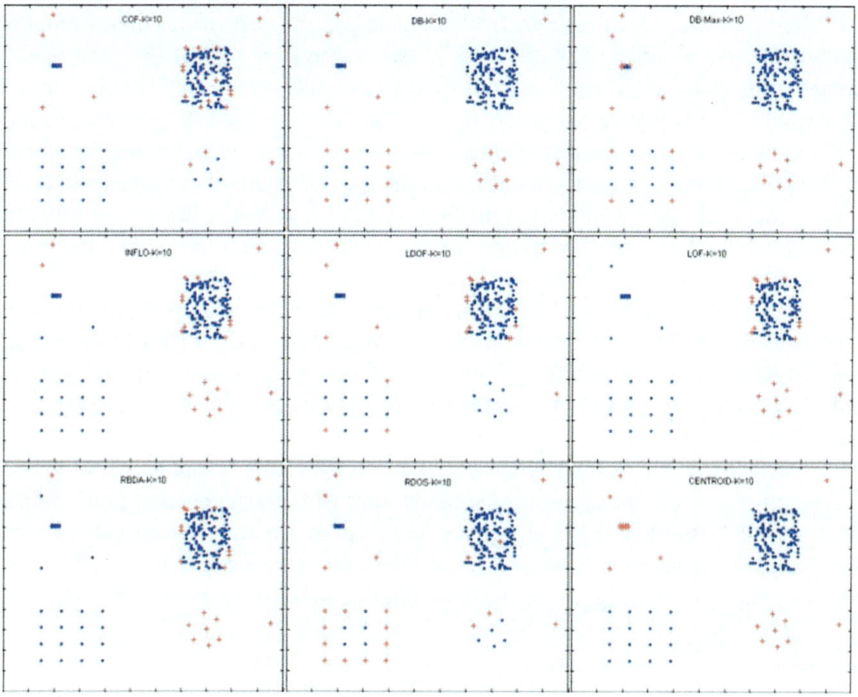

Fig. 4.10 Outlier detecting results for Synthetic Dataset 1 using $k = 10$

are the same, while those of the other two algorithms are slightly different. From Fig. 4.9, we can observe the effectiveness of the proposed detection mechanism for global outlier detection.

When k is set to 10, the two small clusters should also be treated as outliers. The algorithms should return 21 points consisting of two small clusters and 6 global outliers. The experimental results are depicted in Fig. 4.10 in which top 21 outliers returned by each algorithm are marked with red plus.

From Fig. 4.10, it can be seen that only our algorithm successfully identified all outliers as expected. The DB algorithm successfully identifies the sparse cluster with 7 points, but does not recognize the dense cluster with 8 points because of the influence of the sparse normal cluster of size 16. The results of DB-Max and DB are similar, but DB-Max recognizes 4 of the dense cluster of size 8. In this respect, DB-Max achieves better results than DB. For this case, LDOF, COF, and RDOS successfully identify all six global outliers, but fail to identify all the outliers residing in the two outlier clusters. The LOF no longer recognizes all the global outliers when the value of k is changed to 10, but identifies all 7 points in the sparse outlier cluster. Both INFLO and RBDA recognize the sparse outlier cluster, but do not do well in the identification of the other outliers. To summarize, only the proposed kNN-centroid-based outlier detection algorithm identifies all 21 outliers as expected.

To test if our approach can effectively find meaningful outliers in a more complex dataset, we use Synthetic Dataset 2 (as shown in the right part of Fig. 4.8) which contains 155 instances, including six planted global outliers, one small outlier cluster of size eight, and three large clusters of different densities consisting of 25, 16, and 100 instances. To demonstrate the effectiveness of our approach in finding the global outliers, we compare its performance with those of the state-of-the-art methods on this dataset. For these kNN-based methods, we set $k = 3$ and the screen shots in Fig. 4.11 show the results for mining the top 6 outliers which are marked the same way as that for Synthetic Dataset 1.

As can be seen from Fig. 4.11, all six preset global outliers *A*, *B*, *C*, *D*, *E*, and *F* have been successfully detected with DB, DB-Max, and our CENTROID algorithm. Unfortunately, other competing algorithms do not do well on this dataset. Both RBDA and RDOS detect *C* and *F*, but miss other global outliers. LOF, COF, INFLO, and LDOF miss all.

To further test our method, when $k = 9$, the small cluster should also be treated as an outlier cluster. The algorithms should return 14 outliers consisting of 8 points of the small cluster and 6 global outliers. In this case, we are more concerned with the success of detecting all 14 outliers, so all outliers are marked in red plus. The mining results of each algorithm are shown in Fig. 4.12.

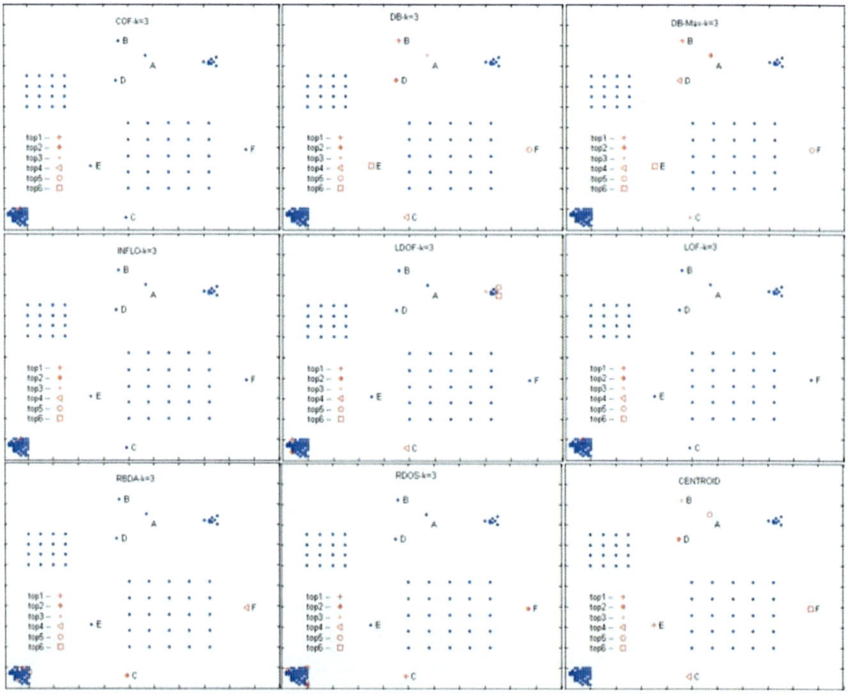

Fig. 4.11 Outlier detecting results for Synthetic Dataset 2 for $k = 3$

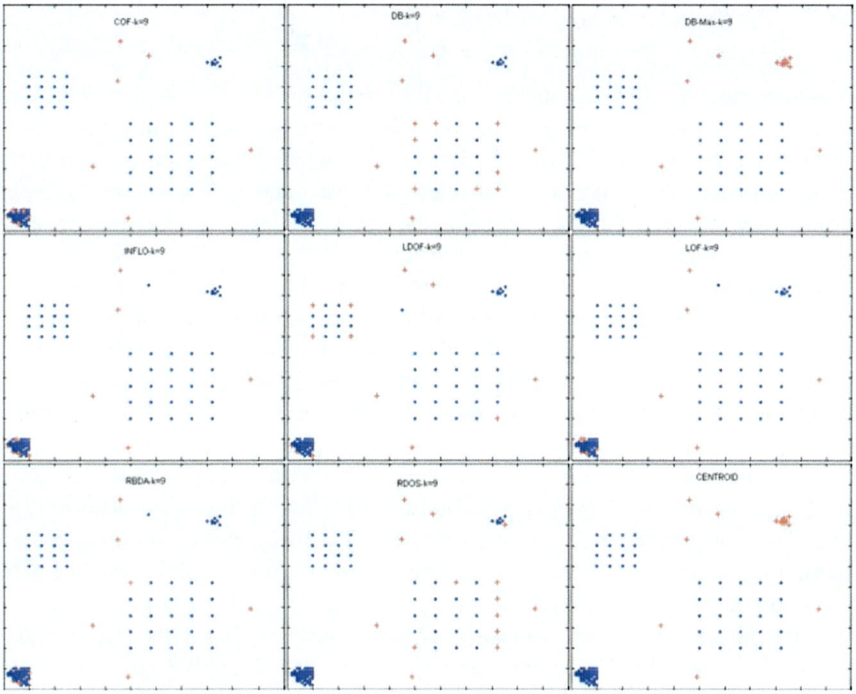

Fig. 4.12 Outlier detecting results for Synthetic Dataset 2 for $k = 9$

From Fig. 4.12, it can be seen that only DB-Max and our algorithm successfully identified all outliers as expected. The DB algorithm fails to identify the dense cluster containing of 8 outlying points because of the influence of the nearby large sparse cluster. INFLO, LOF, and RBDA are all affected by the largest dense cluster and, therefore, only identify 5 global outliers but fail to identify 8 outliers of the small outlier cluster. LDOF and RDOS are affected by two sparse clusters. In this case, RDOS identifies all global outliers, but fails to identify outliers in the small outlier cluster. To conclude, our CENTROID algorithm identifies all 14 outliers as expected.

From Figs. 4.9, 4.10, 4.11, 4.12, the advantage of our outlier detection factors is very evident on this comparably low-dimensional data set having not only global outliers but also outlier clusters. The experimental results show that our method is suitable to detect outliers in data with clusters of different densities and shapes. To summarize, it can be observed that our method has no problems detecting all outliers and clearly offers the best ranking in two synthetic datasets while all other methods do not perform competently with detecting all the outliers one way or the other. The advantage of our outlier detection factors is very evident from these 2-dimensional datasets.

4.4.2 Performance on Real Datasets

As pointed out by Aggarwal and Yu, one way to test how well an outlier detection algorithm works is to run the method on a dataset and test the percentage of points which belong to the rare classes [37]. To evaluate the effectiveness of our proposed method on multi-dimensional real datasets, we compare the algorithms by their performance on detecting rare classes in a number of datasets of different sizes and dimensions, which are downloaded from UCI.

As summarized in Table 4.6, these datasets contain different numbers of classes, so they differ in the processing of detecting outliers. Among these datasets, Arrhythmia, IAD, sonar, wdbc, dermatology, ecoli, and glass identification contain a small number of rare classes which can be selected as outliers. Other datasets do not have rare clusters and can be used for outlier detection purpose by the process of selecting one of these clusters and randomly removing its members until the cluster becomes rare and can be used for outlier detection purpose. As for the parameters used in the algorithms, cut-thred is calculated by Eq. (4.11) for all the experiments in this subsection using our method (i.e., CENTROID). The parameter *k's* for all the other algorithms in this set of experiments are determined by error and trial so that the best performances can be achieved for them.

In the following, the performances of the algorithms on all the datasets are presented by using the ROC graphs. The horizontal axis of the ROC graph is false-positive rate (FPR), and the vertical axis is true-positive rate (TPR). TPR represents the sensitivity and reflects the coverage degree of a positive class. The definition for true positive rate is the ratio of number of positive samples correctly identified to all positive samples in the test set. FPR represents specificity and is the ratio of the number of negative samples correctly identified as positive samples to all negative samples in the test set. On the ROC graph, the corresponding curve of a better result should be above the straight line with a slope of 1. That is to say, the larger the area under curve (AUC) is, the better the algorithm represented by the curve performs. When the curve approaches a straight line with a slope of 1 (AUC $\approx$ 0.5), it means that the outlier detection algorithm has little effect. By plotting the ROC curves for different algorithms on a certain dataset on the same graph, the effects of the algorithms on outlier detection can be compared.

The heart2 dataset has 2 classes, containing 150 and 120 data objects, respectively. We use all the data objects in class '1' and randomly select 20 data objects from class '2' as outliers to constitute a new testing dataset containing 170 data objects. Experiments with different parameter value settings (i.e., $k = 5, 10, 15$) are conducted to compare the results of different algorithms. When $k = 10$, all the other algorithms achieve the best outlier detection results. The algorithms involved in the comparison include DB, DB-Max, LOF, INFLO, COF, RBDA, LDOF, and RDOS. To make the ROC curve more concise, the results of ours and three best performing algorithms, that is, DB, RBDA, and LDOF, are plotted in Fig. 4.13.

From Fig. 4.13, it can be seen that the green curve (our CENTROID algorithm) is entangled with the curves of the other competing algorithms, sometimes on top

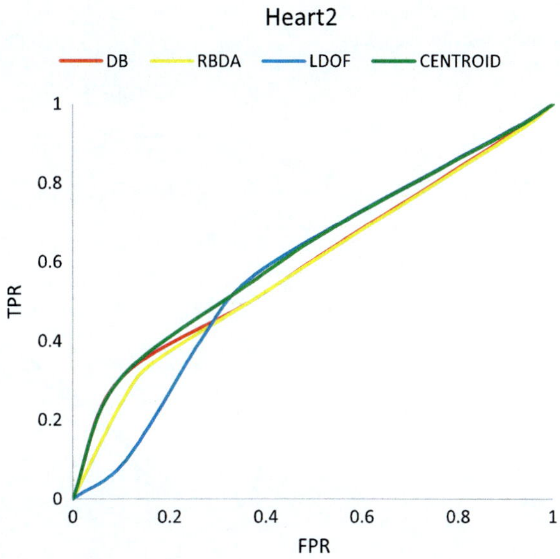

Fig. 4.13 ROC curves of some results on data set heart2

of the other curves and sometimes being coincident with the other curves (i.e., with those of the DB and RBDA algorithms in the range of 0–0.3 and with that of the LDOF algorithm in the range of 0.3–1.0). However, if the AUC of the curve is compared, all the curves are almost the same, except that only the blue curve (LDOF) is obviously below the other curves in the range of 0–0.3 where our algorithm is similar to DB (red curve) and RBDA (yellow curve). Overall, these ROC curves are close to a straight line with a slope of 1, which means that the outlier detection results of all algorithms on the heart2 dataset are not good.

The horse colic dataset has two classes, with class '1' having 99 data objects and class '2' having 201 data objects. The 99 data objects of class '1' are used as outliers. Experiments with different parameter value settings (i.e., $k = 20, 25, 30$) are conducted to compare the results of different algorithms under different parameters. All algorithms achieve the best outlier detection results when $k = 20$. In the experiments on this dataset, the results of the various algorithms are similar, so all the ROC curves are presented in Fig. 4.14.

As you can see from Fig. 4.14, the green curve (our CENTROID algorithm) is always above the other curves in the range 0.0–0.4 or coincides with other curves in the range 0.4–0.6. The only algorithm that has a curve above that of our algorithms is the reddish brown curve (RDOS). When comparing the AUC of the curves, our algorithm dominates. That is to say, our algorithm achieves better outlier detection effect than all competing algorithms. But in fact, overall, all ROC curves are close to a straight line with a slope of 1. That is to say, on the horse colic dataset, all the algorithms in this experiment do not play the role of outlier detection well.

The tae dataset has three classes, each of which contains about 50 data objects. We randomly select 10 data objects in class '1' to be the outliers which are combined

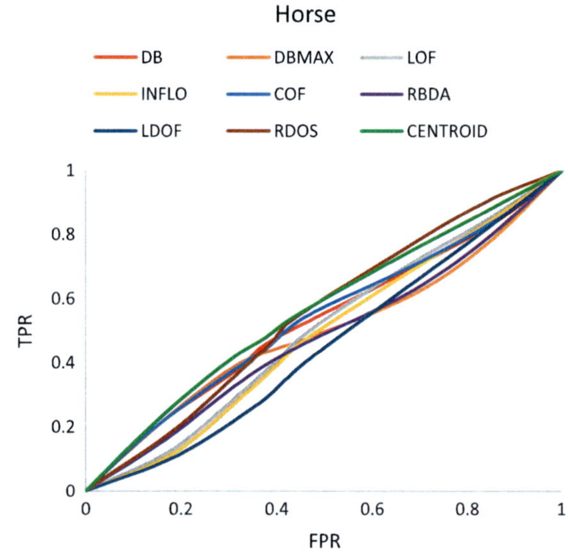

Fig. 4.14 ROC curves of some results on data set horse

with data objects in the other two classes to form a new test dataset consisting of a total of 112 data objects. Experiments with different parameter value settings (i.e., $k = 5$, 10, 15) are conducted to compare the results of different algorithms under different parameters. When $k = 5$, all algorithms achieve the best outlier detection results. The results of the three best competing algorithms, that is, DB, COF, and RDOS, and the results of our algorithm are plotted in Fig. 4.15 in terms of the ROC

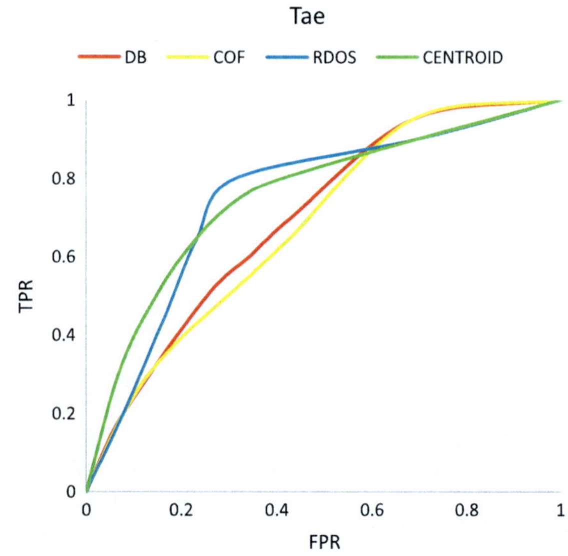

Fig. 4.15 ROC curves of some results on data set tae

graph.

As shown in Fig. 4.15, the green curve (our CENTROID algorithm) and the blue curve (RDOS) are very close with the green curve being above in the range of 0–0.2, being below in the range of 0.2–0.6, and coinciding with the blue curve in the range of 0.6–1.0. When comparing the AUC of the curves, the two curves are almost the same. Similarly, when comparing the green curve to the red curve (DB) and the yellow curve (COF), the green curve is above in the range 0–0.6 and below in the range 0.6–1.0. When the AUC of the curves is compared, the green curve is larger. That is to say, on the Australian dataset, our CENTROID algorithm and RDOS have obtained similar outlier detection performances, which is better than other comparison algorithms.

The sonar dataset contains two classes, of which class '*M*' has 11 data items and the other class contains 97 data items. We can test the performances of outlier detection algorithms by using all the data points in class '*M*' as outliers. Experiments with different parameter value settings (i.e., $k = 5, 10, 15$) are conducted to compare the results of different algorithms under different parameters. All algorithms achieve the best outlier detection results when $k = 15$. In addition to our CENTROID algorithm, the best performing algorithms are DB, LOF, and COF and the results are shown in Fig. 4.16 in terms of ROC curves.

As can be seen from Fig. 4.16, the green curve (our CENTROID algorithm) is above all the other curves in the range of 0–0.35. The green curve is above the yellow curve (LOF) in the range 0–0.7. The green curve and the yellow curve coincide in the range of 0.7–1.0. So in general, the green curve has an AUC greater than the yellow curve. As to the red curve (DB), it is below the green curve in the range of 0–0.35 and above the green curve in the range of 0.35–0.8. When the AUC of the curves is compared, the green curve is larger than the red curve. As for the blue curve

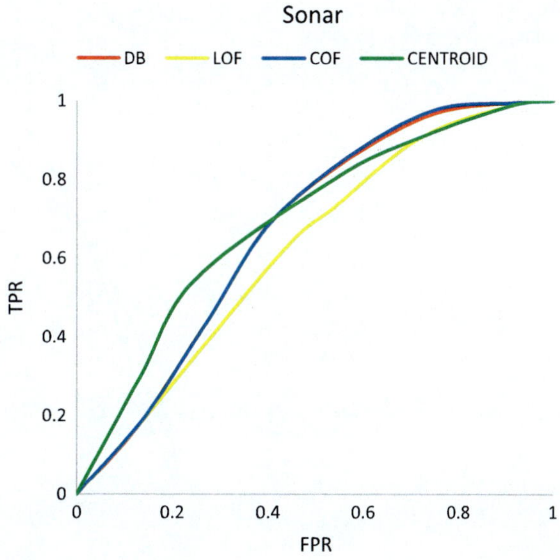

Fig. 4.16 ROC curves of some results on data set sonar

(COF), it is below the green curve in the range of 0–0.35, but above the green curve in the subsequent range. When comparing the AUC of the curves, the green curve and the blue curve are of not much difference. So in general, on the sonar dataset, our CENTROID algorithm has similar efficiency with some algorithms and achieves better outlier detection than most other algorithms.

There are 7 classes in the glass identification dataset. The class '6' has 9 data items, which is the smallest cluster of all classes. Other classes have at least 13 data items. Therefore, the 9 data items of class '6' are treated as outliers, which is appropriate in the case of a total number of data points 214. Experiments with different parameter value settings (i.e., $k = 5, 10, 15$) are performed to compare the results of different algorithms under different parameters. The three competing algorithms with the best results are DB, INFLO, and COF. The results of these three algorithms together with our algorithm are plotted in Fig. 4.17.

As can be seen from Fig. 4.17, the yellow curve (INFLO) is clearly below the other curves in the range of 0.2–1.0, which means the results obtained are relatively poor. The green curve (our algorithm) is above the blue curve (COF) in the range 0–0.2, above the red curve (DB) in the range 0–0.35, and below the other two curves at other times. When the AUC of all the curves is compared, our algorithm is not much different from the other two curves (DB and COF). In general, the outlier detection results obtained by our CENTROID algorithm on the glass identification data set are close to those of DB and COF and better than other competing algorithms.

The australian data set has 2 classes, containing 383 data items and 307 data items, respectively. Class '0' and 30 data items selected randomly from class '1' to be used as outliers form a new test dataset containing 413 data. Experiments with different parameter value settings (i.e., $k = 5, 10, 15$) are conducted to compare the results of

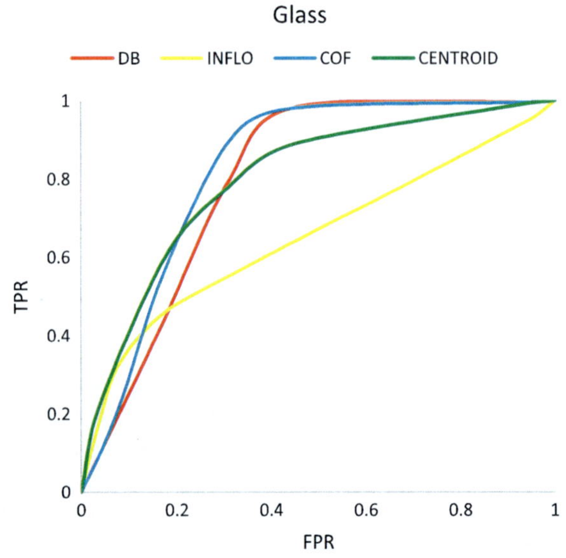

Fig. 4.17 ROC curves of some results on data set glass

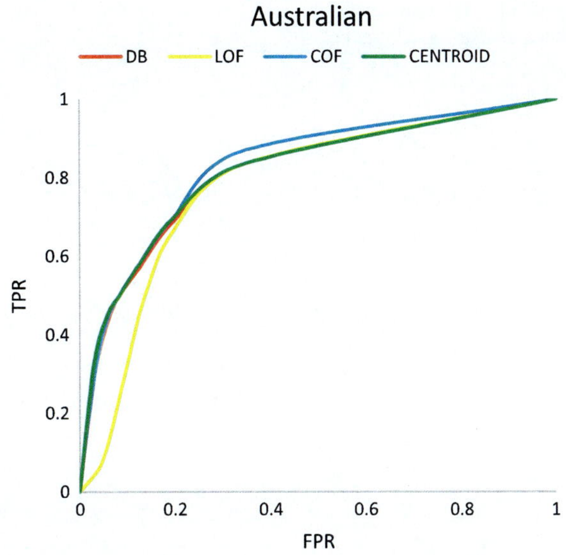

Fig. 4.18 ROC curves of some results on data set Australian

different algorithms under different parameters. When $k = 10$, all algorithms achieve the best outlier detection results. The best three results of the competing algorithms, that is, DB, LOF and COF, are presented together with the results of our algorithm in Fig. 4.18.

As shown in Fig. 4.18, the green curve (our CENTROID algorithm) and the red curve (DB) almost coincide. The green curve is above the yellow curves (LOF) in the range of 0–0.2, and the two are substantially coincident in the range of 0.2–1.0. When compared with the blue curve (COF), the two curves coincide in the range of 0–0.3, and the green curve is below in the range of 0.2–1. In other words, the COF algorithm has achieved better results. However, the AUCs of the two curves are of not much difference. In general, on the australian dataset, our algorithm and the DB and COF algorithms have achieved close outlier detection performances and are better than the other outlier detection algorithms.

The ecoli dataset has 8 classes, of which '5', '7', '8' have a total of 9 data items while the other classes have at least 35 data. Therefore, it is appropriate to use the 9 data items of classes '5', '7', and '8' as outliers. Experiments with different parameter value settings (i.e., $k = 5, 10, 15$) are done to compare the results of different algorithms under different parameters. All algorithms achieve the best results when $k = 10$. On this dataset, the performances of several algorithms are relatively close, so all are drawn in Fig. 4.19. The results of DB-Max and our centroid algorithm are exactly the same, so only ours is drawn in Fig. 4.19.

As shown in Fig. 4.19, the yellow curve (LOF) is either below or coincides with other curves and therefore performs the worst among these competing algorithms. The red curve (DB) and the yellow curve (LOF) almost coincide, and the difference is very small. The green curve (our CENTROID algorithm) is clearly above the other

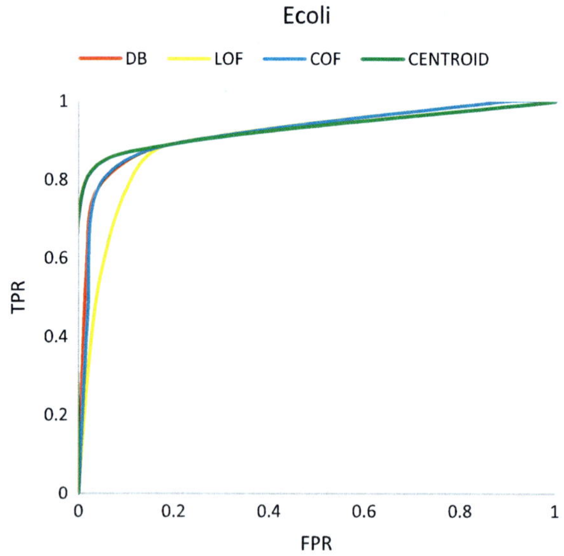

Fig. 4.19 ROC curves of some results on data set ecoli

curves in the range of 0–0.2 and below the blue curve (COF) slightly in the range of 0.2–1. When the AUC of the curves is compared, the green curve has a larger one than the blue curve. As a result, our CENTROID algorithm is better than COF. Taken together, our algorithm has an outlier detection result close to those of LOF and COF on the ecoli dataset and is better than the other competing algorithms.

The hepatitis dataset has two classes, with 85 data objects and 70 data objects, respectively. We use 85 data objects of class '1' and randomly select 10 data objects from class '2' as outliers to form a new test dataset containing 95 data objects. Experiments with different parameter value settings (i.e., $k = 15, 20, 25$) are conducted to compare the results of different algorithms under different parameters. When $k = 15$, all algorithms achieve the best outlier detection results, and the three competing algorithms with the best results are DB, LOF, and DOS. The results of these three algorithms and our algorithm are plotted in Fig. 4.20.

From the experimental results shown in Fig. 4.20, it can be seen that the curves of the three competing algorithms (DB, LOF, and RDOS) are below the other curves within a certain range. The yellow curve (LOF) has achieved the largest value, and the red curve (DB) has achieved the smallest value when the AUCs of their curves are compared. The blue curve (RDOS) is between them. The green curve (our CENTROID algorithm) is either always above the other curves or coincides with other curves, and is only slightly above the yellow curve (LOF) in the range of 0–0.2. That is to say, our CENTROID algorithm has achieved the best outlier detection results on the hepatitis dataset.

Spect heart contains two classes, containing 148 and 119 data items, respectively. We construct a new test dataset with all the data objects in class '1' and 10 data objects randomly selected from class '0' to be considered as outliers, resulting in

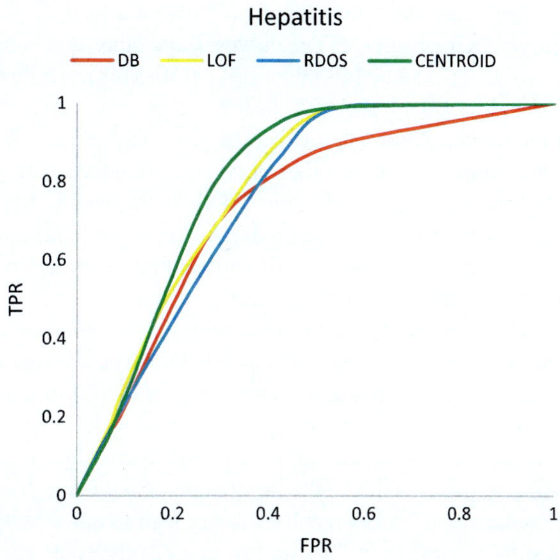

Fig. 4.20 ROC curves of some results on data set hepatitis

a dataset containing 129 data objects. Experiments with different parameter value settings (i.e., $k = 15, 20, 30$) are performed to compare the results of different algorithms under different parameters. When $k = 15$, all algorithms achieve the best outlier detection results. The four algorithms with the best results in the comparison with our algorithm are DB, DB-Max, COF and RBDA. The results of these three algorithms and the results of our algorithm are plotted in Fig. 4.21.

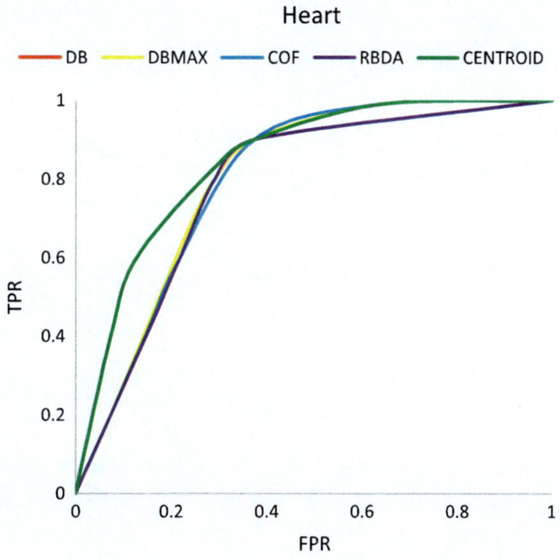

Fig. 4.21 ROC curves of some results on data set heart

From the experimental results shown in Fig. 4.21, the red curve (DB) and the blue curve (COF) coincide. The curves of the other two competing algorithms, the purple curve (RBDA) and the yellow curve (DB-Max), are above all others in a certain range. When comparing the AUC of the curves, the purple curve and the yellow curve are of not much difference. As for the green curve (our CENTROID algorithm), it either always precedes over other curves or coincides with other curves, while it is only slightly below the blue curve (COF) in the range of 0.4–0.6. However, as far as the AUC of the curves is concerned, the green curve is significantly larger than the other curves. In other words, our centroid algorithm achieves the best outlier detection results on the spect heart dataset.

The dermatology dataset contains six classes, of which class '6' has 20 data and the other classes contain at least 49 data. The data set has a total of 366 data points, and therefore, it is appropriate to select the 20 data points in class '6' as outliers. Experiments with different parameter value settings (i.e., $k = 5, 10, 20, 30, 35$) are conducted to compare the results of different algorithms under different parameters. All algorithms achieve the best results when $k = 30$. Our CENTROID algorithm on this dataset achieves results that are significantly better than other algorithms. The results are shown in Fig. 4.22 in terms of the ROC curves of all algorithms.

In Fig. 4.22, the green curve (our CENTROID algorithm) is above all the other curves all the time. In addition, it should be pointed out that, since the DB-Max algorithm achieves the same results as our algorithm, the ROC curve corresponding to DB-Max is not drawn. In general, on the dermatology dataset, our algorithm and DB-Max have achieved the same outlier detection results, which is better than all the other outlier detection algorithms.

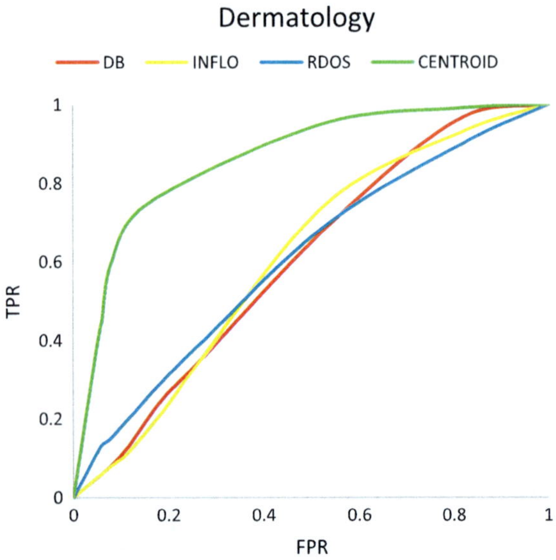

Fig. 4.22 ROC curves of some results on data set heart

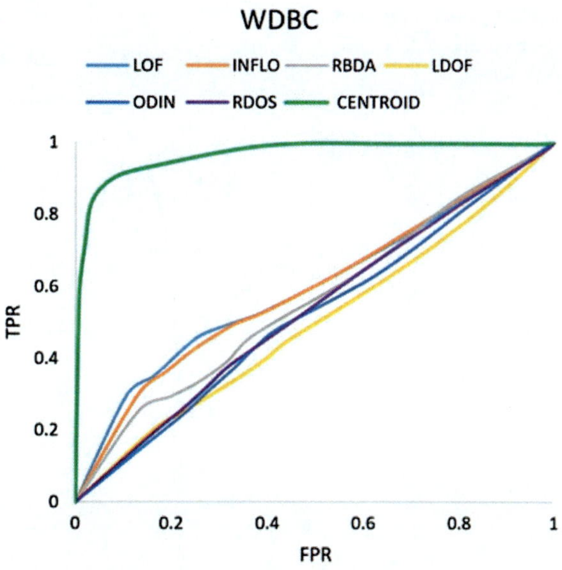

Fig. 4.23 ROC curves of some results on data set wdbc

The class '*M*' in the Breast Cancer Wisconsin (wdbc) dataset contains 212 data objects and is considered as outliers in the performance evaluation tests. Experiments with different parameter values (i.e., $k = 10, 20, 30$) are done to compare the results of different algorithms under different parameters. All algorithms achieve the best outlier detection results when $k = 10$. In addition to our CENTROID algorithm, the best performing algorithms are DB, DB-Max, and COF. And the performances of these three competing algorithms and our CENTROID algorithm are almost the same and, therefore, are not shown together with the results of the other algorithms in Fig. 4.23.

From Fig. 4.23, it can be seen that the green curve (our CENTROID algorithm) is significantly above the other curves. When the green curve is around FPR of 0.1, the TPR reaches 0.9. At FPR of 0.4, its TPR is close to 1.0. That is to say, our algorithm performs quite good on the Breast Cancer Wisconsin (wdbc) dataset. The other curves are close to the straight line with a slope of 1, which means that the algorithms represented by these curves are not good at outlier detection task performed on this dataset and the random guess is almost impossible to detect outliers. Overall, our algorithm has achieved significantly better results on this dataset and is much better than most competing algorithms.

The abalone dataset contains 29 classes, each with a different number of data points. The class that contains most of the data points is class '9', which has 689 data objects. The smallest classes, that is, classes '1', '2', '24', '25', '26', '27', and '29', have a total of 9 data objects. Other classes contain at least 6 data. In comparison with the total number of 4177 data points of this dataset, these 9 data points from different classes can be regarded as outliers in the following outlier detection tests. Experiments with different parameter value settings (i.e., $k = 5, 10, 15$) are done to

Fig. 4.24 ROC curves of some results on data set Australian

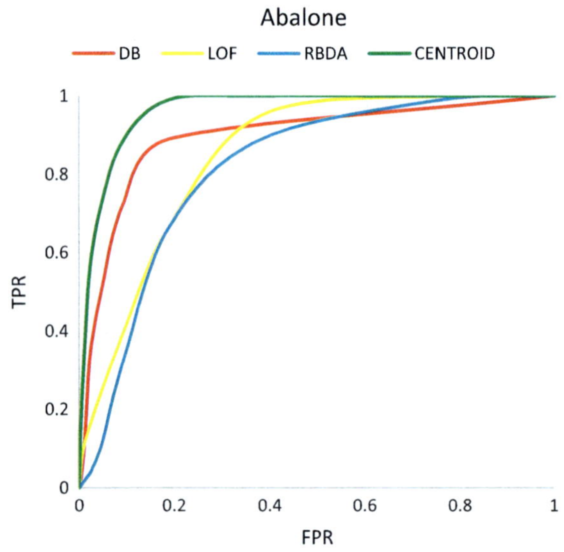

compare the results of different algorithms under different parameters. All algorithms achieve the best results when $k = 5$. The best performing algorithms in the comparison include LOF, RBDA, DB, DB-Max, and COF. The results of DB, DB-Max, and COF are the same, so only the results of the LOF, RBDA, and DB algorithms are drawn in the ROC graph of Fig. 4.24.

As can be seen from Fig. 4.24, the yellow curve (LOF) is either always above or coincides with blue curve (RBDA). In other words, LOF has achieved better results than RBDA. In comparison with the red curve (DB) and the yellow curve (LOF), they all exceed one and another in different ranges. But when comparing the AUC of these two curves, the red curve has a larger one. So, the result of the DB algorithm on this data set is the best among all the competing algorithms. As to the green curve (our CENTROID algorithm), it is always above the other curves. So, on the whole, our algorithm has achieved the best outlier detection results on the abalone dataset.

To summarize, we have conducted experiments on several datasets of different sizes and dimensions using state-of-the-art outlier detection algorithms. The results of the experiments are different. However, overall, our algorithm is better than all other competing algorithms on most datasets in most cases. In short, our algorithm is at least no worse than all other algorithms. From this set of experiments, we come to a conclusion that our method can outperform the state-of-the-art outlier detection techniques in classification accuracy in real situations.

4.4.3 Performance on High-Dimensional Real Datasets

To test the scalability of our algorithm with regard to dimensionality, we next conduct experiments on three higher-dimensional datasets to evaluate our proposed outlier detection method. The three data sets used in this set of experiments, Arrhythmia, Internet Advertisements (IAD), and Musk, are obtained from the UCI and briefly summarized in Table 4.6.

The Arrhythmia dataset comes from H. Altay Guvenir's study of cardiac arrhythmia, which contains 16 classes. While class '1' is a large cluster with 245 data objects, all other classes contain data points ranging from 2 to 50. We choose data objects in categories '5' and '9' (which contain a total of 22 data objects (13 and 9 data objects, respectively)) as outliers. Different k values (i.e., $k = 10, 20, 30$) are set in the experiments to make a performance evaluation among all the algorithms. Most algorithms achieve the best results when $k = 20$. The experimental results are presented in Fig. 4.25 in the form of the ROC plots.

The best performing algorithms are INFLO, COF, and RBDA. In Fig. 4.25, the curve for our CENTROID algorithm is shown in green. As can be seen from Fig. 4.25, the green curve is below the yellow curve (COF) and blue curve (RBDA) in the FPR range of about 0.05–0.25. In other areas, the green curve is above all other algorithms. When comparing the green curve with the yellow curve (COF), except for the range of 0.05–0.25, the blue curve is above or coincides with the yellow curve. Moreover, it can be clearly seen that the AUC of the green curve is significantly larger than that of the yellow curve. When comparing the green curve with respect to the blue curve (RBDA), the green curve is above the blue curve except for the range of 0.05–0.25. The AUC of the green curve is significantly larger than that of the blue curve. As

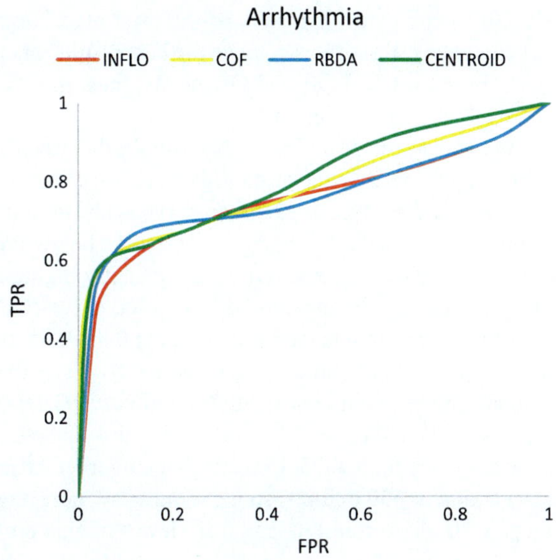

Fig. 4.25 ROC curves of some results on data set Arrhythmia

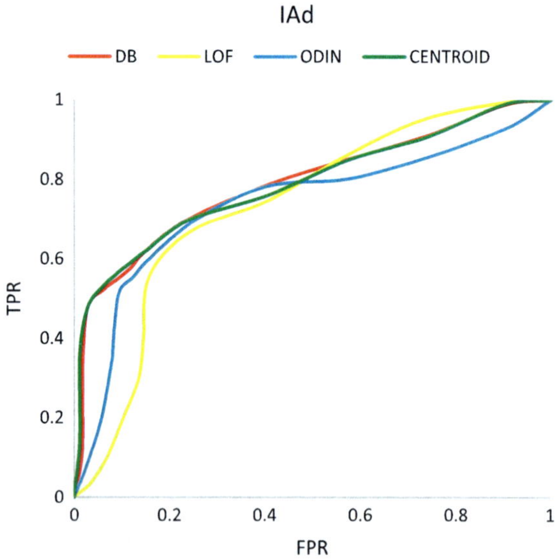

Fig. 4.26 ROC curves of some results on data set IAD

for the red curve (INFLO), it is always below or coincides with the green curve. Therefore, represented by the green curve, our CENTROID algorithm has achieved the best effectiveness on the Arrhythmia dataset than all the competing algorithms.

The Internet Advertisements (IAD) dataset represents a set of possible advertisements on Internet pages. Since there are 459 data items in 'ad' class, which is a minority in comparison with all the other classes (which contain 2820 data items in total), these data points can be regarded as outliers. Similar to the above tests, the experiments are performed under different parameter value settings (i.e., $k = 5$, 10, 20) to compare the results of different algorithms under different parameters. All competing algorithms achieve the best results when $k = 10$. The best competing algorithms are DB, LOF, and ODIN. We show the results of these three algorithms and that of our algorithm in Fig. 4.26.

As can be seen from Fig. 4.26, overall, the green curve (our CENTROID algorithm) is not always above the other curves. When comparing the green curve to the yellow curve (LOF), the green curve is above in the range of about 0–0.43. But in the range of 0.43–0.9, the green curve is below. When the AUCs of the curves are compared, the green curve is significantly higher than the yellow curve. When the green curve is compared with the blue curve (ODIN), the green curve is only below the blue curve in the range of about 0.35–0.55, but the green curve has a much larger value than the blue curve when the AUCs of the curves are compared. When comparing the green curve with the red curve (DB), the green curve mostly coincides with the red curve. The green curve is only below the red curve in the range of approximately 0.25–0.75. In fact, for the Internet Advertisements (IAD) dataset, we have a total of 459 outliers, so we should pay more attention to the curve in the FPR range of 0–0.3. From this point of view, the blue curve is slightly higher than the

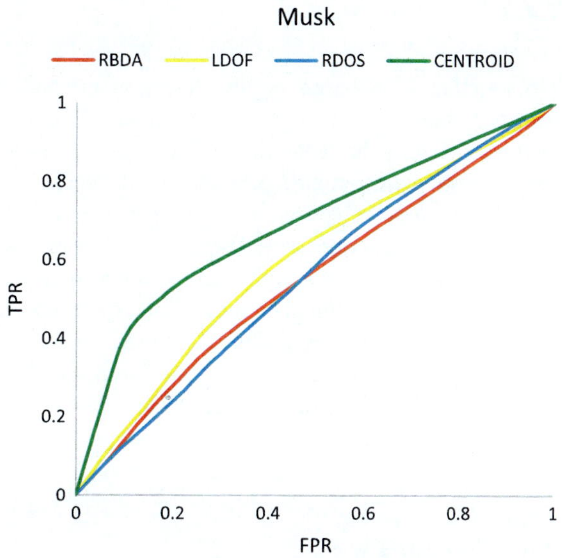

Fig. 4.27 ROC curves of some results on data set heart2

green curve in the range of 0–0.25. Between 0.25 and 0.3, the blue curve is below the green curve. So in general, on the Internet Advertisements (IAD) dataset, our outlier detection algorithm is better than other comparison algorithms.

Musk (version 2) has 2 classes, in which class '1' has 1017 data objects. Since the dataset has a total of 6598 data points, the data points in class '1' are considered as outliers in the tests. Experiments with different parameter values (i.e., $k = 10$, 20, 30) are conducted to compare the results of different algorithms under different parameters. All algorithms achieve the best outlier detection results when $k = 10$. Among them, the three best competing algorithms for this outlier detection task are RBDA, LDOF, and RDOS. Their ROC curves are presented in Fig. 4.27 together with that of ours.

As can be seen from Fig. 4.27, the red curve (RBDA) is always below the yellow curve (LDOF), which means that the result of RBDA is not as good as LDOF. While the blue curve (RDOS) is compared to the yellow curve, the blue curve is always below the yellow curve in the range of 0–0.8. The blue curve slightly exceeds the yellow curve only in the range of 0.8–1.0. In other words, LDOF is the best of the three competing algorithms. The green curve (our CENTROID algorithm) has been on top of other curves all the time. Overall, the three competing algorithms are close to a straight line with a slope of 1, which means that the outlier detection results of these three algorithms on this data set are not good. In other words, our algorithm achieves the best outlier detection results, while other algorithms have poor results.

4.4.4 Discussion

The existing kNN-based outlier detection methods are either distance-based or density-based. We propose a theoretical kNN-centroid-based outlier detection method and study the capabilities of this and other schemes on datasets with different characteristics. In comparing these methods, we find that the density-based schemes are effective in an environment where patterns possess sufficiently higher densities than outliers, while ours works better for relatively isolated outliers. Our study also shows that we should not view one method as being superior to others in all aspects but use it as a compliment to instead of as a replacement of others in applications with different requirements. This is intuitive because, in reality, it is usually difficult to detect all the outliers that fit users' intuitions. Thus, it is probably meaningful to incorporate our proposed outlier factors as a component into the current outlier detection framework.

4.5 Conclusions

In this chapter, we have proposed an effective kNN-centroid-based outlier detection method that can detect both global and local outliers. In addition to being compatible with the traditional distance-based outlier detection methods, our approach also performs better in identifying local outliers that deviate from the main patterns in a given dataset. Candidate outliers are ranked based on the distance-based and density-based outlier scores that are assigned to each data point. Both theoretical justification and empirical validation of the effectiveness of the method have been conducted. To demonstrate the utility of our proposed outlier factors, we performed a detailed comparison with a number of state-of-the-art methods. Through a thorough evaluation, we demonstrate the ability of our new approach to rank the best candidates for being an outlier with high precision and recall. Experiments on different datasets demonstrate the effectiveness of the new algorithm. Compared with the classical and state-of-the-art outlier detection algorithms, we prove that our algorithm has higher detection efficiency.

In addition to the basic approach, we propose to use different numbers of nearest neighbors for different data points in the determination of outlier scores, which are obtained by minimizing the distance between each data point and its kNN-based centroid. It has been shown that a small number of top global and local outliers can be detected effectively by our methods together with outlier indicators. Furthermore, the idea of using our kNN-centroid-based outlier scores can be effective on mining high-dimensional data. For future work, it would be interesting to conduct a more extensive study on larger and higher-dimensional datasets and compare the performance of outlier detection approaches especially designed for high-dimensional data with the ones proposed in this chapter.

References

1. Eskin, E., Arnold, A., Prerau, M., Portnoy, L., & Stolfo, S. (2002). A geometric framework for unsupervised anomaly detection: Detecting intrusions in unlabeled data. In D. Barbará & S. Jajodia (Eds.), *Applications of Data Mining in Computer Security. Advances in Information Security* (Vol. 6, pp. 77–101).
2. Lane, T., & Brodley, C. E. (1998). Temporal sequence learning and data reduction for anomaly detection. In *Proceedings of the 1998 5th ACM Conference on Computer and Communications Security* (CCS-5), San Francisco, CA, USA, pp. 150–158.
3. Bolton, R. J., & David, J. H. (2002). Unsupervised profiling methods for fraud detection. *Statistical Science, 17*(3), 235–255.
4. Wong, W., Moore, A., Cooper, G., & Wagner, M. (2002). Rule-based anomaly pattern detection for detecting disease outbreaks. In *Proceedings of the 18th National Conference on Artificial Intelligence*, Edmonton, Alta., Canada, pp. 217–223.
5. Sheng, B., Li, Q., Mao, W. and Jin, W. (2007). Outlier detection in sensor networks. In P*roceedings of ACM International Symposium on Mobile Ad Hoc Networking and Computing*, pp.219–228.
6. Hodge, V. J., & Austin, J. (2004). A survey of outlier detection methodologies. *Artificial Intelligence Review, 22*(2), 85–126.
7. Chandola, V., Banerjee, A., & Kumar, V. (2009). Anomaly detection: A survey. *ACM Computing Surveys, 41*(3), 15.1–15.58.
8. Knorr, E. M., & Ng, R. T. (1998). Algorithms for mining distance-based outliers in large datasets. In *Proceedings of the International Conference on Very Large Data Bases* (VLDB'98), New York, pp. 392–403.
9. Breuning, M.M., Kriegel, H.P., Ng, R.T. and Sander, J. (2000). LOF: Identifying density-based local outliers. In *Proceedings of the 2000 ACM SIGMOD International Conference on Management of Data* (SIGMOD'00), Dallas, TX, United states, pp. 93–104.
10. MacQueen, J. (1965). Some methods for classification and analysis of multivariate observations. In *Proceedings of the 5th Berkeley Symposium on Mathematical Statistics and Probability*, pp. 281–297.
11. Kaufman, L., & Rousseeuw, P. (1990). Finding groups in data: an introduction to cluster analysis. John Wiley & Sons.
12. Jiang, M. F., Tseng, S. S., & Su, C. M. (2001). Two-phase clustering process for outliers detection. *Pattern Recognition Letters, 22*(6–7), 691–700.
13. He, Z., Xu, X., & Deng, S. (2003). Discovering cluster-based local outliers. *Pattern Recognition Letters, 24*(9–10), 1641–1650.
14. Hautamäki, V., Cherednichenko, S., Kärkkäinen, I., Kinnunen, T., Fränti, P. (2005). Improving *K*-means by outlier removal. In *Proceedings of the 14th Scandinavian Conference on Image Analysis* (SCIA'05), Joensuu, Finland, pp. 978–987.
15. Rehm, F., Klawonn, F., & Kruse, R. (2007). A novel approach to noise clustering for outlier detection. *Soft Computing, 11*(5), 489–494.
16. Zhou, Y., Yu, H., & Cai, X. (2009). A novel k-means algorithm for clustering and outlier detection. In *Proceedings of the 2nd International Conference on Future Information Technology and Management Engineering* (FITME'09), Sanya, China, pp. 476–480.
17. Zhang, K., Hutter, M., & Jin, H. (2009). A new local distance-based outlier detection approach for scattered real-world data. In *Proceedings of the 13th Pacific-Asia Conference on Advances in Knowledge Discovery and Data Mining* (PAKDD '09), Bangkok, Thailand, pp. 813–822.
18. Pamula, R., Deka, J., & Nandi, S. (2011). An outlier detection method based on clustering. In *Proceedings of the 2nd International Conference on Emerging Applications of Information Technology* (EAIT'11), pp. 253–256.
19. Ahmed, M., & Naser, A. (2013). A novel approach for outlier detection and clustering improvement. In *Proceedings of the 8th IEEE Conference on Industrial Electronics and Applications* (ICIEA'13), Melbourne, VIC, Australia, pp. 577–582.

20. Chawla, S., & Gionis, A. (2013). K-means: a unified approach to clustering and outlier detection. In *Proceedings of the 13th SIAM International Conference on Data Mining* (SDM'13), Austin, TX, United states, pp. 189–197.
21. Whang, J., Dhillon, I.S., & Gleich, D. (2015). Non-exhaustive, overlapping K-means. In *Proceedings of the SIAM International Conference on Data Mining* (SDM'15), Vancouver, BC, Canada, pp. 936–944.
22. Yu, Q., Luo, Y., Chen, C., & Ding, X. (2016). Outlier-eliminated K-means clustering algorithm based on differential privacy preservation. *Applied Intelligence, 45*(4), 1179–1191.
23. Aparna, K., Nair, M. K. (2016). Effect of outlier detection on clustering accuracy and computation time of CHB *K*-means algorithm. In *Proceedings of the 2nd International Conference on Computational Intelligence in Data Mining* (ICCIDM'15), Berhampur, OR, India, (Vol. 411(1), pp. 25–35).
24. Barai, A., & Dey, L. (2017). Outlier detection and removal algorithm in *K*-means and hierarchical clustering. *World Journal of Computer Application and Technology, 5,* 24–29.
25. Diez-Olivan, A., Pagan, J. A., Sanz, R., & Sierra, B. Data-driven prognostics using a combination of constrained K-means clustering, fuzzy modeling and LOF-based Score. *Neurocomputing, 241,* 97–107.
26. Gan, G., & Ng, M.K.-P. (2017). K-means clustering with outlier removal. *Pattern Recognition Letters, 90,* 8–14.
27. Zhao, S., Li, W., & Cao, J. (2018). A user-adaptive algorithm for activity recognition based on K-means clustering, local outlier factor and multivariate Gaussian distribution. *Sensors, 18*(6), 1850.
28. Karczmarek, P., Kiersztyn, A., Pedrycz, W., & Al, E. (2020). K-means-based isolation forest. *Knowledge-Based Systems, 195.* https://doi.org/10.1016/j.knosys.2020.105659
29. Ramaswamy, S., Rastogi, R., & Shim, K. (2000). Efficient algorithms for mining outliers from large data sets. In *Proceedings of the 2000 ACM International Conference on Management of Data* (SIGMOD'00), Dallas, pp. 427–438.
30. Angiulli, F., & Pizzuti, C. (2002). Fast outlier detection in high dimensional spaces. In *Proceedings of the 6th European Conference on Principles of Data Mining and Knowledge Discovery* (PKDD'02), Helsinki, pp. 15–26.
31. Tang, J., Chen, Z., Fu, A. W. C., Cheung, D. W. (2002). Enhancing effectiveness of outlier detections for low density patterns. In *Proceedings of the 6th Pacific-Asia Conference on Knowledge Discovery and Data Mining (PAKDD'02)*, Taipei, Taiwan, pp. 535–548.
32. Jin, W., Tung, A. K. H., Han, J., Wang, W. (2006). Ranking outliers using symmetric neighborhood relationship. In *Proceedings of the 10th Pacific-Asia Conference on Knowledge Discovery and Data Mining (PAKDD'06)*, Singapore, pp. 577–593.
33. Huang, H., Mehrotra, K., & Mohan, C. K. (2013). Rank-based outlier detection. *Journal of Statistical Computation and Simulation, 83*(3), 518–531.
34. Tang, B., & He, H. (2017). A local density-based approach for outlier detection. *Neurocomputing, 241,* 171–180.
35. Li, X., Wu, X., Lv, J., He, J., Guo, J., & Li, M. (2017). Automatic detection of boundary points based on local geometrical measures. *Soft Computing, 22*(11), 3663–3674.
36. UCI: The UCI KDD Archive. https://kdd.ics.uci.edu/. Irvine, CA: University of California.
37. Aggarwal, C., & Yu, P. (2001). Outlier detection for high-dimensional data. In Proceedings of *the 2001 ACM International Conference on Management of Data* (SIGMOD'01), Santa Barbara, CA, USA, pp. 37–46.

Chapter 5
A Minimum Spanning Tree Clustering-Inspired Outlier Detection Technique

Abstract Due to its important applications in data mining, many techniques have been developed for outlier detection. In contrast to k-nearest neighbor-based outlier detection techniques such as distance-based and density-based algorithms whose results have a relatively strong sensitiveness to the setting of the parameters, clustering-based outlier detection algorithms regard data items in the small groups as outliers and often obtain them as a by-product. Unlike K-means clustering algorithms, minimum spanning tree-based clustering algorithms can find clusters of arbitrary shapes, different sizes, and different densities. However, minimum spanning tree clustering-based outlier detection approaches may incur high computational costs. To partially circumvent these problems, in this chapter, an efficient outlier detection technique is proposed which is inspired by minimum spanning tree-based clustering. Extensive performance evaluations on synthetic as well as real datasets are conducted to show that the proposed approach works well for identifying global as well as local outliers with respect to the state-of-the-art outlier detection methods.

Keywords Distance-based outlier detection · Density-based outlier detection · Clustering-based outlier detection · Minimum spanning tree-based clustering · Minimum spanning tree-based outlier detection

5.1 Introduction

Human species cannot survive without the ability to discover the anomalous patterns in the observed data that do not conform to the expected normal behavior. In statistics, anomalous patterns are often referred to as outliers, although other names such as anomalies, discordant observations, exceptions, faults, defects, aberrations, noise, errors, damage, surprise, novelty, peculiarities, or contaminants are also used in different scientific areas. In a classic definition, 'an outlier is an observation that deviates so much from other observations that it aroused suspicions that it is generated by a different mechanism' [1]. On the one hand, the presence of an outlier indicates some sort of problem. For example, outliers can be unintentionally produced by human either due to recording and measurement errors (such as incorrect data entry and missing values), or due to incorrect distribution assumptions given for an

X. Wang et al., *New Developments in Unsupervised Outlier Detection*,
https://doi.org/10.1007/978-981-15-9519-6_5

unknown data structure (such as applying a single model for a homogeneous set to a rather heterogeneous set of two or more classes). On the other hand, outliers can be the results of the uncontrollable natural variation of population or process and happen when rare events or novel phenomena arise [2]. In this case, the identification of outliers can lead to the discovery of useful knowledge and exploring why outliers exist in a particular domain can benefit us. Actually, many great discoveries in human history can be traced to researchers exploring some outlying or unusual data. Due to their important applications, outlier detection has found immense use in a wide variety of practical domains, such as intrusion detection for cyber-security [3], fraud detection for credit cards, insurance and tax [4], early detection of disease outbreaks in the medical field [5], fault detection in sensor networks for monitoring health, traffic, machine status, weather, pollution, surveillance [6], and so on [7, 8].

As a result, the task of finding outliers in a data set has long been an area of active research, different outlier detection techniques, such as distribution-based, depth-based, distance-based, density-based, and clustering-based approaches, have been developed, and no techniques are completely satisfactory for all the situations. Originating from statistics, distribution-based methods study the outlier detection problem in the context of a given distributional model (e.g., normal, Poisson, etc.) [9–11]. However, for modern large databases, in most cases, the distributional model of a data set is not known a priori, and it is not practical to test which distributional model captures the spread of the data. As a result, they have limited usefulness. Stemming from computational geometry, depth-based methods organize the observations into layers through the computation of k-dimensional convex hulls. Outliers are defined according to a definition of depth under the assumption that the shallow layers are more likely to contain outliers than deep ones. These methods eliminate the need to fit a distribution model to data but are very computationally expensive for more than a few dimensions [12]. Distance-based methods and density-based methods are mainly k-nearest neighbor-based approaches that actually address the detection problems of two different notions of outliers: more globally oriented outliers (the former) and more locally oriented outliers (the latter). Clustering-based approaches obtain outliers as the by-products of clustering; that is, outliers are the data items that reside in the smallest clusters. Other types of outlier detection algorithms have also been developed to look into the problem from different aspects, either trying to mine the outliers in a more efficient way [13] given an outlier definition, or trying to mine the outliers in a more accurate way but with quadratic even cubic running time [14–16]. However, they are beyond the scope of this study and will not be discussed further here.

Unlike K-means clustering algorithms which group data points around some 'centroids' and therefore do not work well when the boundaries of the clusters are irregular, graph theory-based methods, typified by minimum spanning tree (MST)-based clustering algorithms, can find clusters of arbitrary shapes, different sizes, and different densities. In this chapter, we propose an efficient in-memory minimum spanning tree clustering-inspired outlier detection mechanism for mining global as well as local outliers in modern large multi-dimensional datasets. This is promising because, with random access memory getting cheaper, it becomes increasingly affordable to build computers with larger and larger main memories so as to store the whole database in memory for faster system response. Our first contribution in this chapter is that the proposed approach combines the advantages of distanced-based, density-based, and clustering-based outlier detection to give a better intuition to view such outlier detection techniques which can converge without the number of outliers given a priori. Our second contribution is that it can detect both global outliers and local outliers, and can provide accurate ranking with reduced dependency of the results on some man-set parameters. Finally, extensive experimental evaluation on both synthetic and real datasets demonstrates the effectiveness and efficiency of the proposed outlier detection approach.

The rest of this chapter is organized as follows. In Sect. 5.2, we provide a short introduction to minimum spanning tree (MST)-based clustering and review some existing work on MST-clustering-based outlier detection algorithms. We next present our proposed approach in Sect. 5.3. In Sect. 5.4, an empirical study is conducted to evaluate the performances of our algorithm with respect to some state-of-the-art outlier detection algorithms. Finally, conclusions are made in Sect. 5.5.

5.2 Background

There are two parts of the unsupervised learning literature that are related to our study: minimum spanning tree-based clustering and minimum spanning tree clustering-based outlier detection.

5.2.1 Minimum Spanning Tree-Based Clustering

As a very important data mining tool, the main concern of clustering algorithms is to find clusters. Relying on a notion of similarity (Euclidean distance is used in this research), clustering aims to group the objects of a database into clusters by optimizing some criterion, such as maximizing the intra-cluster similarity (minimizing the intra-cluster distance) and minimizing the inter-cluster similarity (maximizing the inter-cluster distance). For example, being a branch of hierarchical clustering techniques (which start by treating each data point as a singleton cluster and then successively merge clusters until all points have been merged into a single remaining cluster), single-linked clustering algorithms merge in each step the two clusters whose two closest members have the smallest distance (or, the two clusters with the smallest minimum pairwise distance), while complete-linked clustering algorithms merge in each step the two clusters whose merger has the smallest diameter (i.e., the two clusters with the smallest maximum pairwise distance). A definition for clustering is given in the following together with a definition for a distance between two clusters.

Definition 1 (*Clustering*) Given a dataset T, a general clustering algorithm attempts to partition the dataset into K clusters, $C_1, C_2, \ldots, C_K$, $C_i \neq \emptyset$, $C_i \cap C_j = \emptyset$, $T = C_1 \cup C_2 \cup \ldots \cup C_K$, $i, j = 1{:}\ K$, $i \neq j$, such that the distance between any pair of points in a cluster is less than the distance between any point in the cluster and any point not in the cluster (or, its nearest cluster).

Definition 2 (*Distance between two clusters*) Let C_i, C_j be the clusters of the dataset T, and the distance between C_i and C_j is defined as $\rho(C_i, C_j) = \min\{\rho(x_i, x_j)|x_i \in C_i, x_j \in C_j\}$.

Classical clustering algorithms, such as K-means algorithm or K-medoids algorithm, rely on grouping the data points around some 'centers.' Therefore, they do not work well when the boundaries of the clusters are irregular. Fortunately, graph theory-based methods, typified by the minimum spanning tree-based clustering algorithms, can find clusters of arbitrary shapes, different sizes, and different densities.

The minimum spanning tree is a graphical analysis of an arbitrary point set of data. In such a graph, two points or vertices can be connected by either a direct edge or a sequence of edges called a path. The length of a path is the number of edges in the path. A loop in a graph is a closed path. A connected graph has one or more paths between any pair of points. A tree is a connected graph with no closed loops. A spanning tree is a tree that contains every point in the dataset. If a value is assigned to each edge in the tree, the tree is called a weighted tree. The weight of a tree is the total sum of all the edge weights in the tree. The minimum spanning tree is the spanning tree that has the minimal total weight among all possible spanning trees for a dataset. A definition for minimum spanning tree is given in the following.

Definition 3 (*Minimum spanning tree*) Given a weighted and connected graph, $G(T) = (V, E)$ with vertex set $V = T$ (i.e., a set of T data points) and edge set $E = \{(x_i, x_j)|x_i, x_j \in T, i \neq j\}$. Each edge $e_{ij} = (x_i, x_j)$ has a weight $\rho(x_i, x_j)$, and a minimum spanning tree (MST) of graph $G(T)$ is an acyclic subset of E that connects all the vertices in V with minimum total weights $W(T) = \sum_{x_i x_j \in T \rho(x_i, x_j)}$ and satisfies the following two conditions [17]:

1. The cut property: the edge with the smallest weight crossing any 2 partitions of the vertex set must belong to an MST.
2. The cycle property: the edge with the largest weight in any cycle in a graph cannot be in an MST.

As a result, when the weight associated with each edge denotes a distance between the two end points, any edge in the minimum spanning tree will be the shortest distance between the two subtrees that are connected by that edge. Therefore, as illustrated in Fig. 5.1, removing the longest edge will theoretically result in a two-cluster grouping. Removing the next longest edge will result in a three-cluster grouping and so on. This corresponds to choosing the breaks where the maximum weights occur in the sorted edges.

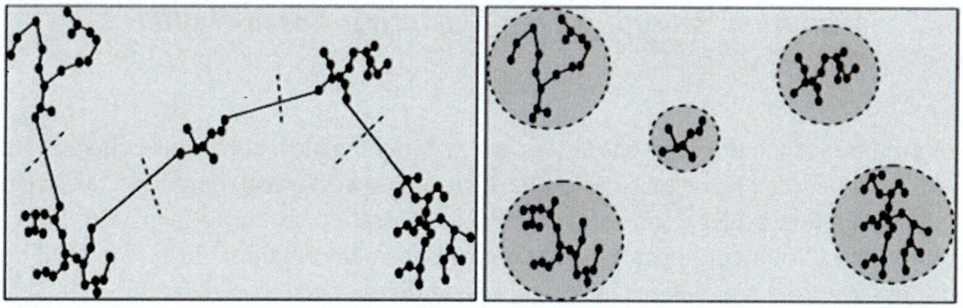

Fig. 5.1 (Left) A minimum spanning tree of a data set and (right) its minimum spanning tree-based clusters

Definition 4 (*Link*) A partition of the nodes of a graph G is a division into two disjoint nonempty subsets (C_i, C_j). A link is any edge in G whose weight is equal to the distance $\rho(C_i, C_j)$.

Theorem 1 *All MST edges are links of some partition in graph G. (The proof is given in* [18]).

Definition 5 (*Minimum spanning tree-based clustering*) Given a dataset T, let P, $Q \subseteq T$, $P \cap Q = \emptyset$, and $P, Q \neq \emptyset$. P, Q are minimum spanning tree-based clusters if they are formed by removing the link whose edge weight (i.e., distance) is significantly larger than the average of nearby edge weights on both sides of the link, that is, $\rho(P, Q) \gg \max\{\rho(P), \rho(Q)\}$, where $\rho(P) = \max\{\rho(x_i, x_j)|x_i, x_j \in P, i \neq j\}$and $\rho(Q) = \max\{\rho(x_i, x_j)|x_i, x_j \in Q, i \neq j\}$.

Based on this finding, MST-based clustering algorithms were initially proposed by Zahn [17] and have so far been extensively studied [19–22]. As shown in Fig. 5.2, one difficult problem associated with MST-based clustering algorithms is the chaining problem.

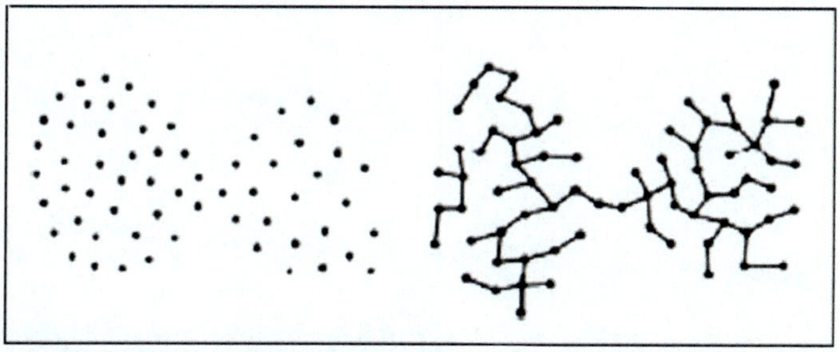

Fig. 5.2 Chaining problem of MST-based clustering

5.2.2 Minimum Spanning Tree Clustering-Based Outlier Detection

For minimum spanning tree-based clustering, under the ideal conditions, clusters are well separated and there exist no outliers. In a real-world task, outliers often exist, which make the longest edges an unreliable indication of cluster separations. As defined in the following, therefore, the data points in the smallest clusters formed by cutting the sorted MST edges may likely be outliers.

Definition 7 (*Minimum spanning tree clustering-based outliers*) Let C_1, C_2, ...,C_K be the clusters of the dataset T discovered by MST in the sequence that $|C_1| \geq |C_2| \geq \cdots \geq |C_K|$. Given parameters α and β, cluster-based outliers are the clusters in C_1 through C_K such that: $|C_1| + |C_2| + \cdots + |C_{i-1}| \geq |T|*\alpha$ and $|C_1| + |C_2| + \cdots + |C_{i-2}| \leq |T|*\alpha$ and $C_{i-1}/C_i > \beta$.

According to the normal distribution, α can take value of 0.3%. Without loss of generality, β can take a minimum value of 3.

Definition 8 (*MST-clustering-based global outliers*) Let C_1, C_2,..., C_K be the clusters of the dataset T discovered by MST and ρ_{MAX} be the largest intra-cluster edge weight, and a cluster C_i of MST-clustering-based outliers is a global one if $\rho(C_i, C_j) \gg \rho_{\text{MAX}}$, where C_j is the nearest neighboring cluster of C_i.

Definition 9 (*MST-clustering-based local outliers*) Let C_1,C_2, ..., C_K be the clusters of the dataset T discovered by MST and ρ_{MAX} be the largest intra-cluster edge weight, and a cluster C_i of MST-clustering-based outliers is a local one if $\rho(C_i, C_j) < \rho_{\text{MAX}}$, but $\rho(C_i, C_j) \gg \rho(C_j)$, where C_j is the nearest neighboring cluster of C_i.

For the sample dataset shown in Fig. 5.3, it can be seen that o_1 (i.e., cluster C_3)

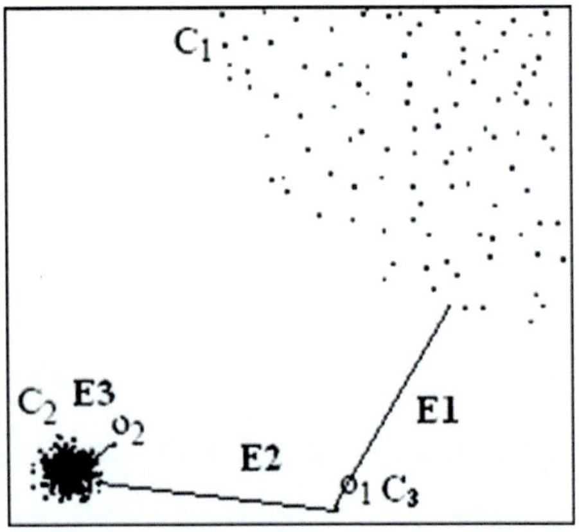

Fig. 5.3 Sample clusters in a 2-D data set, reprinted from Ref. [38], copyright 2015, with permission from Elsevier

is an MST-based global outlier since $E1 \gg \rho_{\text{MAX}}$ (i.e., $\rho(C_1)$ and $E1 < E2$, while o_2 is an MST-based local outlier since $E3 \gg \rho(C2)$ and $E3 < E2$.

Based on these ideas, several MST-based outlier detection techniques have been proposed. In 2008, Lin et al. proposed a new algorithm based on MST clustering to identify spatial outliers [23]. In their approach, basic spatial structure characteristics of spatial objects are maintained by Delaunay triangles and MST clustering is achieved by cutting off several most inconsistent edges. They claim their algorithm is a $O(N\log N)$ algorithm.

In 2011, Peter and Victor proposed an integrated approach using minimum spanning tree-based clustering and density-based approach for detecting outliers [24]. The algorithm works in two phases. The first phase of the algorithm creates an optimal number of clusters using a new cluster validation criterion based on the geometric property of data partitions of the dataset and small clusters are then determined and considered as outliers, whereas the second phase of the algorithm detects the rest of the outliers (if any) in the clusters based on density-based approach. As a continuation of this work, in the same year, Peter proposed another two-phase minimum spanning tree-based clustering algorithm for detecting outliers [25]. In the first phase, the algorithm follows the same procedure as in the previous one to find the proper number of clusters and regards data points in the small clusters as outliers. In the second phase, the rest of the outliers (if any) are then detected in the remaining clusters based on absolute distance between the center point of the current cluster and each data point in the same cluster.

Although MST-based clustering algorithms are very computationally expensive (from $O(N\log N)$ to $O(N^2)$), recently in 2012, Wang et al. propose an efficient three-phase outlier detection technique (referred to as MST + LOF in the following) [26]. First, an efficient construction of a spanning tree very close to a minimum spanning tree of the data set is conducted. Second, the longest edges in the obtained spanning tree are removed to form clusters. Based on the intuition that the data points in small clusters may be most likely all outliers, they are selected and regarded as potential outlier candidates. Finally, density-based outlying factors, LOF, are calculated for potential outlier candidates and accessed to pinpoint the local outliers. The main advantage of the algorithm is its computation efficiency.

In 2013, Daneshgar et al. proposed a graph-based data clustering algorithm which is based on an exact clustering of a minimum spanning tree in terms of a minimum isoperimetry criterion [27]. It was shown that their generalized graph model, which also allowed the use of potentials at vertices, can be used to extract an extra piece of information related to anomalous data patterns and outliers. In this regard, they proposed an algorithm that extracts outliers in parallel to data clustering. Finally, they showed that their basic clustering algorithm runs in $O(N\log N)$ and with a post-processing in almost $O(N\log N)$ (average case) and $O(N^2)$ (worst case) time where N is the size of the dataset.

In 2014, Zhu et al. presented an outlier detection algorithm based on k-neighborhood minimum spanning tree (MST) [28]. Taking density and directional factor into consideration, this algorithm proposes a new dissimilarity measure based on k-neighborhood. Then, a minimum spanning tree (MST) is built based on this k-neighborhood dissimilarity measure. Finally, the tree is progressively constrained to cutting so that the outliers can be found. Being applicable to datasets with clusters of any arbitrary shapes and densities, this algorithm can effectively detect local outliers and local outlying clusters. In the same year, Cipolla and Vella presented a technique to discover spatiotemporal outliers based on the search of the minimum spanning tree that allows to retain the strongest connections among nodes while minimizing a cost function at a global optimization [29]. They characterized the data with a graph that took into account the mutual information contents of neighboring nodes. The main advantage of this work is the early detection of events of particular intensity.

In 2018, Abghari et al. proposed an unsupervised approach for outlier detection in a sequence dataset [30]. The proposed approach combines sequential pattern mining, cluster analysis, and a minimum spanning tree algorithm in order to identify clusters of outliers. Initially, the sequential pattern mining is used to extract frequent sequential patterns. Next, the extracted patterns are clustered into groups of similar patterns. Finally, the minimum spanning tree algorithm is used to find groups of outliers.

5.3 An Improved MST-Clustering-Inspired Outlier Detection Algorithm

Due to the fact that the weight associated with every edge in the minimum spanning tree is the smallest one between the two subtrees that are connected by that edge, the MST-based clustering algorithms can be very useful in detecting the small clusters that can be selected and regarded as outliers. However, all the proposed methods for MST-based clustering techniques require a complete MST be constructed in the first step. In other words, the time complexity of these algorithms is at least $O(N\log N)$. For modern databases typically consisting of millions of high-dimensional data items, constructing an MST for such databases can be very computationally expensive. In this section, we describe a new scheme to facilitate efficient MST-clustering-inspired outlier detection for modern large databases. The design of such a scheme is motivated by the following observation. The MST-based outlier detection algorithms can be more efficient if the longest edges of an MST can be identified quickly before most of the shorter ones are found. This is because, for some MST-clustering-based outlier detection approaches, if we can find the longest edges in the MST corresponding to cluster breakers very quickly, there is no need to compute the exact distance values associated with the shortest ones. For cases where the number of the longest edges that separate the clusters (including the noisy ones) can be much fewer than the number of the shorter ones, this divide-and-conquer approach allows us to save distance computation tremendously.

5.3.1 A Simple Idea

From our study of distance-based, density-based, and clustering-based outlier detection algorithms, we have obtained several observations. Firstly, MST-based clustering criterion favors cutting small sets of isolated nodes in a graph and can give a bad partition when no outliers exist. This fact can be easily manifested from the dataset shown in Fig. 5.2. Therefore, there must be some way to judge whether outliers exist or not. To do so, as a first-degree approximation, the edge weights (i.e., edge distances) within a cluster of an MST can be assumed to follow a uniform distribution and the corresponding mean and standard deviation can thus be calculated. The ratio of the standard deviation over the mean can be used to judge to some degree whether outliers exist or not.

Secondly, density-based outlier detection algorithms may not do well in global outlier detection. For example, consider the data set in Fig. 5.4. Apparently data point A is farthest away from its six closest neighbors and therefore should be identified as a global outlier. However, for $k = 6$, the LOF algorithm assigns a higher outlier score to data point B than to A. This is because that LOF is computed as a ratio and data point B gets a higher ratio for this k. As a result, LOF-based algorithms fail to identify A as the most significant outlier [16]. From an MST point of view as shown on the right of Fig. 5.4, if tree edge weight (i.e., distance) is used as the outlier factor, this will not happen and data A's score remains as a constant for different k's.

Thirdly, according to the cut property, MST-based clustering algorithms can be very useful in detecting small clusters connected to the normal data points by long edges that can be selected and regarded as outliers as illustrated in Fig. 5.4. However, standard MST-clustering-based outlier detection algorithms usually take a quadratic running time to ensure the properties of MSTs to be satisfied. For modern databases typically consisting of millions of high-dimensional data items, constructing an MST for such databases can be very computationally expensive. From our point of view, this is neither efficient nor necessary. In this section, we describe a new scheme to

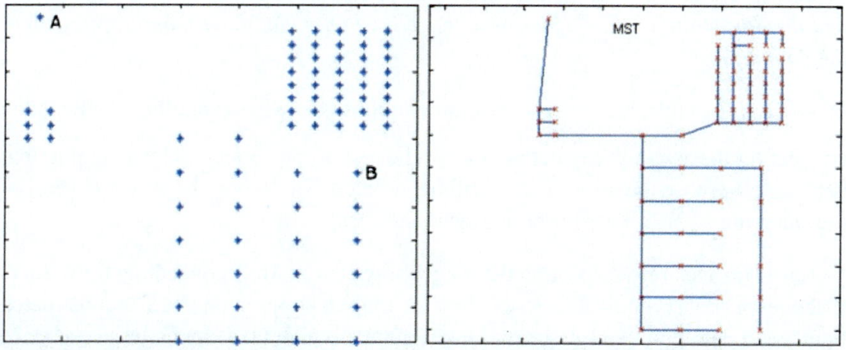

Fig. 5.4 Sample 2-D data set illustrating deficiency of LOF-based outlier detection (left) and the data set (right) its MST, reprinted from Ref. [38], copyright 2015, with permission from Elsevier

facilitate efficient MST-clustering-based outlier detection for modern large databases. The design of such a scheme is motivated by the following observation. The MST-clustering-based outlier detection algorithms can be more efficient if the longest edges of an MST connected to the small clusters of data points can be identified quickly before most of the shorter ones are found.

According to the working principle of MST-based clustering algorithms, a database can be split into partitions by identifying and removing the longest edges in the tree. As a result, the goal of finding MST-based clusters that correspond to small outlier clusters can be achieved efficiently by quickly locating the longest edges in the MST. The point is that, for a database that has several dense clusters and a small number of outlier groups, if the longest edges that are outlier breaks in the MST can be found very quickly, there is no need to find the normal clusters that most data items reside in the MST. By this divide-and-conquer approach, the number of distance computations can be reduced to a value smaller than $O(N^2)$.

Based on these observations, the working ideas behind our efficient MST-inspired outlier detection algorithm are formalized in the following subsection.

5.3.2 Two New Outlier Factors

Since there are only a small number of outliers in a dataset, MST-clustering-based outlier detection algorithms can be more efficient if the longest edges connecting the outlier clusters to the normal data can be correctly identified and quickly located. In other words, some of the longest edges do not correspond to any cluster separations but are associated with the outliers. Basically, this observation for the design of a more efficient scheme is formalized in the following theorem.

Theorem 2 *Given a data set and a minimum spanning tree constructed upon it, for an outlying group of* k *data points, the edge connecting these* k *data points to the rest of data set is the same as the one in a miniMST constructed by these* k *data points and the data point at the other end of this edge in the minimum spanning tree upon the data set.*

Proof To prove this, we use contradiction. Suppose there exists another smaller edge connecting these k outlying data points to the rest of the data set. This is impossible because otherwise the cut property will be violated. Therefore, they must have been the same one. This is true for both global and local outliers.

Based on Theorem 2, locating the longest edges in an MST connecting the outliers to the rest of data is equivalent to locating the longest edges in miniMSTs constructed upon each data point and its k-nearest neighbors, which takes much less time since many efficient kNN computation methods have been proposed [31]. Therefore, we can carry an outlier detection procedure (e.g., an MST-clustering-based) within each data point and its k-nearest neighbors to detect top outliers. On the other hand, since MST-based clustering has been an active research area and many sophisticated algorithms have been developed for it for small datasets, MST-based clustering algorithms can be dwelt upon and tapped for their potential for outlier detection within a miniMST. This is where the combination of kNN structure and the cutting property can be exploited.

In this sense, it will be more meaningful for k if it is defined to be the size of the largest outlying cluster plus 1 for global outliers and plus 2 for the local outliers. To identify the relatively small number of long edges connecting small outlier groups to the majority normal data of an MST, a miniMST for every point and its kNN is first constructed.

To find global outliers, we first search for the edge in these miniMSTs that has the largest edge distance value. If it is significantly larger than the average values of neighboring edges, this longest edge among miniMST can be regarded as the global outlier score for those data items that are in the same partition as the outlier data point itself. In other words, these data points are this longest edge away from the rest data points of the database. As illustrated in Fig. 5.4, it is evident that, compared to DB and DB-Max, these global scores are less sensitive to k.

Further for global outlier detection, we are more interested in those miniMSTs in which the weights of the longest edges are significantly larger than the average values of those of the neighboring edges. To further quantify the significance of an edge being larger than the average values of neighboring edges for global outlier detection, we use the uniform distribution as a first-degree approximation for tree edge weights in miniMSTs.

Definition 12 (*MiniMST-based global outlier indicator*) Let a miniMST be constructed upon a data point and its k-nearest neighbors, and dist[i] denote the i-th edge weight of such miniMST starting at the point; miniMST-based global outlier indicator, SOM_{MST}, is defined to be the ratios of the standard deviation, Std_{MST}, and the mean, Mean_{MST}, of these edge weights (i.e., edge distances) is defined as

$$\text{Mean}_{\text{MST}} = \frac{1}{k}\sum_{i=1}^{k}\text{dist}[i] \tag{5.1}$$

$$\text{Std}_{\text{MST}} = \sqrt{\frac{1}{k}\sum_{i=1}^{k}(\text{dist}[i] - \text{Mean}_{\text{MST}})^2} \tag{5.2}$$

$$\text{SOM}_{\text{MST}} = \frac{\text{Std}_{\text{MST}}}{\text{Mean}_{\text{MST}}} \tag{5.3}$$

SOM_{MST} is a quantitative measure of deviation from normal and can be used as a threshold to rule out the large portion of normal data. Based on these ideas, our k-nearest neighbor-based global outlier score is given in the following definition.

Definition 13 (k*NN miniMST-based global outlier factor*) Let a miniMST be constructed upon a data point and its k-nearest neighbors, dist[i] denote the i-th edge weight of such miniMST, and cut-thred be a user provided measure of the possibility of outlier existence threshold for SOM_{MST}, and an MST-inspired kNN-based global outlier factor is defined as

$$\text{MST-MAX} = \max_{i=1:k}\{\text{dist}[i]\} \tag{5.4}$$

$$\text{and } \text{SOM}_{\text{MST}} \geq \text{cut-thred} \tag{5.5}$$

For local outliers, since they are significantly far away only from their nearest neighboring clusters, the local outlier score is defined as the ratio of the largest edge weight value over the smallest edge weight value in each miniMST.

Definition 14 (k*NN miniMST-based local outlier factor*) Let a miniMST be constructed upon a data point and its k-nearest neighbors and dist[i] denote the ith edge weight of such miniMST, and the MST-inspired kNN-based local outlier factor is defined as

$$\text{MST-MAX-MIN} = \frac{\max_{i=1:k}\{\text{dist}[i]\}}{\min_{i=1:k}\{\text{dist}[i]\}} \tag{5.6}$$

The local outlier detection can be assumed to be over when the ratio drops below a threshold (say 3). Finally, the outlier scores are used to assign the returned data points a degree of being an outlier.

5.3.3 The Proposed MST-Clustering-Inspired Outlier Detection Algorithm

We combine the above three factors to create our kNN miniMST-based outlier detection algorithm.

1. Set k to be the largest outlying cluster size plus 1.
2. Sequentially read the data set in, and find k-nearest neighbors for each data item.
3. Compute three 1-dimensional arrays of the estimated outlying scores for each data item, and sort them in a nonincreasing order.
4. For data items with top global outlying scores, check its SOM_{MST}, return it if the SOM_{MST} is larger than a threshold (say 1.5), then if $i = n$, the global outlier detection is completed, otherwise, $i = i + 1$, and go to 4.

5. For data items with top local outlying scores, return it if the score is larger than a threshold (say 3), then if $i = n$, the local outlier detection is completed, otherwise, $i = i + 1$, and go to 6.
6. Steps 4, 5, and 6 are repeated until all the outlying scores above the global and the local thresholds are mined.

Since the number of outliers is expected to be relatively small, the number of distance computations consumed is expected to be relatively small as well. Physically, the resource consumed by our algorithm includes the space to hold the whole data set in memory, space to store their k neighbors, and some temporary space for miniMSTs. The numerical parameters our algorithm needs from the user include the number of nearest neighbors (i.e., k), the global and local outlier thresholds, while the outputs include a set of ranked outliers from the dataset. To improve the readability, our MST-clustering-inspired kNN-based outlier detection algorithm is presented in a pseudo-code format in Table 5.1.

5.3.4 *Time Complexity Analysis*

Since most of the data points are normal, the goal of finding top n outliers can be achieved by first quickly finding a good estimate of the outlying score for each data item and then focusing on top $m \geq n$ ones. By removing all the inliers among them, the required top n outliers show up. In d-dimensional space, the time complexity of the kNN computation is upper bounded by $O(\mathrm{d}N^2)$. To be more efficient, we are particularly interested in an approximate nearest neighbor search facility, called the divisive hierarchical clustering algorithm (DHCA) [18]. Essentially, the DHCA recursively partitions the dataset into K subclusters and forms at most K^{p} partitions for the p-th stage. The procedure continues until the number of elements in a partition is below $K + 2$, at which time, a nearest neighbor search among all the data items in that partition is conducted. Such a strategy ensures that points that are close to each other in space are likely to be collected in the same partition, and multiple runs of DHCA greatly enhance such possibilities. For a balanced tree, the tree height l is $\log_K N$. The best case time complexity of a DHCA is $\mathrm{O}(dNK\log_K N + N)$. A more detailed demonstration and proof of the effectiveness of DHCA on approximate nearest neighbors' search have been given in Chap. 3 and will not be repeated here. After several iterations, exact kNNs and the correspondingly scores are computed for top outliers. The number of top outliers is small, and thus the computation time is fast.

Table 5.1 Our MST-inspired outlier detection algorithm

Input:			
data	A set of N data points		
k	The number of *NN*s of a data item		
K	The number of clusters at each step		
SOM-TH	Global outlier termination threshold		
MAX-MIN-TH	Local outlier termination threshold		
Output:			
A set of ranked global outliers *GO*, and a set of ranked local outliers *LO*			
Begin			
	set k to be the largest outlying cluster size plus 1, find k nearest neighbors for each data item		
	compute two 1-dimensional arrays of the estimated outlying scores and sort both arrays in a non-increasing order;		
	for each $i = 2{:}k + 1$		
	{		
		compute three 1-dimensional arrays of the estimated outlying scores for *iNN*, namely, *MST-MAX*, *MST-MAX-MIN* and SOM_{MST}, and sort the first two in a non-increasing order and the orders are remembered in *MST-MAX-INDEX* and *MST-MAX-MIN-INDEX*, respectively;	
		for j = 1:N	
		{	
			find true *iNN* for *MST-MAX-INDEX*[j];
			recomputed the global score and SOM_{MST};
			if(SOM_{MST}[*MST-MAX-INDEX*[j]]>*SOM-TH*

(continued)

Table 5.1 (continued)

			{	
				GO_T = [*GO_T* *MST-MAX-INDEX*[j]];
			}	
			Sort *GO_T* according their scores;	
			GO = [*GO GO_T*];	
		}		
		for j = 1:*N*		
		{		
			find *iNN* for *MST-MAX-MIN-INDEX*[j];	
			recomputed the local score;	
			if(*MST-MAX-MIN*[j]]> *MAX-MIN-TH*	
			{	
				LO_T = [*LO_T* *MST-MAX-MIN-INDEX*[j]];
			}	
			Sort *LO_T* according their scores;	
			LO = [*LO LO_T*];	
		}		
	}			
End				

5.4 A Performance Study

In this section, we present the results of two sets of experiments performed to evaluate our MST-clustering-inspired outlier detection algorithm. In Experiment 1, three 2-dimensional synthetic data sets are used to show that our MST-clustering-inspired outlier detection method can outperform classic outlier detection algorithms in classification accuracy. In Experiment 2, five real multi-dimensional data sets obtained from the UCI Machine Learning Repository [32] are used to check the technical robustness of this study and to illustrate the effectiveness of our method in real-world situation by comparing the proposed outlier detection algorithm with the state-of-the-art outlier detection methods, that is, DB-Max [10], DB [11], LOF [12], COF [33], INFLO [34], LDOF [35], RBDA [16], MST + LOF [26], and PBOD [14]. All the data sets are briefly summarized in Table 5.2.

All the algorithms are implemented in C/C++ and run on a computer with Intel Core 2 Duo Processor E6550 2.33 GHz CPU and 2 GB RAM. The operating system running on this computer is Windows XP. The results show that, overall, our MST-inspired outlier detection algorithm is superior to other state-of-the-art outlier detection algorithms in classification accuracy. For consistency, we only use the parameters k to represent the neighborhood size in the investigation of these three methods.

5.4.1 *Performance on Synthetic Datasets*

In this subsection, we use three synthetic data sets to show that the proposed method can efficiently identify local and global outliers in various scenarios. In each dataset, there are multiple clusters and six outliers (A, B, C, D, E, and F) are planted in the vicinities of the clusters. A particular challenging feature of these data sets is that clusters are of different sizes and have different densities. We exclusively use 2-dimensional data sets due to their convenience for visual inspection.

To study the relative effects of our proposed outlier detection algorithm to parameter k, in comparison with DB, DB-Max, LOF, COF, INFLO, LDOF, RBDA, MST

Table 5.2 Sets of data

Data set	Size (N)	Dimensionality
Dataset1	515	2
Dataset2	78	2
Dataset3	473	2
Shuttle	14,500	9
Lymphography	148	18
Ionosphere	351	34
Optical digits	5620	64
Multiple features	2000	649

+ LOF, and PBOD, we first decide the largest number of data points for an outlying cluster, next calculate the outlier scores according to each outlier definition, sort them in nonincreasing order, and return top ones.

Synthetic Dataset 1 consists of 515 instances, including six planted outliers, one large normally distributed cluster, and two small uniform clusters. This is a global outlier detection task. For our approach, if the size of the largest outlying group is set to 2, k is set to 3 for all the algorithms in this case. The experimental results are depicted in Fig. 5.5 where 6 top outliers returned are marked with red plus, red star, red cross, red triangle, red circle, and red square in order, respectively.

From the figures, it can be seen that DB, DB-Max, LOF, INFLO, MST + LOF, and our method (i.e., MST-MAX) correctly detect all the outliers, though the rankings are

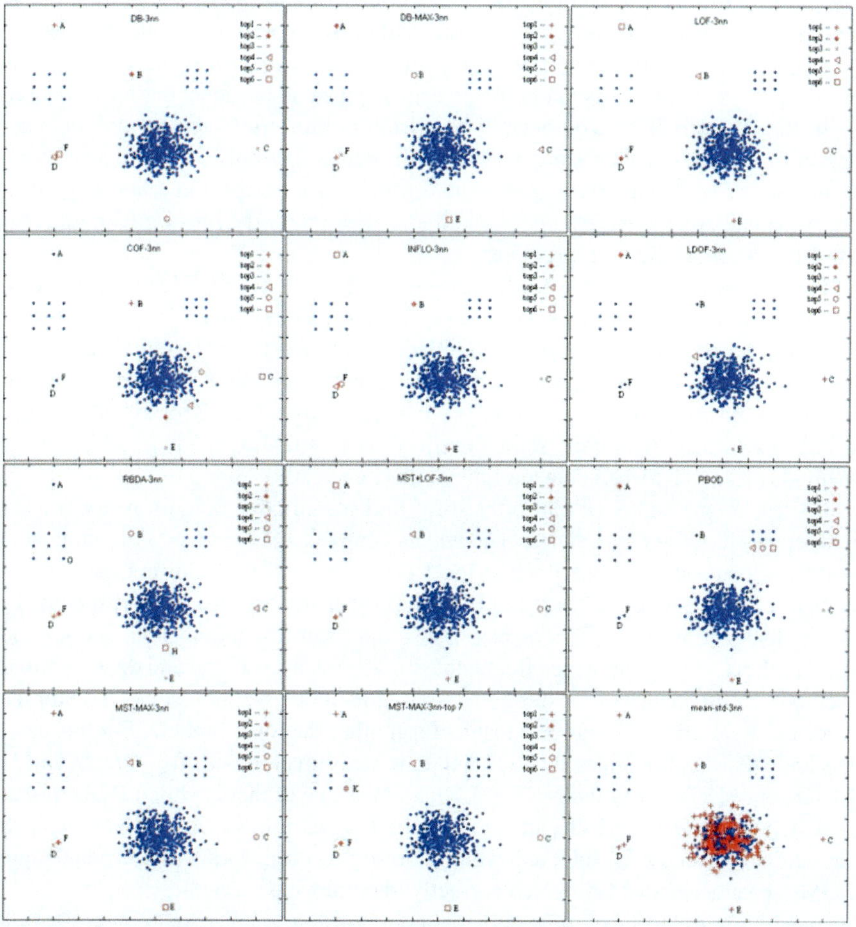

Fig. 5.5 Outlier detecting results on Synthetic Dataset 1 for $k = 3$, reprinted from Ref. [38], copyright 2015, with permission from Elsevier

slightly different for different methods. Unfortunately, COF does not do very well and misses A, D, F. Since the average of k-nearest distances and that of the $k(k-1)$ pairwise inner distances are both large for B, D, F, their ratios are near 1. As a result, they are missed by LDOF. RBDA misses one global outlier, (i.e., A) in Fig. 5.5. This phenomenon is due to the fact that the rank of A in its 3 nearest neighbors is relatively lower than the ranks of H in its 3 nearest neighbors. PBOD detects 3 right for top 6 outliers.

So far, it is implicitly assumed that the number of outliers is given beforehand and it is not clear when the detection process should come to a close. As an improvement, SOM_{MST} can be a first-degree approximation for such purpose.

To test our termination mechanism, we show top 7 outliers detected using our algorithm for $k = 3$ in the middle plot on the last line of Fig. 5.5. From the figure, we can observe the limitation of our global outlier factor; that is, data point K is mistakenly detected by our global outlier definition. Fortunately, our termination mechanism for global outlier detection identifies data object K to have a very low SOM_{MST} value (which is actually 0) and declares the end of global outlier detection.

In addition to 6 planted outliers, there exist three clusters. Correspondingly, if the size of the largest outlying group is set to 9, the dataset has only a major data pattern. Depicted in Fig. 5.6 are the experimental results for each method when k is set to 10, at which case, there are totally 24 outliers. To simplify the illustration, totally 24 top outliers are marked with red plus.

From the figure, it can be seen DB misses one for $k = 10$. LOF identifies only 14 out of 24 correctly. COF does not do very well, either, and misclassifies 17 out of 24. INFLO detects only 9 out of 24. LDOF does not detect two clusters of size 9 and therefore misses 18 out of 24. RBDA misses 9 points in one of the two clusters of size 9. PBOD works a little better for this case and only misses 2 for top 24 outliers. LOCI misses 2 but with quite some false positives. DB-Max, MST + LOF, and our method (i.e., MST-MAX) detect all the outliers correctly.

To test if our approach can effectively find meaningful outliers in a complex dataset, we use Synthetic Dataset 2 which contains 78 instances, including five planted global outliers, one planted local outlier B, and four clusters of different densities consisting of 36, 8, 12, and 16 uniformly distributed instances, respectively.

To demonstrate the effectiveness of our approach in finding both global and local outliers, we compare the effectiveness of the distance-based and density-based methods on this dataset. For these kNN-based methods, we first set $k = 3$ and the screen shots in Fig. 5.7 show the results for mining the top 6 outliers. For this case, DB and DB-Max both miss B and C, and have the same rankings for A, D, E, and F. LDOF, our MST-MAX, and MST + LOF all miss one of them, while PBOD misses 5. RBDA, INFLO, COF, LOF, and our local outlier factor MST-MAX-MIN detect all six outliers correctly. The plot at lower bottom-right corner shows the corresponding SOM_{MST} values above 1.5, which correctly identifies the six outliers.

In addition to the six planted outliers, there is also a small cluster consisting of 8 points. If it is also regarded as an outlying group, we set k to 9. For this case, there are 14 outliers. The detection results are shown in Fig. 5.8. From the figure, it can be observed that DB-Max and MST-MAX do not perform well when local outlier

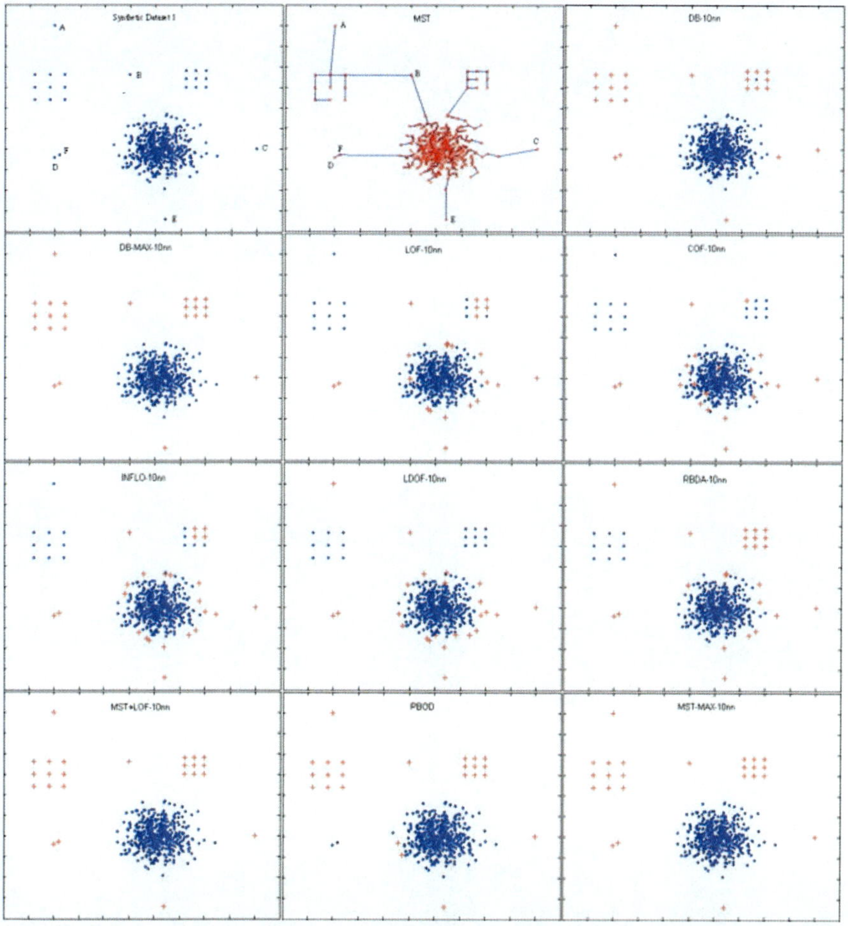

Fig. 5.6 Outlier detecting results on Synthetic Dataset 1 for $k = 10$, reprinted from Ref. [38], copyright 2015, with permission from Elsevier

exists. DB misses 10, DB-Max misses 11, LOF misses 3, COF misses 6, INFLO misses 4, LDOF misses 8, RBDA misses 6, and PBOD misses 12. LOCI misses 11. MST + LOF and our local outlier detector, MST-MAX-MIN, both only miss 2.

From Figs. 5.7 and 5.8, the advantage of our outlier detection factor, MST-MAX-MIN, is very evident on this comparably low-dimensional data set having not only global outliers, but also local outliers. The experiment shows that our method is suitable to mixed type data in this sense.

To test our method as a whole, Synthetic Dataset 3, which has 473 points and consists of 5 clusters with different densities and 6 outliers, is used. A particular challenging feature of this data set is that three denser clusters are buried into one sparse cluster on the upper-right corner. Since the largest outlying group has 2 data points (i.e., E and F), k is first set to 3. The performance of outlier detection methods

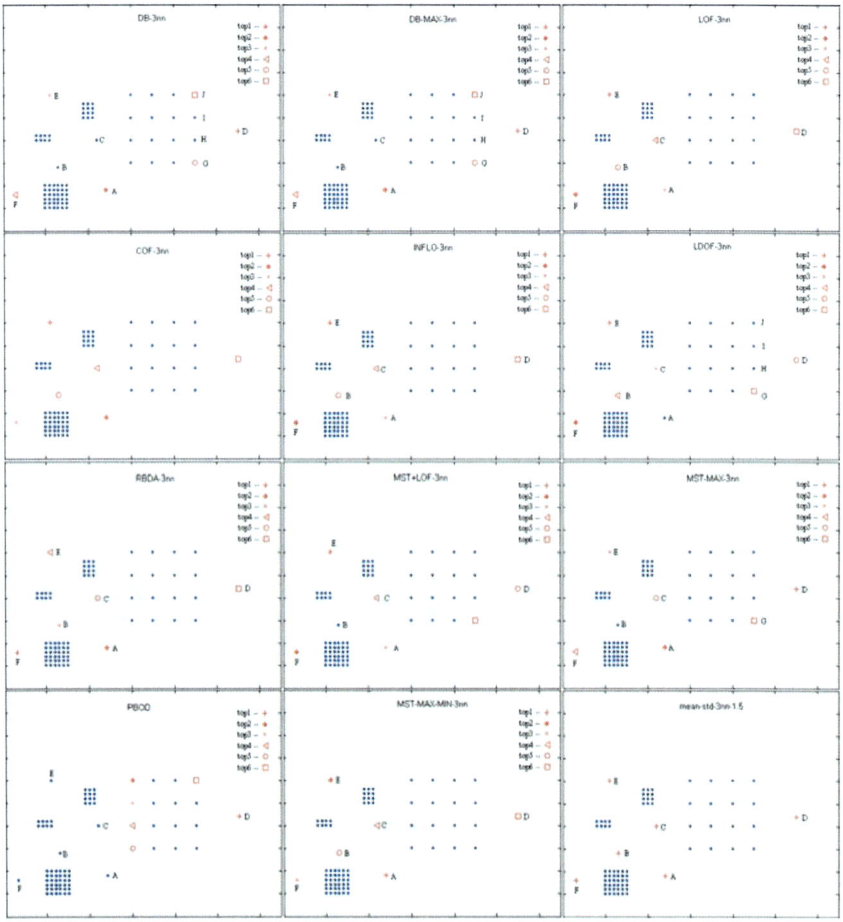

Fig. 5.7 Outlier detecting results for Synthetic Dataset 2 for $k = 3$, reprinted from Ref. [38], copyright 2015, with permission from Elsevier

is shown in Fig. 5.9. The same conventions are employed for the plots. From the figure, it is easily seen that this is also a global outlier detection situation with the difference from Synthetic Dataset 1 being that the detection process is disturbed by the immediate connection of clusters with different densities.

For detecting top 6 outliers, COF misses *B* and *C*, LDOF misses *E* and *F*, RBDA misses *C*, and PBOD misses all. The good news is that DB, DB-Max, LOF, INFLO, MST + LOG, and our method detect all six outliers correctly but with different rankings. If the sparse cluster (i.e., the one consisting of 7 data points) is also regarded as an outlying group, *k* is set to 8. For this situation, both DB-Max and LOF miss 1, COF misses 6, INFLO misses 2, LDOF misses 8, RBDA misses 6, PBOD misses 6, LOCI misses 8, while DB, MST + LOF, and our outlier detector get all right, as shown in Fig. 5.10.

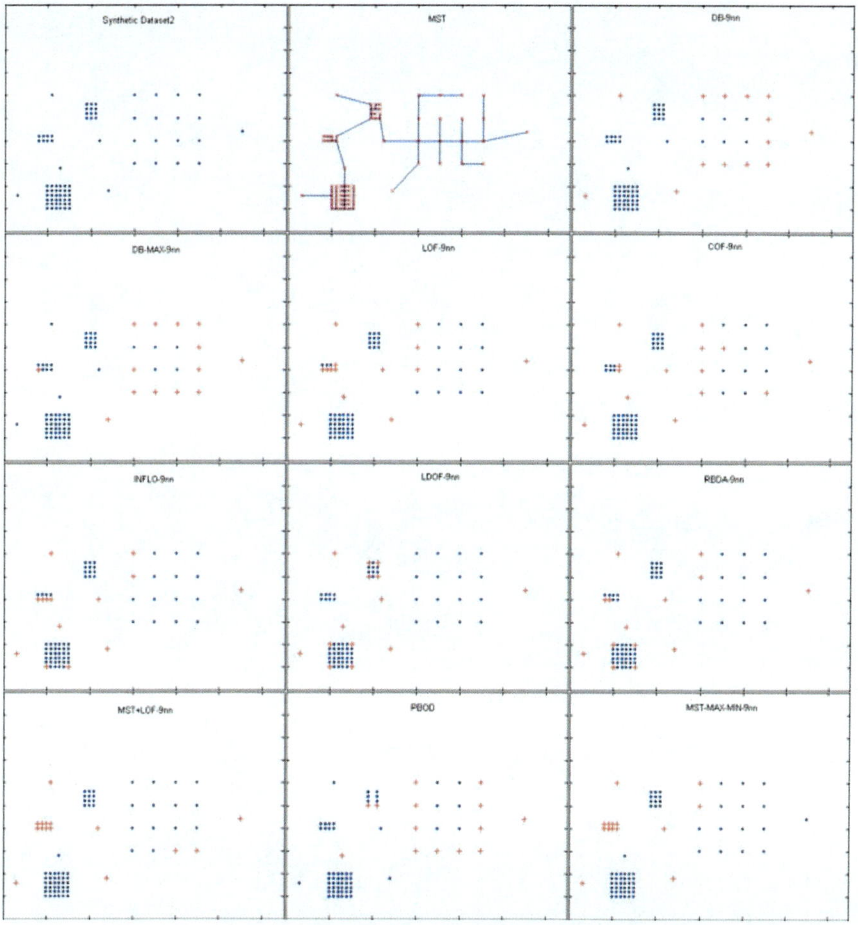

Fig. 5.8 Outlier detecting results for Synthetic Dataset 2 for $k = 9$, reprinted from Ref. [38], copyright 2015, with permission from Elsevier

To summarize, it can be observed from Figs. 5.5 through 5.10 that our method has no problems detecting all outliers and clearly offers the best ranking for three synthetic datasets while all other methods do not perform competently with detecting all the outliers for both k's one way or the other. In other words, the advantage of our outlier detection factors is very evident on these 2-dimensional data sets.

5.4.2 Performance on Multi-dimensional Real Datasets

To evaluate the effectiveness and accuracy of our proposed method on multi-dimensional real data, we compare the algorithms by their performance on detecting

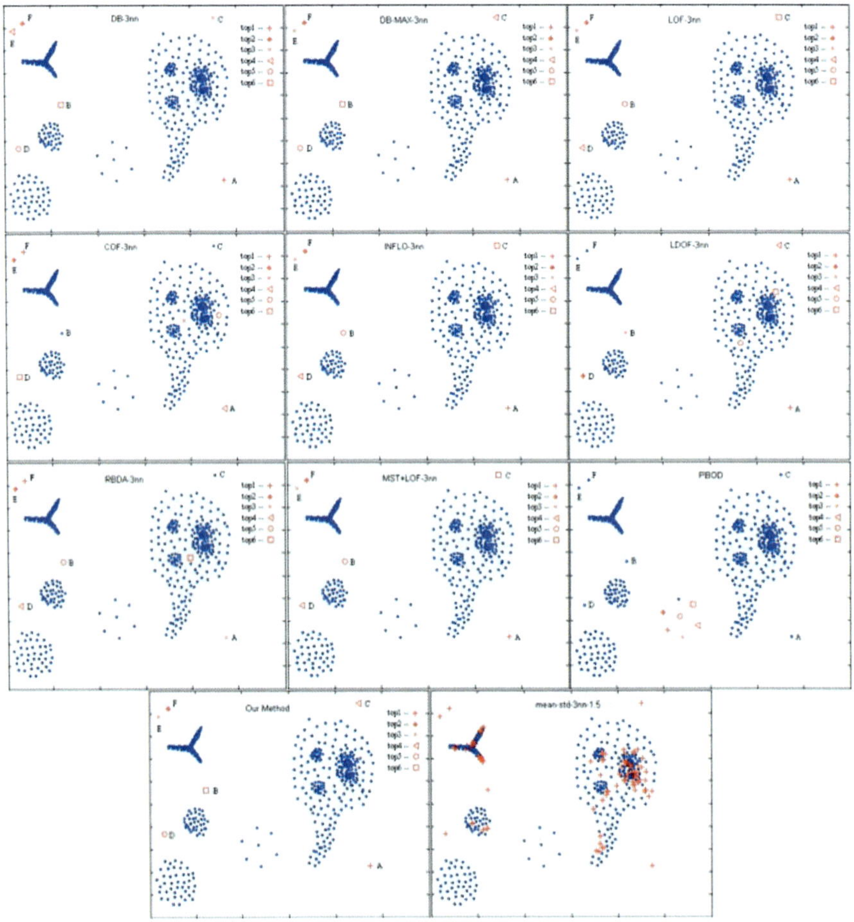

Fig. 5.9 Outlier detecting results for Synthetic Dataset 3 using $k = 3$, reprinted from Ref. [38], copyright 2015, with permission from Elsevier

rare classes in five real datasets, namely the Lymphography, Shuttle, Ionosphere, Optical Digits and Multiple Features datasets, which are downloaded from UCI [32]. As pointed out by Aggarwal and Yu, one way to test how well an outlier detection algorithm works is to run the method on a dataset and test the percentage of points which belong to rare classes [36].

To quantitatively measure the performance of an outlier detection scheme, three popular metrics, namely precision, recall, and rank power, are employed here. Assume that a dataset $D = D_o \cup D_n$ where D_o denotes the set of all outliers and D_n denotes the set of all nonoutliers. Given any integer $m \geq 1$, if O_m denotes the set of outliers among the objects in the top m positions returned by an outlier detection scheme, precision and recall are defined as

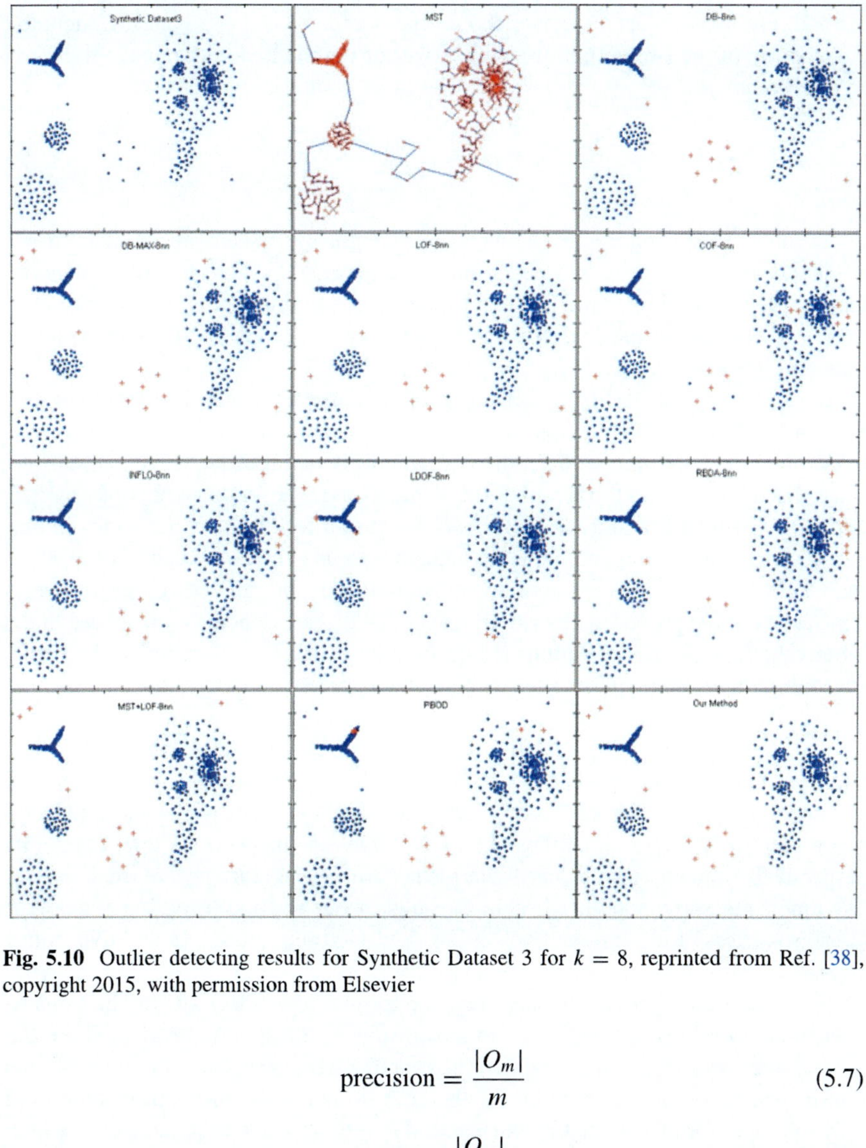

Fig. 5.10 Outlier detecting results for Synthetic Dataset 3 for $k = 8$, reprinted from Ref. [38], copyright 2015, with permission from Elsevier

$$\text{precision} = \frac{|O_m|}{m} \tag{5.7}$$

$$\text{recall} = \frac{|O_m|}{|D_o|} \tag{5.8}$$

Therefore, precision measures the percentage of outliers among top m ranked objects returned by a method, while recall measures the percentage of total outlier set included in top m ranked objects. Usually, users are not only interested in how many true outliers being returned by a method, but also in where they are placed. Rank power is a metric that considers both the placements and the number of outliers returned by a method. Here, the rank power given in [37] is used. Suppose that a method returns m objects, n of which are true outliers. For $1 \leq i \leq n$, if L_i denotes the position of the i-th outlier, the rank power of the method with respect to m can be defined as

$$\text{Rank Power} = \frac{n(n+1)}{2\sum_{(i=1)}^{n} L_i} \tag{5.9}$$

As can be seen from Eq. (5.9), rank power weighs the placements of the returned outliers heavily. An outlier placed earlier in the returned list adds less to the denominator of the rank power (and thus contributes more to the rank power metric) than those placed later in the list. A value of 1 indicates the best performance, and 0 indicates the worst.

To measure the capability of each algorithm to retrieve the most likely outliers and to compare the quality of the ranking provided by each algorithm, in the following, the performance is measured with the three metrics of recall, precision, and rank power. The value of m indicates top m ranked records returned by an outlier detection scheme. *SOM-TH* is set to 1.5, and *MAX-MIN-TH* is set to 3 for all the experiments in this subsection using our method. In the figures, the x-axis denotes the values of m while the y-axis denotes the corresponding precision, recall, and rank power, respectively. The results of the proposed algorithm, that is, 'OUR,' are denoted by solid red lines, while other competing algorithms are denoted by dotted line, dashdot, dashed lines together with different colors such as blue, black, green, and magenta.

Lymphography Data

The Lymphography dataset has 148 instances with 18 attributes. The dataset contains a total of 4 classes. Classes 2 and 3 have the largest number of instances (81 and 61, respectively). The remaining two classes have 6 instances in total (2 and 4, respectively) and are regarded as outliers (i.e., rare classes) for they are small in size. We report the corresponding detecting results of our method, PBOD method, and three best cases of the rest methods in Fig. 5.11 for four values of k and five values of m.

From the figures, it can be seen that, for k equal to 7, PBOD is the first one to mine all six outliers within top 10 ranked instances. LOF and our method are the second one. For k equal to 30, LOF, INFLO, and RBDA perform better than PBOD. The detection performances of our method are similar with other four methods and stay the same for all k's; that is, our method is less sensitive to parameter k, which agrees with our expectation.

Space Shuttle Data

The Shuttle dataset contains 14,500 objects, in which each object has 9 real-valued features and an integer label (1 ~ 7), and has 7 clusters. We regard those objects with labels 2, 6, and 7 (with 13, 4, and 2, objects respectively) as outliers, and the rest 4 classes as normal data (i.e., classes 1, 3, 4, and 5, with 11,478, 39, 2155, and 809 instances, respectively).

We show in Fig. 5.12 the experimental results of our method, PBOD method, and

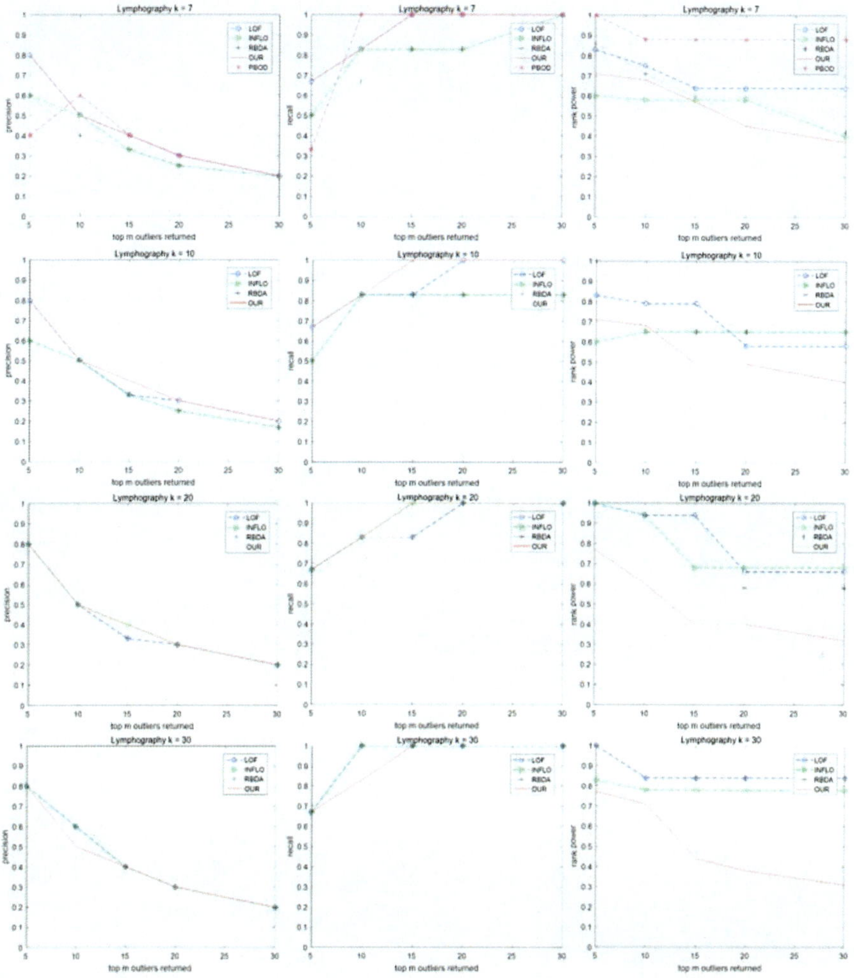

Fig. 5.11 Outlier detecting results for lymphography data for $k = 7, 10, 20, 30$

three best cases of the rest methods for four different values of k's and seven values of m. As reflected in the figure, this is not a very easy task for all other techniques. Our method performs significantly better in comparison with others for all k's and all m's. More specifically, our method has the highest precision and recall and is less sensitive to k as in the case of Lymphography data, which again manifests the intuition that using tree edge weights in miniMST or MST as indication of outlying degree is less sensitive to the different values of k and, therefore, more reliable.

Ionosphere Data

The Johns Hopkins University Ionosphere dataset contains 351 instances with 34 attributes, of which 225 are labeled as good instances while the rest 126 are labeled

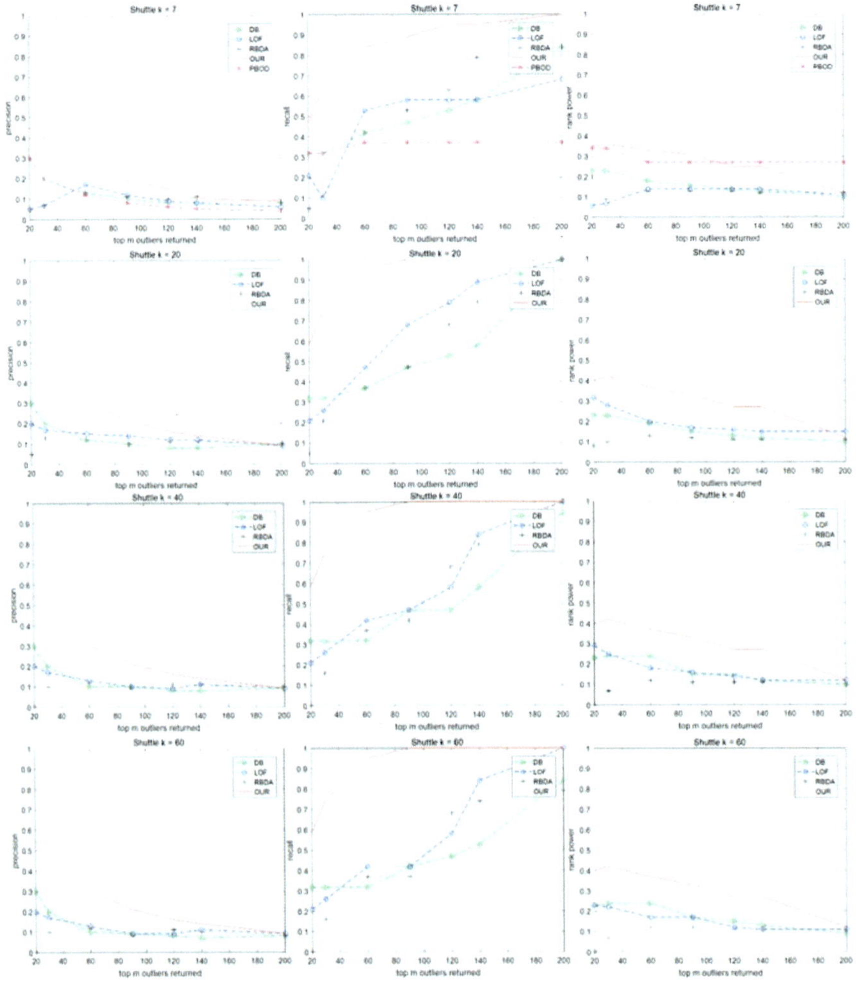

Fig. 5.12 Outlier detecting results for Shuttle data for $k = 7, 20, 40, 60$

as bad instances. All attributes are normalized in the range from 0 to 1, and there are no duplicate instances in the dataset. Instead of randomly selecting sample instances from the bad class to form the rare class, we use all 351 instances and compare the performances of all the methods for five different k's and eight different m's. The experimental results obtained by PBOD, our method, and three best cases from the rest methods are presented in Fig. 5.13.

From the figures, it can be clearly seen that, overall, our method performs the best, and DB method performs the next. RBDA performs a little better than MST+LOF. PBOD performs the worst. For this test dataset, all the methods are less sensitive to k than they are in the previous experiments.

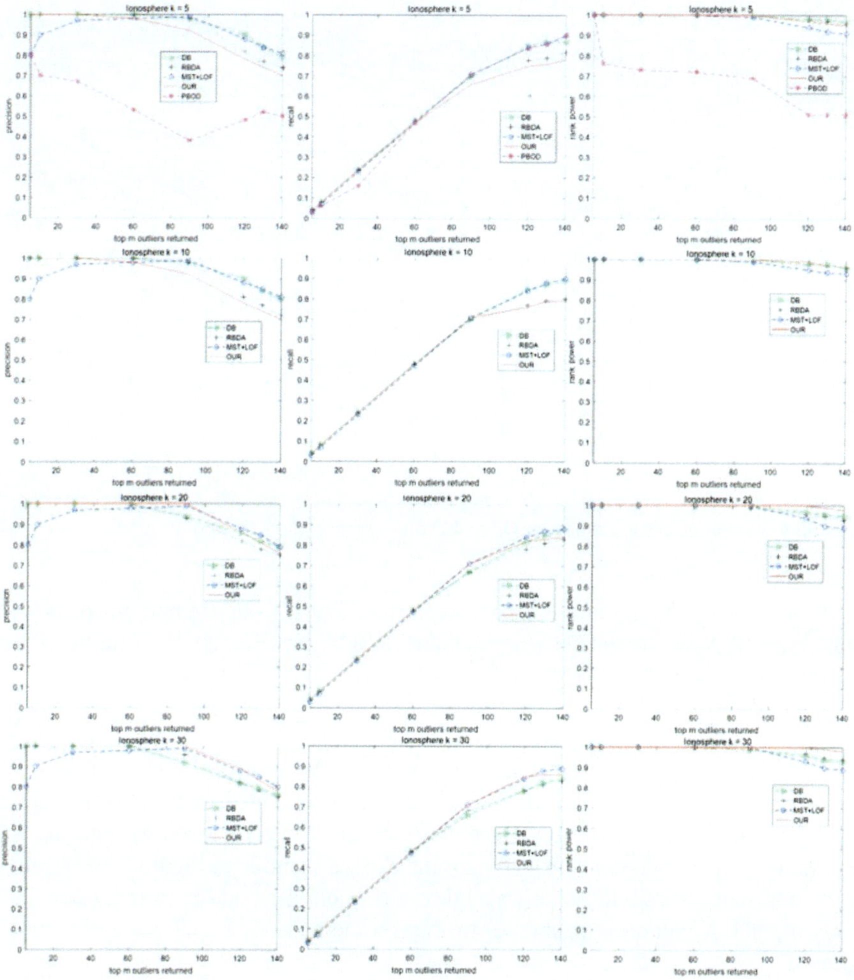

Fig. 5.13 Outlier detecting results for Ionosphere data for $k = 5, 10, 20, 30$

Optical Digits

For real high-dimensional data sets, we pick two datasets (Multiple Features and Optical Digits). The Optical Digits dataset consists of the data of handwritten numbers ('0'–'9') and contains 5620 instances with 64 attributes. Since there are no small rare classes to be regarded as outliers, we randomly pick 5 data points from one of the classes as outliers. We show in Fig. 5.14 the experimental results of our method for two classes (i.e., '2' and '0') and four best cases of the rest methods for $k = 7$ and eight values of m's.

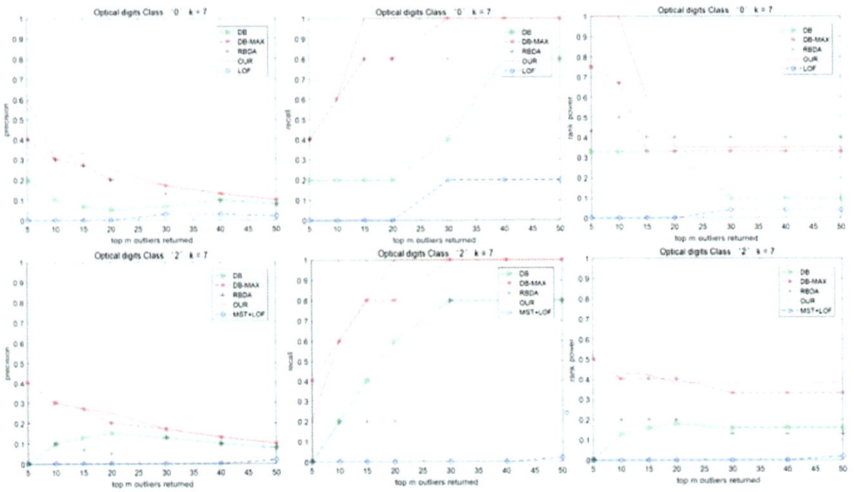

Fig. 5.14 Outlier detection results for Optical Digits data of classes '0' and '2' for $k = 7$

From the figures, it can be clearly seen that, overall, our method performs the best, and DB-Max method performs the next. RBDA performs a little better than DB method.

Multiple Features

The Multiple Features dataset consists of the data of handwritten numbers ('0'–'9') as well and contains 2000 instances with 649 attributes. Since all the classes contain 200 data points each and there are no rare classes to be regarded as outliers, we randomly pick 5 data points from one of the classes as outliers. Figure 5.15 summarizes our findings from the experimental results of our method for three classes (i.e., '9', '6', and '4') and four best cases of the rest methods for $k = 7$ and eight values of m's.

From the figures, it can be clearly seen that, overall, our method is the only one that performs best for class '9' for which others perform much worse. For the other two cases, our method and DB-Max method perform similarly. DB performs well for classes '6' and '4'. RBDA performs well for class '4'. Therefore, overall, our method performs the best.

From this set of experiments, we come to a conclusion that our method can outperform state-of-the-art outlier detection techniques in classification accuracy in real situations.

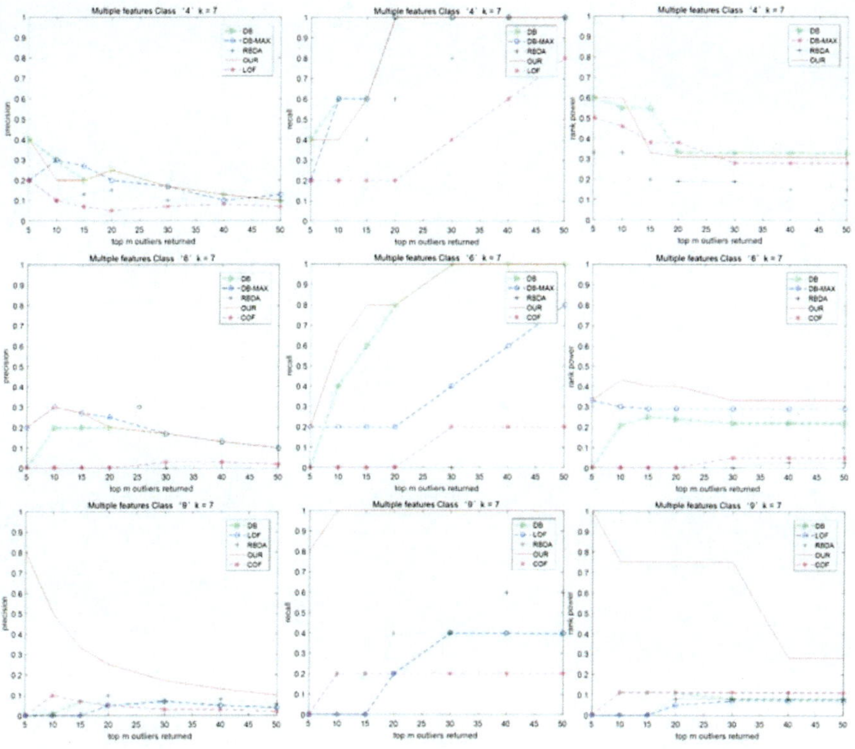

Fig. 5.15 Outlier detection results for Multiple Features data of classes '4', '6', and '9' for $k = 7$

5.4.3 *Performance of the Proposed Algorithm with Varying* SOM-TH

In the above experiments, we keep the two thresholds, SOM-TH and MAX-MIN-TH, fixed to be 1.5 and 3.0, respectively. In this subsection, we conduct experiments to study the detecting effectiveness of the measures on the behavior of the proposed method by varying the SOM-TH in the range [1.0, 1.5, 2.0, 2.5, 3.0] and the results for two datasets, Ionosphere and Shuttle, are shown in Figs. 5.16 and 5.17, respectively.

In the figures, the x-axis denotes the values of m while the y-axis denotes for five values of SOM the corresponding precision, recall, and rank power, respectively. The dashed green line, the dashed blue line, the dotted black line, the solid red line, and the dashed magenta line denote the results with SOM being set to be 1.0, 1.5, 2.0, 2.5, and 3.0, respectively. From the figures, we can see that our methods can obtain better results when SOM-TH is set to take value of 1.0.

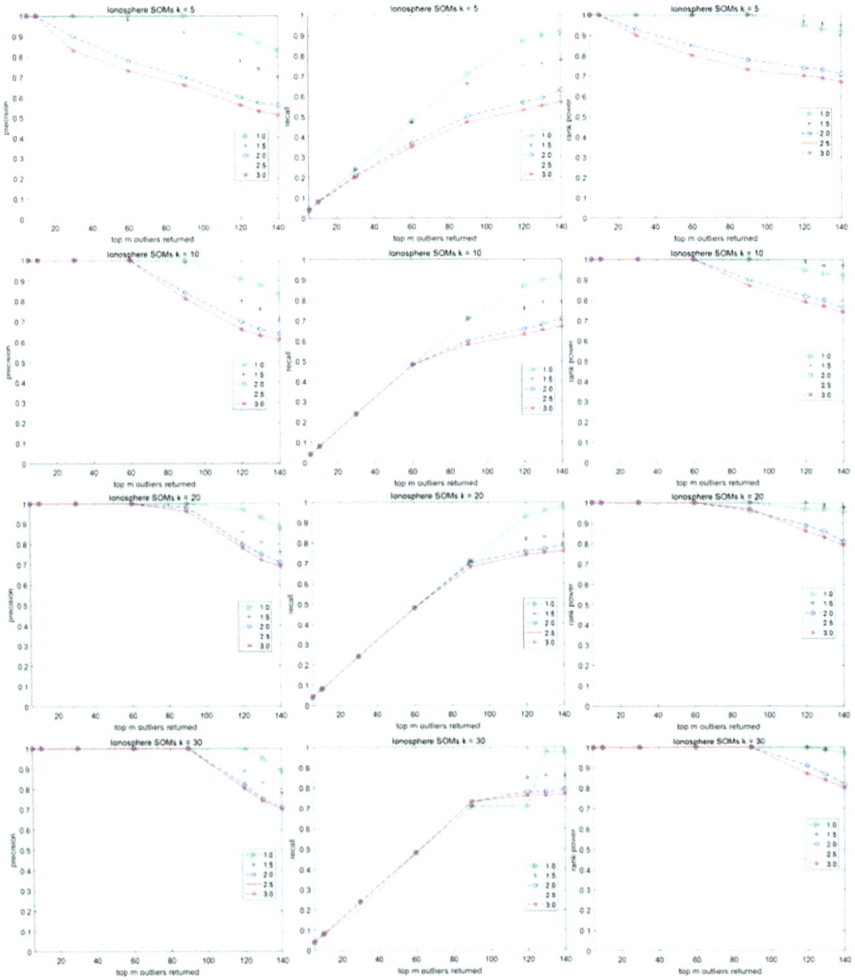

Fig. 5.16 Outlier detection results for Ionosphere data for k = 5, 10, 20, 30

5.5 Concluding Remarks

In this chapter, we have proposed an efficient MST-clustering-inspired *k*NN-based outlier detection method that can detect both global and local outliers. In addition to being compatible with the traditional distance-based outlier detection methods, our approach also performs better in identifying local outliers that deviate from the main patterns in a given dataset. Candidate outliers are ranked based on the miniMST-based outlier scores that are assigned to each data point. To demonstrate the utility of our proposed outlier factors, we have performed a detailed comparison of its performance with a number of state-of-the-art outlier detection methods. Through

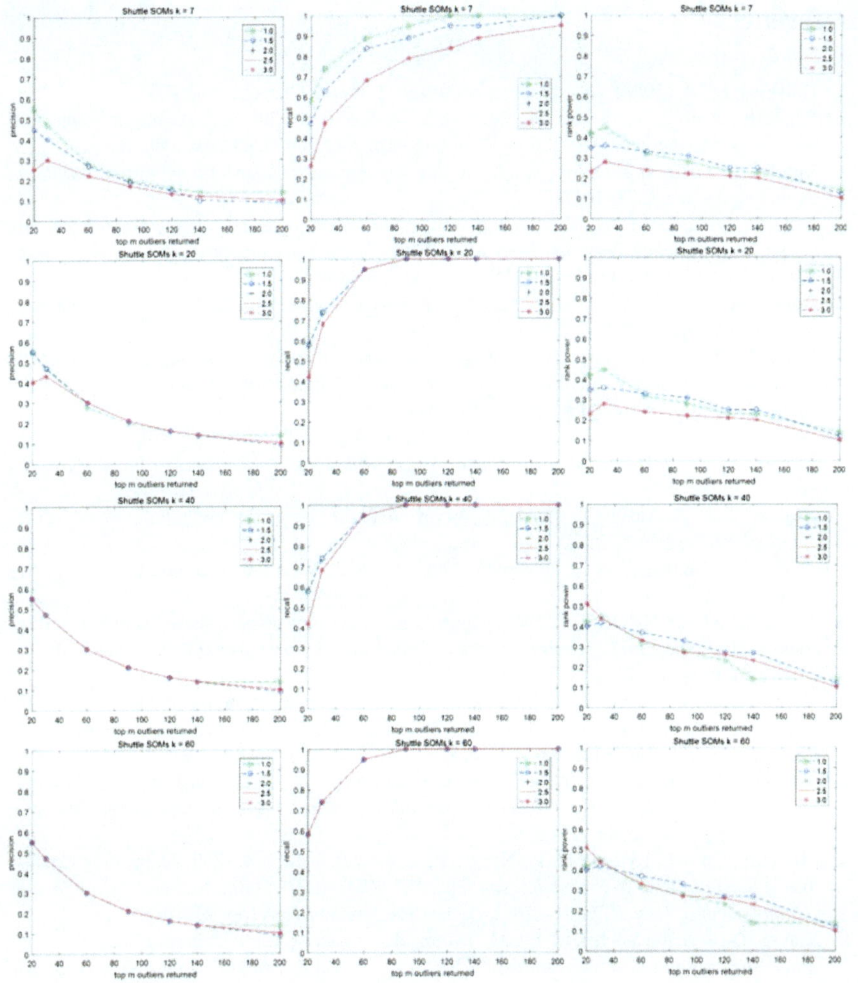

Fig. 5.17 Outlier detection results for Shuttle data for $k = 7, 20, 40, 60$

a thorough evaluation, our method manifests its ability to rank the best candidates for being an outlier with high precision and recall. Both theoretical justification and empirical validation show the effectiveness of the proposed method.

Acknowledgements This chapter was modified from the paper published by our group in Information Systems [38]. The related contents are reused with permission.

References

1. Hawkins, D. M. (1980). *Identification of outliers*. London: Chapman and Hall.
2. Eskin, E., Arnold, A., Prerau, M., Portnoy, L., & Stolfo, S. (2002). A geometric framework for unsupervised anomaly detection: Detecting intrusions in unlabeled data. In: Barbará, D., & Jajodia, S. (Eds.), *Applications of data mining in computer security. Advances in Information Security* (Vol. 6, pp. 77–101).
3. Lane, T., & Brodley, C. E. (1998). Temporal sequence learning and data reduction for anomaly detection. In *Proceedings of the 1998 5th ACM Conference on Computer and Communications Security* (CCS-5), San Francisco, CA, USA (pp. 150–158).
4. Bolton, R. J., & David, J. H. (2002). Unsupervised profiling methods for fraud detection. *Statistical Science, 17*(3), 235–255.
5. Wong, W., Moore, A., Cooper, G., & Wagner, M. (2002). Rule-based anomaly pattern detection for detecting disease outbreaks. In *Proceedings of the 18th National Conference on Artificial Intelligence*, Edmonton, Alta., Canada (pp. 217–223).
6. Sheng, B., Li, Q., Mao, W., & Jin, W. (2007). Outlier detection in sensor networks. In *Proceedings of ACM International Symposium on Mobile Ad Hoc Networking and Computing* (pp. 219–228).
7. Hodge, V. J., & Austin, J. (2004). A survey of outlier detection methodologies. *Artificial Intelligence Review, 22*(2), 85–126.
8. Chandola, V., Banerjee, A., & Kumar, V. (2009). Anomaly detection: A survey. *ACM Computing Surveys, 41*(3), 15.1–15.58.
9. Knorr, E. M., & Ng, R. T. (1998). Algorithms for mining distance-based outliers in large datasets. In *Proceedings of the International Conference on Very Large Data Bases* (VLDB'98), New York (pp. 392–403).
10. Ramaswamy, S., Rastogi, R., & Shim, K. (2000). Efficient algorithms for mining outliers from large data sets. In *Proceedings of the 2000 ACM International Conference on Management of Data* (SIGMOD'00), Dallas (pp. 427–438).
11. Angiulli, F., & Pizzuti, C. (2002). Fast outlier detection in high dimensional spaces. In *Proceedings of the 6th European Conference on Principles of Data Mining and Knowledge Discovery* (PKDD'02), Helsinki (pp. 15–26).
12. Breuning, M. M., Kriegel, H. P., Ng, R. T., & Sander, J. (2000). LOF: Identifying density-based local outliers. In *Proceedings of the 2000 ACM SIGMOD International Conference on Management of Data* (SIGMOD'00), Dallas, TX, United States (pp. 93–104).
13. Bay, S. D., & Schwabacher, M. (2003). Mining distance-based outliers in near linear time with randomization and a simple pruning rule. In *Proceedings of the 9th ACM SIGKDD International Conference on Knowledge Discovery and Data Mining (KDD '03)*, Washington, DC, United States (pp. 29–38).
14. Aggarwal, C., & Yu, P. (2005). An effective and efficient algorithm for high-dimensional outlier detection. *The VLDB Journal, 14*(2), 211–221.
15. Kriegel, H.-P., Schubert, M., & Zimek, A. (2008). Angle-based outlier detection in high-dimensional data. In *Proceedings of the 14th ACM SIGKDD International Conference on Knowledge Discovery and Data Mining (KDD'08)*, Las Vegas, Nevada, USA (pp. 444–452).
16. Huang, H., Mehrotra, K., & Mohan, C. K. (2013). Rank-based outlier detection. *Journal of Statistical Computation and Simulation, 83*(3), 518–531.
17. Zahn, C. T. (1971). Graph-theoretical methods for detecting and describing gestalt clusters. *IEEE Transactions on Computers, C-20*(1), 64–82.
18. Wang, X., Wang, X. L., & Wilkes, D. M. (2009). A divide-and-conquer approach for minimum spanning tree-based clustering. *IEEE Transactions on Knowledge and Data Engineering, 21*(7), 945–958.
19. Zhong, C., Miao, D., & Wang, R. (2010). A graph-theoretical clustering method based on two rounds of minimum spanning trees. *Pattern Recognition, 43*(3), 752–766.

20. Luo, T., & Zhong, C. (2010). A neighborhood density estimation clustering algorithm based on minimum spanning tree. In *Proceedings of the 5th International Conference on Rough Set and Knowledge Technology (RSKT'10)*, Beijing, China (pp. 557–565).
21. Luo, T., Zhong, C., Li, H., & Sun, X. (2010). A multi-prototype clustering algorithm based on minimum spanning tree. In *Proceedings of 2010 7th International Conference on Fuzzy Systems and Knowledge Discovery (FSKD'10)* (pp. 1602–1607).
22. Zhong, C., Miao, D., & Franti, P. (2011). Minimum spanning tree based split-and-merge: A hierarchical clustering method. *Information Sciences, 181*(16), 3397–3410.
23. Lin, J., Ye, D., Chen, C., & Gao, M. (2008). Minimum spanning tree based spatial outlier mining and its applications. In *Proceedings of the 3rd International Conference on Rough Sets and Knowledge Technology (RSKT'08)*, Chengdu, China, LNAI 5009 (pp. 508–515).
24. John Peter, S., & Victor, S. P. (2011). An integrated approach for local outlier detection using dynamic minimum spanning tree. *Journal of Discrete Mathematical Sciences and Cryptography, 14*(1), 89–106.
25. John Peter, S. (2011). Minimum spanning tree based clustering for outlier detection. *Journal of Discrete Mathematical Sciences and Cryptography, 14*(2), 149–166.
26. Daneshgar, A., Javadi, R., & Shariat Razavi, S. B. (2013). Clustering and outlier detection using isoperimetric number of trees. *Pattern Recognition, 46*(12), 3371–3382.
27. Wang, X., Wang, X. L., & Wilkes, D. M., A spanning tree-inspired clustering based outlier detection technique. In *Proceedings of the 12th Industry Conference on Data Mining*, Berlin, Germany (pp. 209–223)
28. Zhu, Q., Fan, X., & Feng, J. (2014). Outlier detection based on K-Neighborhood MST. In *Proceedings of the 2014 IEEE 15th International Conference on Information Reuse and Integration (IEEE IRI'14)*, San Francisco, CA, United States (pp. 718–724).
29. Cipolla, E., & Vella, F. (2014). Identification of spatio-temporal outliers through Minimum Spanning Tree. In *Proceedings of the 10th International Conference on Signal-Image Technology and Internet-Based Systems (SITIS'14)*, Marrakech, Morocco (pp. 248–255).
30. Abghari, S., Boeva, V., Lavesson, N., Grahn, H., Ickin, S., & Gustafsson, J. (2018). A minimum spanning tree clustering approach for outlier detection in event sequences. In *Proceedings of the 2018 17th IEEE International Conference on Machine Learning and Applications (ICMLA'18)*, Orlando, FL, United States (pp. 1123–1130).
31. Wang, X., Wang, X.L., & Wilkes, D. M. (2012). Modifying iDistance for a fast CHAMELEON with application to patch based image segmentation. In *Proceedings of the 9th IASTED International Conference on Signal Processing, Pattern Recognition and Applications (SPPRA 2012)*, Crete, Greece (pp. 107–114).
32. UCI: The UCI KDD Archive. [http://kdd.ics.uci.edu/]. Irvine, CA: University of California.
33. Tang, J., Chen, Z., Fu, A. W. C., & Cheung, D. W. (2002). Enhancing effectiveness of outlier detections for low density patterns. In *Proceedings of the 6th Pacific-Asia Conference on Knowledge Discovery and Data Mining (PAKDD'02)*, Taipei, Taiwan (pp. 535–548).
34. Jin, W., Tung, A.K.H., Han, J., & Wang, W. (2006). Ranking outliers using symmetric neighborhood relationship. In *Proceedings of the 10th Pacific-Asia Conference on Knowledge Discovery and Data Mining (PAKDD'06)*, Singapore (pp. 577–593).
35. Zhang, K., Hutter, M., & Jin, H. (2009). A new local distance-based outlier detection approach for scattered real-world data. In *Proceedings of the 13th Pacific-Asia Conference on Advances in Knowledge Discovery and Data Mining (PAKDD '09)*, Bangkok, Thailand (pp. 813–822).
36. Aggarwal, C., & Yu, P. (2001). Outlier detection for high-dimensional data. In *Proceedings of the 2001 ACM International Conference on Management of Data (SIGMOD'01)*, Santa Barbara, CA, USA (pp. 37–46).
37. Meng, X., & Chen, Z. (2004). On user-oriented measurements of effectiveness of web information retrieval systems. In *Proceedings of the International Conference on Internet Computing (ICIC'04)*, LasVegas, Nevada, USA (vol. 1, pp. 527–533).
38. Wang, X., Wang, X. L., Ma, Y., & Wilkes, D. M. (2015). A fast MST-inspired *k*NN-based outlier detection method. *Information Systems, 48,* 89–112.

Chapter 6
A *k*-Nearest Neighbour Spectral Clustering-Based Outlier Detection Technique

Abstract The problem of outlier analysis is an important one because of its applicability to a variety of problem domains such as intrusion detection, fraud detection, discovery of criminal activities in electronic commerce, and so on. Many models have been developed for outlier detection, including probabilistic models, distance-based models, density-based models, and clustering models. These models extract various indicators (e.g., frequencies of certain values) from the amount of available data, which are useful in understanding the behaviors of the outliers. Spectral clustering receives much attention as a competitive clustering algorithms emerging in recent years. However, it is not very well scalable to modern large datasets. To partially circumvent this drawback, in this chapter, we propose a new outlier detection method inspired by spectral clustering. Our algorithm combines the concept of *k*-nearest neighbors and spectral clustering techniques to obtain the abnormal data as outliers by using the information of eigenvalues in the feature space statistically. We compare the performance of the proposed method with state-of-the-art outlier detection methods. Experimental results show the effectiveness of our algorithm for identifying outliers.

Keywords Outlier detection · Distribution-based outlier detection · Clustering-based outlier detection · Spectral clustering · Eigenvalues

6.1 Introduction

With the rapid development of information technology, data mining technologies come into being consequently to find important and useful information from massive and high-dimensional data. Outlier detection is an important data mining technique, which focuses on discovering the small portion of data objects in a dataset that, being inconsistent with the conventional pattern of the majority of the data set, however, may imply important information. In recent years, there have been a broad range of definitions for outliers and researchers have developed many types of algorithms for outlier detection, mainly classified into distribution-based outlier detection algorithms, depth-based outlier detection algorithms, distance-based outlier detection

X. Wang et al., *New Developments in Unsupervised Outlier Detection*,
https://doi.org/10.1007/978-981-15-9519-6_6

algorithms, density-based outlier detection algorithms, and clustering-based outlier detection algorithms [1, 2].

Distribution-based methods (also known as model-based and statistical methods) generally assume that normal data objects are generated by a statistical (stochastic) model and data not following the model are outliers. If a given set of data points fit a Gaussian distribution gD, where for location x in the data space, gD(x) gives the probability density at x. Thus, Gaussian distribution gD can be used to model the normal data, that is, most of the data points in the data set. For each outlying object y, gD(y) can be estimated to be the probability that this point fits the Gaussian distribution. Because gD(y) is very low, y is unlikely generated by the Gaussian model and thus is an outlier with a probability of $1 - \text{gD}(y)$. Data involving only one attribute or variable are called univariate data. For simplicity, we often choose to assume that univariate data are generated from a normal distribution. Data involving two or more attributes or variables are called multivariate data. Many univariate outlier detection methods can be extended to handle multivariate data. The central idea is to transform the multivariate outlier detection task into a univariate outlier detection problem.

An unsupervised outlier detection method expects that the normal objects follow a pattern far more frequently than outliers. Clustering is, after all, about the determination of *frequent groups* of data points, whereas outliers are about determination of *rare groups* of data points. There are three general approaches to clustering-based outlier detection. If an object does not belong to any cluster, it is identified as an outlier. If there is a large distance between the object and the cluster to which it is closest, it is an outlier. This approach considers the distance between an object and the cluster to which it is closest to. If the distance is large, then the object is likely an outlier with respect to the cluster. Thus, this approach detects individual outliers with respect to clusters. If an object is part of a small or sparse cluster, all the objects in that cluster are outliers.

As a rule, statistical learning methods find it much easier to determine frequent characteristics than rare characteristics of a data set. This problem is further magnified in high dimensionality. Recently, spectral clustering has been widely applied to pattern recognition and data mining because it can obtain global optimal solution on sample spaces of arbitrary shape [3, 4]. Meanwhile, since it is only related to the number of data points, not the dimensionality, spectral clustering can help solve the curse of dimensionality suffered by distance-based methods and density-based methods in high-dimensional data space. However, it is not very well scalable to modern large datasets. In this chapter, we propose a spectral clustering-based outlier mining approach, which, when applied to a local setting in terms of k-nearest neighborhood of a data point, manifests its effectiveness and efficiency.

The rest of the chapter is organized as follows. In Sect. 6.2, we present some preliminaries of spectral clustering and review some existing work on spectral clustering-based outlier detection. In Sect. 6.3, the proposed spectral clustering-based outlier mining approach is introduced. In Sect. 6.4, we present the results of our experiments conducted to evaluate the performance of the proposed algorithm. Finally, conclusions are given in Sect. 6.5.

6.2 Spectral Clustering and Its Application to Outlier Detection

6.2.1 *Preliminaries*

Clustering is the process to divide points in a dataset into a number of clusters so that the similarity between two points within the same cluster and the dissimilarity between two points belonging to respectively two different clusters are as high as possible. Cluster analysis plays an important part in data mining. Spectral clustering receives much attention as a competitive clustering algorithms emerging in recent years, which is mainly applied to image segmentation. Basically, possessing numerous advantages, spectral clustering treats points in a dataset as the vertices of a weighted undirected graph and the similarity between two vertexes as the weights of edges in the graph, and converts the problem of clustering into an optimal partitioning problem of a graph. The goal is to find a graph partitioning methodology to make the weights of edges connecting two different subgraphs as large as possible and the weights of edges within a subgraph as small as possible. The most common graph-dividing criteria proposed include Mini Cut [5], Normalized Cut [6, 7], and Ratio Cut [8]. Mini Cut produces a good result on image segmentation but is prone to skew-segmentation, while Normalized Cut and Ratio Cut take both the minimum sum of the weights of the cutting edges and the balance of the paritioning into account.

In this chapter, we utilize the *k*-way Normalized Cut partitioning approach, for a given data set which contains *N* points, $x_1, x_2, \ldots, x_N$, the goal of clustering is to divide this *N* points into *K*-clusters so that the similarities between two points belonging to the same cluster are maximized and those between two points belonging to different clusters are minimized. Treating data point x_i in the dataset *S* as a vertex v_i in a graph and the similarity between x_i and x_j, W_{ij}, as the weight of the edge connecting x_i and x_j, we obtain a weighted undirected graph $G = (V, E)$ with *V* denoting the set of all the vertexes and *E* denoting the set of all the edges in the graph *G*. Suppose that the multi-way spectral clustering partitions the graph *G* into *K*-subgraphs, A_1, $A_2, \ldots, A_K$, then the target function of Normalized Cut (*N*cut) that makes the optimal partitioning can be described as:

$$N\text{cut}(A_1, A_2, \cdots, A_K) = \sum_{i=1}^{K} \frac{\text{cut}\left(A_i, \bar{A_i}\right)}{\text{vol}(A_i)} \tag{6.1}$$

where

$$\text{cut}(A, B) = \sum_{i \in A, j \in B} W_{ij}, \underset{i}{\bar{A}} = \left\{v_p \in V, v_p \notin A_i\right\}, \text{vol}(A_i)$$

$$= \sum_{p \in A_i, q \in V} W_{pq}.$$

After mathematic transformations, the solution for the above Ncut problem can be finally converted to a problem of finding eigenvalues (and eigenvectors) of a Laplacian matrix and the smallest series of eigenvalues corresponds to the optimal partitioning of the graph. Thus, the discrete clustering problem becomes to find the eigenvectors on a contiguous data space.

In the general frame of the spectral clustering, the following three Laplacian matrixes are commonly used,

$$L = D - W \tag{6.2}$$

$$L_{\text{sys}} = D^{-\frac{1}{2}} L D^{-\frac{1}{2}} = I - D^{-\frac{1}{2}} W D^{-\frac{1}{2}} \tag{6.3}$$

$$L_{\text{rw}} = D^{-1} L = I - D^{-1} W \tag{6.4}$$

where W is the similarity matrix, D is the corresponding diagonal matrix, $D_{ij} = \sum^{N}{}_{j=1} W_{ij}$, and I is the identify matrix. In this chapter, we adopt the second form of Laplacian matrix, and notice that $L_{\text{sys}} = I - D^{-1/2} W D^{-1/2}$, hence the sum of eigenvalues of $D^{-1/2} W D^{-1/2}$ and the corresponding eigenvalues of L_{sys} are one and the corresponding eigenvectors are equal. Therefore, finding the eigenvectors of $D^{-\frac{1}{2}} W D^{-\frac{1}{2}}$ which correspond to the K-largest eigenvalues is equivalent to finding the eigenvectors of L_{sys} which correspond to the K-smallest eigenvalues. With these ideas in mind, the general frame of spectral clustering works as following,

1. Given a set of data points, $x_1, x_2, \ldots, x_N$, calculate the similarity between every pair of points according to a certain kind of similarity definition (e.g., the Gaussian kernel function) and construct the similarity matrix $\boldsymbol{W}$;
2. Compute the corresponding Laplacian matrix L_{sys}, find its K-eigenvectors corresponding to the K-smallest eigenvalues and use these K-eigenvectors as columns to construct a feature matrix in a K-dimensional space, $H \in R^{N \times K;}$
3. Treat each row of the feature matrix, H, as a data point in K-dimensional space and perform K-means clustering on these points. Each original d-dimensional data point is assigned the same cluster number as the K-dimensional feature vector of the corresponding row in the feature matrix H.

From the general framework of spectral clustering given in the above, it can be seen that a similarity definition should first be provided to establish the similarity matrix. The most commonly used one is the Gaussian kernel function, which has the following form,

$$W_{ij} = e^{-\frac{\|x_i - x_j\|^2}{2\sigma^2}} \tag{6.5}$$

where $\|x_i - x_j\|$ denotes the Euclidean distance between two points x_i and x_j, and σ is the width parameter, controlling the radial function scope of the Gaussian kernel function. Obviously, the similarity is closer to 1 if two points are very close; otherwise, the similarity is closer to 0 if two points are far away from each other.

With respect to parameter σ, in the original NJW algorithm, it was proposed to perform spectral clustering, respectively, using several pre-set values and choose the one that yields the best clustering results. To improve the running time performance, some researchers suggested to determine σ by empirical formulas. However, all these practices require a combination of the domain knowledge and are not applied to all the situations. Therefore, researchers start to focus on how to automatically determine this parameter according to the dataset itself. Under this line, Zelnik-Manor and Perona proposed the 'Self-Tuning' algorithm [9]. Based on "Local Scale" idea, the algorithm constructs a self-adaptive parameter $\sigma_i = \|x_i - x_p\|$ (where x_p is the k-th nearest neighbor of x_i, usually $k = 7$ is used) for each data point using its own neighborhood information. The similarity between two points is then replaced by the following form,

$$W_{ij} = e^{-\frac{\|x_i - x_j\|^2}{\sigma_i \sigma_j}} \tag{6.6}$$

Empirical evidence shows that this latter approach can address those data sets with a multi-scale nature where a universal σ can not do, and isolate accurately the dense clusters embedded in a sparse background.

In recent years, spectral clustering is one of the most prominent clustering approaches that can capture hidden cluster structures in the data and has wide applications in data mining because it can obtain global optima solution and adapt to sample spaces with any shape. However, it is highly sensitive to noisy input data. In 2017, Bojchevski et al. proposed a robust spectral clustering technique that is able to handle such scenarios [10]. To achieve this goal, they proposed a sparse and latent decomposition of the similarity graph used in spectral clustering. In their model, the spectral embedding as well as the corrupted data can be jointly learned, thus enhancing the clustering performance overall. In 2018, to cope with the problem that spectral clustering methods do not scale well to large-scale problems due to its quadratic complexity in constructing similarity graphs and computing subsequent eigen-decomposition, Wu et al. proposed a novel scalable spectral clustering method using Random Binning features (RB) to simultaneously accelerate both similarity graph construction and the eigen-decomposition [11]. The key step of this spectral clustering is learning the affinity matrix to measure the similarity among data points. Therefore, also in the same year, Tan et al. proposed a new spectral clustering method, which used mutual k-nearest neighbors to obtain the affinity matrix by removing the influence of noise [12]. Then, the characteristics of high-dimensional data can be self-represented to ensure local important information of data by using affinity matrix in standardized processing. Furthermore, they also used the normalization method to further improve the performance of clustering. In 2019, Tong et al. proposed a new spectral clustering method based on local principal component analysis (PCA) and connected graph decomposition [13]. Specifically, their method randomly selected centroids of the dataset to take global structure of data points into consideration and then used local PCA to preserve the local structure of data points for constructing the similarity matrix. Furthermore, their method employed the connected graph decomposition to partition the resulting similarity matrix to group data points into clusters. In 2020, to overcome the issues that the spectral clustering methods are not suitable for large-scale problems due to its high computational complexity and that the neighborhood weighted graph is constructed by the Gaussian kernel so that more work is required to tune the heat-kernel parameter, Yang et al. proposed a novel spectral clustering based on hierarchical bipartite graph (SCHBG) approach by exploring multiple-layer anchors with a pyramid-style structure [14]. Classical spectral clustering algorithms consist of two separate stages: (1) solving a relaxed continuous optimization problem to obtain a real matrix followed by (2) applying K-means or spectral rotation to round the real matrix (i.e., continuous clustering result) into a binary matrix called the cluster indicator matrix. Such a separate scheme is not guaranteed to achieve jointly optimal result because of the loss of useful information. Therefore, in the same year, Pang et al. proposed a joint model to simultaneously compute the optimal real matrix and binary matrix so as to obtain a better clustering result [15]. The existing joint model adopts an orthonormal real matrix to approximate the orthogonal but nonorthonormal cluster indicator matrix. It is noted that only in a very special case (i.e., all clusters have the same number of samples), the cluster indicator matrix is an orthonormal matrix multiplied by a real number. The error of approximating a nonorthonormal matrix is inevitably large. To overcome the drawback, they further proposed to replace the nonorthonormal cluster indicator matrix with a scaled cluster indicator matrix which is an orthonormal matrix. Their method is capable of obtaining better performance because it is easy to minimize the difference between two orthonormal matrices.

6.2.2 Spectral Clustering-Based Outlier Detection

Outlier detection is widely used for many areas such as credit card fraud detection, discovery of criminal activities in electronic commerce, weather prediction, and marketing. As mentioned in [16, 17], clustering-based outlier detection techniques regard small clusters as outliers or identify outliers by removing clusters from the original dataset. Thus, clustering algorithms can detect outliers as a by-product of the clustering process. Spectral clustering has wide applications in pattern recognition and data mining because it can obtain global optima solution and adapt to sample spaces with any shape [18, 19]. It is simple to implement and can be solved efficiently by standard linear algebra software [20, 21]. In 2008, Yang and Huang proposed a spectral clustering-based outlier detection algorithm [22]. Through spectral method, the information of feature space with eigenvectors rather than that of the whole dataset can be used to obtain K-stable clusters. Then a cluster-based local outlier factor (CBLOF) is introduced to identify and find the outliers in a dataset [23].

Finally, the top n outliers are output according to their CBLOF on the K-clusters. In the same year, Yang and Huang also proposed a new spectral clustering algorithm and attempted to use it for outlier detection in a dataset [24]. To construct a spectral graph, the proposed spectral clustering algorithm took the number of neighborhoods shared by the objects as the similarity measure, which could help to isolate outliers as well as construct a sparse matrix. Before defining the shared neighborhoods, they calculated the neighborhoods of a given object in the adjacent matrix and limited the number of neighborhoods to a set of nodes with high similarity values by using a threshold to neglect low similarity neighborhoods. The selection of threshold T was important for achieving an appropriate adjacent matrix and higher outlier detection rate. Let X be the set of Gaussian similarity values. Let μ and δ be the mean and standard deviation of X, respectively. The suitable value of T will be inside the interval $[\mu - \delta, \mu)$. Then, for every object in the dataset, the value of CBLOF was computed with Definition 3 in [23]. In 2012, Lin and Zhu proposed a dataset structure analysis and outlier detection process to solve the problem that, in real applications with spectral clustering algorithm, the number of clusters of a dataset is not given as a prior [25]. The proposed process was on the basis of spectral clustering algorithm and consisted of four steps. The first step applied some algorithm, such as DBSCAN, which does not need the number of clusters as input to cluster the dataset to get an approximation of the number of clusters. The second step used the approximation obtained from the first step to get the upper bound of the number of the clusters. The third step used the upper bound obtained from the second step to get the optimal value of the number of the clusters, and output the optimal clustering result. The last step applied the LOF algorithm with the result from the third step to find the data objects with the largest LOF values to be outliers.

In 2018, motivated by the low number of outlier detection algorithms focusing on weighted undirected graphs, Tyuryukanov et al. proposed an efficient and effective method to detect loosely connected graph clusters below a certain number of nodes [26]. The essence of the method lied in the efficient examination of the spectral minimal spanning tree of the input graph. Compared to outlier detection in Euclidean point cloud data, there are noticeably less methods to find outliers in graphs [27]. The majority of existing graph outlier detection methods focuses either on community outliers [28] (i.e., residual nodes loosely coupled to the identified network communities), or on identifying anomalous elements in graph matrices via matrix factorization techniques [29]. These methods are not tailored to find outlier clusters (i.e., loosely connected groups of two or more nodes). Additionally, many state-of-the-art outlier detection methods have a higher time complexity than spectral clustering [29, 30], which may result in a computational bottleneck. The main idea of the proposed method is to embed the graph in a low-dimensional Euclidean space and detect outlier clusters by analysing the minimal spanning tree (MST) of the embedded graph. The graph embedding is constructed in a well-established manner by finding several largest eigenvectors of the normalized adjacency matrix. However, an output of graph clustering may not reveal all of the outliers as they may mask each other. That is, some outliers may look similar to the normal data in the presence of a substantially more severe outlier [27]. For this reason, the proposed method searched the entire graph for loosely connected nodes and clusters below a certain size to return a collection of such nodes and clusters together with their severity ranking for outlier detection in the dataset.

6.3 The Proposed Spectral Clustering-Based Outlier Mining Algorithm

In the above section, spectral clustering and spectral clustering-based outlier detection methods are presented. In view of the curse of dimensionality problem suffered by distance-based outlier detection algorithms [31, 32] and density-based outlier detection algorithms [33] in high-dimensional data space [34], spectral clustering shields this problem by computing similarity between two data objects using the Euclidean distance but suffers from high computation cost and high memory storage cost. For a given dataset with N number of objects, the size of the similarity matrix used by spectral clustering is $N \times N$. When N is very large, the size of the similarity matrix will become too large to fit into the main memory.

6.3.1 A Simple Idea

It is generally believed outliers comprise a small portion of the whole dataset and reside in small clusters in sparse region and behave differently relative to the majority of the normal data. To take both the dimension and the number of data objects in a database into consideration for clustering-based outlier detection, we propose a preprocess to find for each data point its k-nearest neighbors (kNN) and then perform a spectral clustering process on these $k + 1$ data point. By this way, it is more easily to examine the abnormal behavior of a small number of data points by taking advantage of a small neighborhood of each data point and its kNN while keeping the tempo- and spatio-computational cost as low as possible. The result of this pre-processing step is N new small datasets consisting of each data point and its k-nearest neighbors. Next, we perform spectral clustering on every new dataset, which results in N new sets of $(k + 1)$-dimensional feature data in the eigen-space. For these $(k + 1)$-dimensional feature data in the eigen-space, the smallest eigenvalues are zeros for all the data. Other eigenvalues denote the occurrence frequency of patterns. Our spectral clustering-based outlier detection algorithm is based on the observation that the values of the second smallest eigenvalues associated with outliers have lowest frequency of occurrence. From a statistical point of view, we select those points (e.g., 15~20% of the total number) as outlier candidates whose corresponding second smallest eigenvalues appear least frequently. From our experience, there are some inliers among the outlier candidates. To remove inliers, we then compute the distance of each outlier candidate to its k-th nearest neighbor as its outlier index to rank the

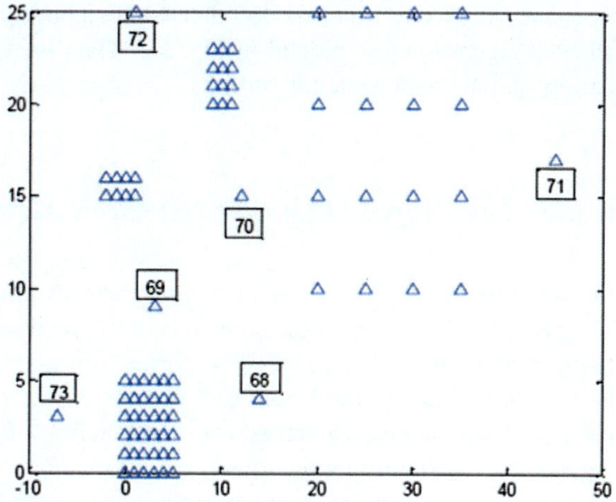

Fig. 6.1 Synthetic dataset for illustration of our working idea. Reprinted from Ref. [35], copyright 2016, with permission from Springer

outlier candidates (i.e., the distance-based outlier score for the DB-Max method). Finally, top n ranked outlier candidates with the largest outlier indices will be returned as outliers in the database.

To illustrate this observation, a synthetic two-dimensional dataset is plotted in Fig. 6.1. It consists of 73 data points and, in our preprocess stage, produces 73 size-reduced mini datasets. After performing spectral clustering process on these 73 new groups of data, we plot the second smallest eigenvalues for each new data set in Fig. 6.2. From the plot, it can be clearly seen that six points, labeled by 68, 69, 70, 71, 72, 73, have their second smallest eigenvalues occurring with the lowest

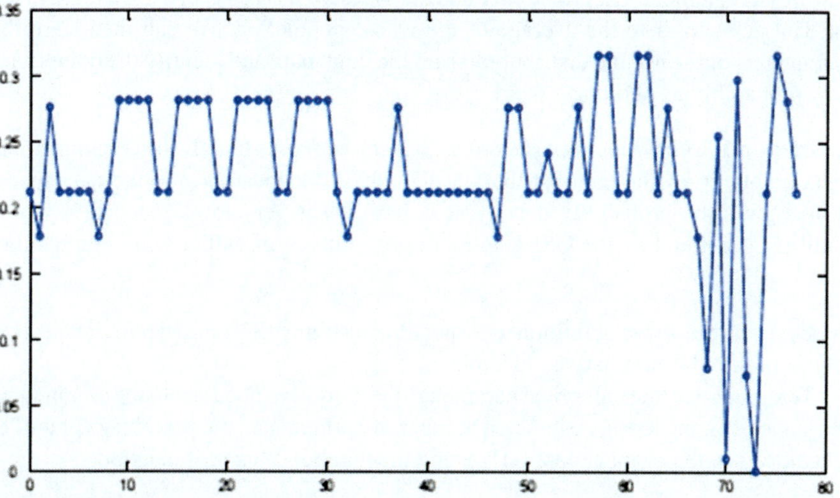

Fig. 6.2 Distribution of the second smallest eigenvalues of synthetic dataset Reprinted from Ref. [35], copyright 2016, with permission from Springer

frequency (i.e., one time, while other values appear more than one time) among all such eigenvalues as plotted in Fig. 6.2. Then back to Fig. 6.1, we can see they correspond to six outstanding outliers.

6.3.2 The Proposed Outlier Detection Algorithm

Statistical methods for outlier detection make assumptions about data normality. They assume that normal objects in a data set are generated by a stochastic process (a generative model). Consequently, normal objects occur in regions of high probability for the stochastic model, and objects in the regions of low probability are outliers. The general idea behind statistical methods for outlier detection is to learn a generative model fitting the given data set and then identify those objects in low-probability regions of the model as outliers. However, there are many different ways to learn generative models. In general, statistical methods for outlier detection can be divided into two major categories: parametric methods and nonparametric methods, according to how the models are specified and learned. A parametric method assumes that the normal data objects are generated by a parametric distribution with parameter Θ. The probability density function of the parametric distribution $f(x, \Theta)$ gives the probability that object x is generated by the distribution. The smaller this value, the more likely x is an outlier. Data involving only one attribute or variable are called univariate data. There are many kinds of statistical models. The effectiveness of statistical methods highly depends on whether the assumptions made for the statistical model hold true for the given data. For simplicity, we often choose to assume that data are generated from a normal distribution. This is because, according to the law of large numbers, in probability and statistics, as a sample size grows, the mean of data gets closer to the average of the whole population. We can then learn the parameters of the normal distribution from the input data and identify the points with low probability as outliers.

Definition 1 (*Statistical-based Outlier Detection Model*) If a given set of univariate data points fit a Gaussian distribution gD, where for location x in the data space, gD(x) gives the probability density at x. If gD(y) is very low for location y, y is unlikely generated by the Gaussian model and thus is an outlier with a probability of $1 - \text{gD}(y)$.

Based on the above definition, our spectral clustering (SC)-based outlier detection algorithm can be summarized in Table 6.1.

To summarize, the numerical parameters the algorithm needs from the user include the data set, S, the loosely estimated number of outliers (i.e., the percentage of outlier candidates in the original data set), p, and the number of nearest neighbors, k.

Table 6.1 A spectral clustering-based outlier detection algorithm

Input:	S: a set of N data objects; p: a loosely estimated percentage of the number of outliers; k: the number of nearest neighbors
Output:	m: a desired number of ranked outliers
Begin:	
1:	Compute k-nearest neighbors for each data point; generate N mini datasets $\{S_1, S_2, \ldots, S_N\}$ consisting of each data point and its kNN
2:	Perform spectral clustering on each mini dataset $S_i, 1 \le i \le N$; collect all the corresponding second smallest eigenvalues $\lambda_{i2}, 1 \le i \le N$
3:	Add top p data objects with the least occurring frequency of the second smallest eigenvalues to outlier candidates C
4:	For each outlier candidate in C, calculate its distance to the k-th nearest neighbor as the outlier index
5:	Rank all the outlier candidates in a nonincreasing order according to their outlier index and return top m ones with the biggest outlier indices as the final outliers
End	

6.3.3 Complexity Analysis

From the description in the previous subsections, it can be seen that our algorithm mainly consists of three steps, an initial k-nearest neighbor search for each of N data points, the spectral clustering on N mini datasets each of size $k + 1$ and the calculation of the outlier scores, and finally, top outliers' mining by one-dimensional sorting. Therefore, most of the calculation time is spent on the first two steps, k-nearest neighbor search and N spectral clustering. As a result, the total time complexity for the average case can be $O(N^2 + Nk^2 + \log N)$ or $O(N\log N + Nk^2 + \log N)$ if some index structure can be used. Since the number of outliers is small, the third step takes a nearly logarithm time on average. Physically, the resource consumed by our algorithm includes the space for hold the whole data set in memory, space to store their k-nearest neighbors, and some space for spectral clustering on mini dataset of size $k + 1$ and some index structure if used.

6.4 Experimental Results

In this section, experiments are conducted to evaluate the performance of our proposed algorithms in comparison with those of state-of-the-art outlier detection methods on different datasets. First, we select two two-dimensional outlier detection problems to show that our spectral clustering-based algorithm can outperform classic outlier detection algorithms in the detection accuracy. We then evaluate our algorithm on a number of multi-dimensional real datasets with no assumptions made on the data distribution, which are downloaded from the UCI machine learning

Table 6.2 Sets of data

Data Name	Data size	Dimension	# of clusters	# of outliers
syn_ Data1	89	2	1	11
syn_ Data2	78	2	4	6
connectionist-mines-vs-rocks	208	60	2	97
habermans-survival	306	3	2	81
heart-disease-processed-hungarian	294	13	2	106
hepatitis	155	19	2	32
hepatitis_withoutdupl_05_v06	70	19	2	3
hepatitis_withoutdupl_05_v07	70	19	2	3
hepatitis_withoutdupl_norm_10_v08	74	19	2	7
parkinson's	195	22	2	48
phoneme	5404	5	2	1586
seismic-bumps	2584	18	2	170
statlog-australian-credit	690	14	2	307
statlog-german-credit	1000	20	2	300
statlog-german-credit-numeric	1000	24	2	300
wholescale-region	440	7	3	47

repository [36], to check the technical soundness of this study. All the data sets are briefly summarized in Table 6.2.

We implement all the algorithms in Python and perform all the experiments on a computer with AMD A6-4400 M Processor 2.70 GHz CPU and 4.00G RAM. The operating system running on this computer is Windows 7. In our evaluation, we focus on the outlier detection accuracy rate of the outlier detection algorithms on different data sets. The results show that, overall, our spectral clustering-based outlier detection algorithm is superior over other state-of-the-art outlier detection algorithms.

6.4.1 *Performance of Our Algorithm on Synthetic Data*

In this subsection, we use two synthetic datasets to show that the proposed spectral clustering-based outlier detection method can more effectively identify local and global outliers in various scenarios than classic outlier detection algorithms, namely the DB method [32], the DB-Max method [33], and the LOF method [34].

The two synthetic datasets, syn_Data1 and syn_Data2, are shown in Fig. 6.3. The first synthetic dataset, syn_Data1, consists of 89 instances, including one large uniformly distributed cluster surrounded by eleven planted easily-identified outliers denoted by red stars. This is a global outlier detection task. The best experimental results are obtained with the parameter k's being determined by error and trial to

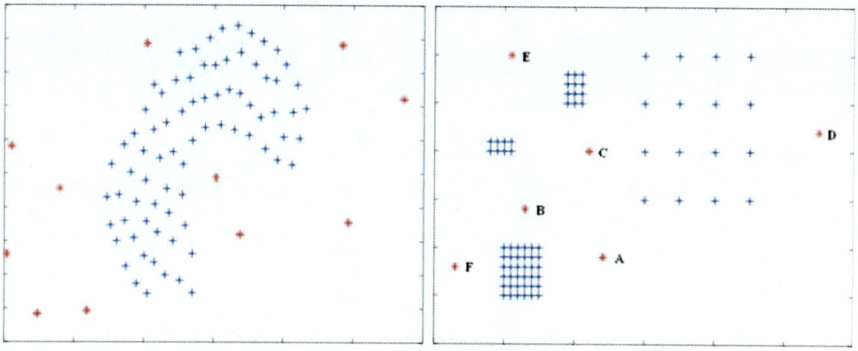

Fig. 6.3 Synthetic Datasets (left) syn_Data1, (right) syn_Data2

be 5, 5, 5, 14 for DB, DB-Max, LOF and our method, respectively and depicted in Fig. 6.4. From the figures, it can be seen that, the LOF method and the proposed method can correctly detect all the outliers (denoted by red circles) while the DB and DB-Max methods both miss one.

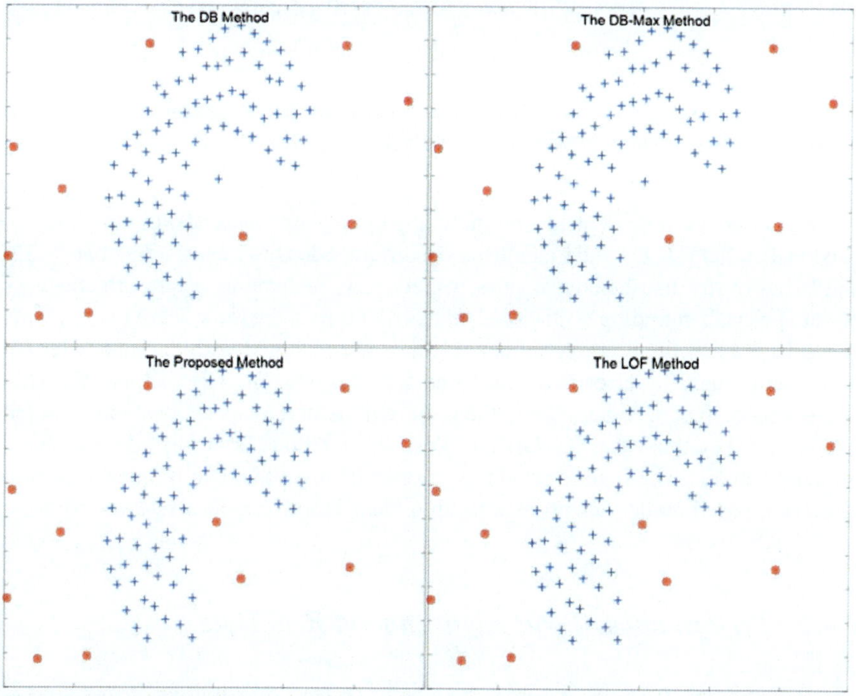

Fig. 6.4 Experimental results for syn_Data1 (upper left) the DB method, (upper right) the DB-Max, (lower left) the LOF method, (lower right) our method

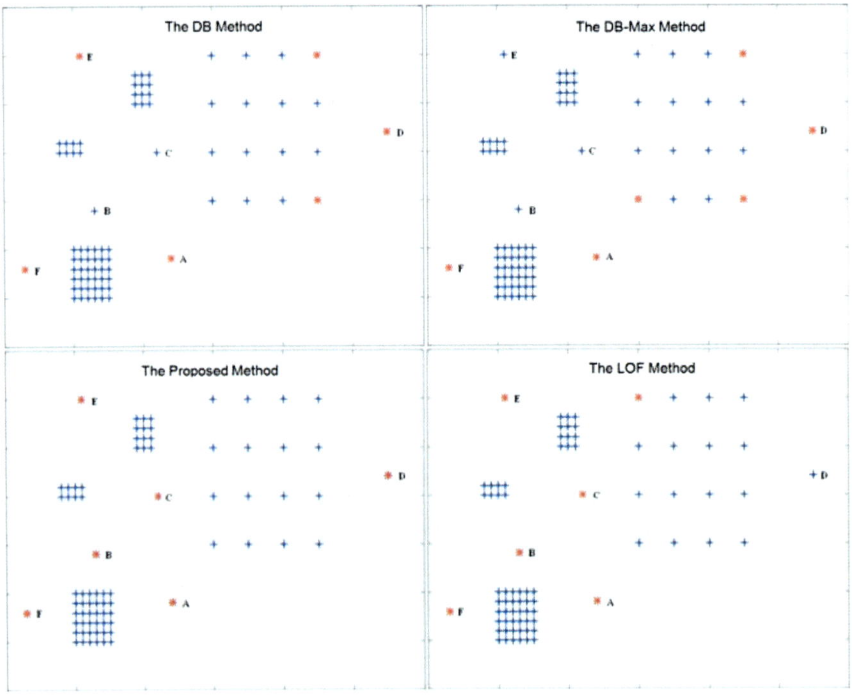

Fig. 6.5 Experimental results for syn_Data2 (upper left) the DB method, (upper right) the DB-Max, (lower left) the LOF method, (lower right) our method

syn_Data2 contains 78 instances, including three planted global outliers (*A*, *D*, *E*), two local outliers (*B*, *C*), and four clusters of different densities consisting of 36, 8, 12, and 16 uniformly distributed instances, respectively. To demonstrate the effectiveness of our approach in finding both global and local outliers, the same set of experiments is conducted and the best experimental results are obtained with the parameter *k*'s being determined by error and trial to be 5, 5, 5, 5 for the DB method, the DB-Max method, the LOF method and the proposed method, respectively, and depicted in Fig. 6.5. For this case, the DB and DB-Max methods both miss the two local outliers, *B* and *D*. The LOF method misses one global outlier. Our proposed spectral clustering-based outlier detection algorithm identifies all six outliers correctly.

6.4.2 Performance of Our Algorithm on Real Data

As pointed out by Aggarwal and Yu, one way to test how well the outlier detection algorithm works is to run the method on the dataset and test the percentage of points which belongs to the rare classes [37]. To evaluate the effectiveness and accuracy of the proposed method, we compare the algorithms by their performance on detecting rare classes in some real multi-dimensional datasets, which are downloaded from UCI [36]. To quantitatively measure the performance of an outlier detection scheme, two metrics called recall and precision are employed here. Assuming that a dataset $D = D_o \cup D_n$ where D_o denotes the set of all outliers and D_n denotes the set of all normal data. Given any integer $m \geq 1$, if O_m denotes the set of outliers among objects in the top m positions returned by an outlier detection scheme, precision and recall are defined as,

$$\text{Precision} = \frac{|O_m|}{m} \tag{6.7}$$

$$\text{Recall} = \frac{|O_m|}{D_o} \tag{6.8}$$

To easily visualize the performances of the proposed method in comparison with several state-of-the-art outlier detection methods, namely the KNN method [32, 33], the LOF method [34], the SOS method [38], the MOGAAL method [39], the CFAR method [40], the RDOS method [41], the ABOD method [30], and the OCSVM method [42], we use the precision-recall plots to manifest the results of top outlier detection methods for all data sets. In all the plots, the red solid curve is used to manifest the results of the proposed method.

6.4.2.1 Connectionist-Mines-Versus-Rocks Data

The Connectionist-mines-vs-rocks dataset consists of two clusters, of which class '1' contains 111 patterns (obtained by bouncing sonar signals off a metal cylinder at various angles and under various conditions) and class '2' contains 97 data points (obtained from rocks under similar conditions). We choose the points in the second cluster as outliers. The experimental results of the proposed algorithm (denoted by PESA), and three best cases of the rest methods are presented in Fig. 6.6 for $k = 10$, 20, 30 from left to right. From the figures, it can be seen that the LOF method can achieve the best performance when k is set to 20. For $k = 10$ and $k = 30$, the SOS method performs the best at beginning while the LOF method performs the best at the

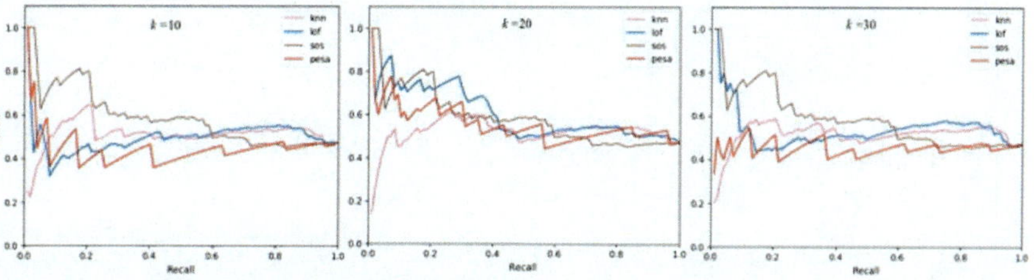

Fig. 6.6 Experimental results by the precision-recall plot for connectionist-mines-versus-rocks data

end. When k is set to 10 and 20, the precision of the proposed method can achieve to 1. However, the proposed method performs well on the first 60% outliers relatively for $k = 10$ and $k = 20$ but does not perform very well when $k = 30$. The KNN algorithm does not perform well at beginning, but outperforms the SOS algorithm and our algorithm (i.e., PESA) for the last 40% outliers.

6.4.2.2 Habermans-Survival Data

The Habermans-survival dataset consists of 2 classes that describe the survival and dead patients who had undergone surgery of breast cancer, respectively. Class '1' is composed of 225 patterns and denotes the number of patients who survived more than 5 years after the surgery, while class '2' is composed of 81 patterns and denotes the number of patients who died within 5 years after the surgery. We select class '2' as outliers. The experimental results of the proposed algorithm (denoted by PESA) and three best cases of the rest methods are presented in Fig. 6.7 for three values of k's, that is, 10, 20, 30. It can be seen that our algorithm performs best among all of the algorithms on the first 20% outliers while KNN and ABOD methods perform better than ours in the last 80% outliers. Unfortunately, MOGAAL does not perform well when $k = 30$. Therefore, we choose the SOS method in comparison with our PESA algorithm for $k = 30$. Further, it can be seen that the precisions of ABOD and MOGAAL can achieve 1 when k is equal to 10 and 20 at the beginning.

6.4.2.3 Heart-Disease-Processed-Hungarian Data

The Heart-disease-processed-hungarian dataset consists of two classes concerning heart disease diagnosis, with 188 and 106 patterns, respectively. We select class '2' as outliers. The experimental results of the proposed algorithm (denoted by PESA) and three best cases of the rest methods are presented in Fig. 6.8 for three values of k's, that is, 10, 20, 30. It is evident that our method outperforms all the other methods for $k = 20$ and 30 and can retain the good for the last 50% outliers for $k = 10$. The KNN algorithm performs the next on this data set for $k = 20$ and $k = 30$. Unfortunately, the MOGAAL and RDOS methods do not performs well on this

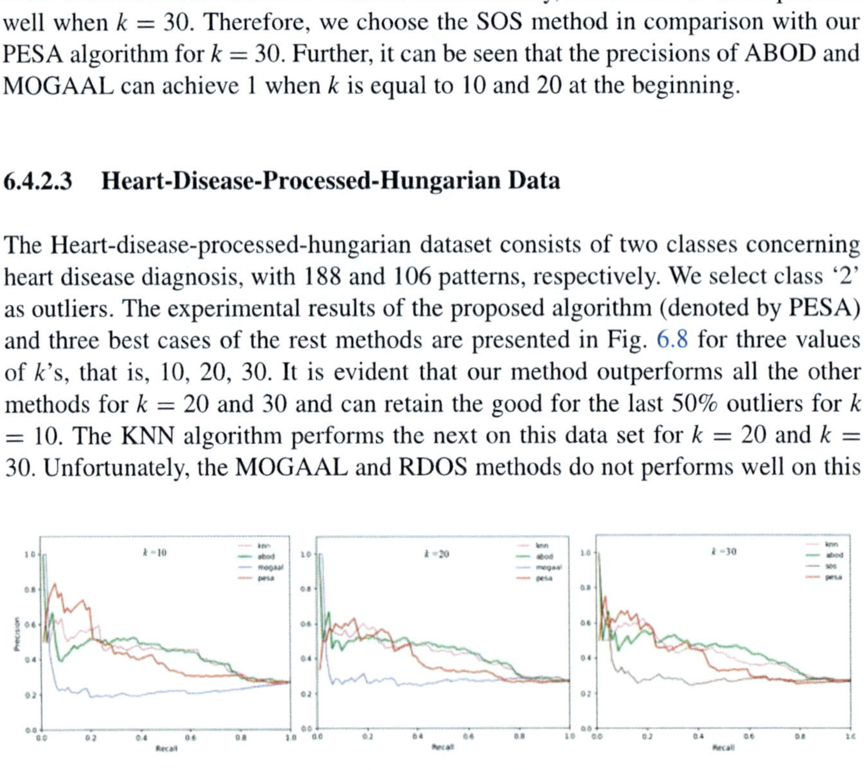

Fig. 6.7 Experimental results by the precision-recall plot for Habermans-survival data

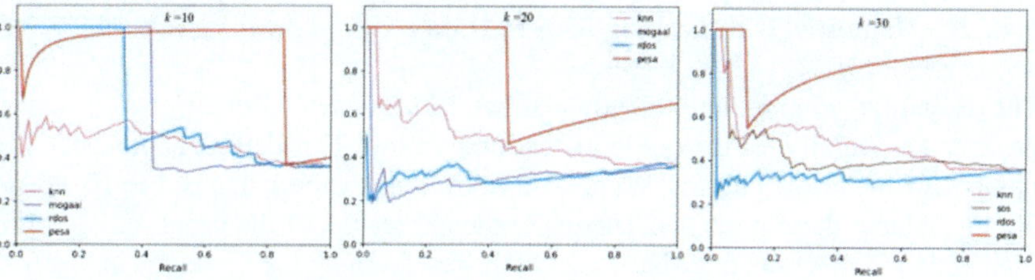

Fig. 6.8 Experimental results by the precision-recall plot for Heart-disease-processed-hungarian data

dataset for larger values of k. Since the MOGAAL method performs worse when k is equal to 30, the results of the SOS method are shown as a comparison to our method for $k = 30$.

6.4.2.4 Hepatitis Data

There are two classes (concerning die and live) in the Hepatitis dataset which consists of 155 data points. Class '1' and class '2' contain 32 and 123 patterns, respectively. We choose class '1' as outliers. The results of our algorithm (i.e., PESA) and three best cases of the rest methods for $k = 30$ are presented in Fig. 6.9. Our method outperforms others at beginning (the precision can achieve 1) and performs relatively better at the end as well. MOGAAL performs better than our method in the middle. RDOS does not perform well on this data set.

Fig. 6.9 Experimental results by the precision-recall plot for Hepatitis data

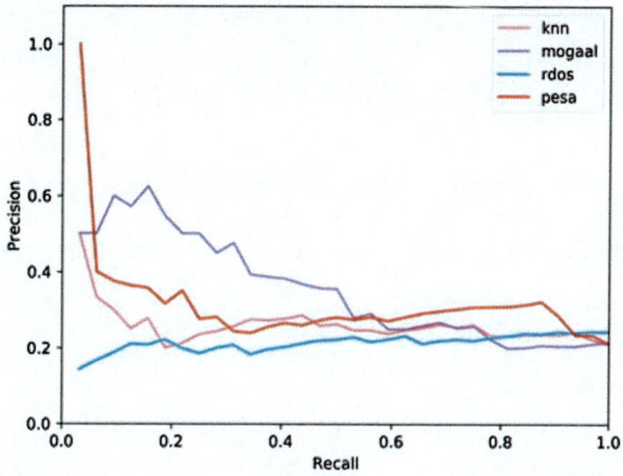

6.4.2.5 Hepatitis_withoutdupl_05_v06 Data

The Hepatitis_withoutdupl_05_v06 dataset is a subset of the Hepatitis dataset produced especially for the outlier detection purpose. This dataset contains 70 data points, of which 3 are outliers. We choose 3 different k values, that is, $k = 10, 20$, and 30, to evaluate these methods. The experimental results of the proposed algorithm (denoted by PESA) and three best cases of the rest methods are presented in Fig. 6.10. It can be seen that our method performs significantly better than others when k is set to 20 and 30. All other algorithms do not performs well on the first 60% outliers while our method perform ineffectively for $k = 10$, for which the ABOD method performs best. The performances of the KNN and OCSVM methods are almost unaffected by k.

6.4.2.6 Hepatitis_withoutdupl_05_v07 Data

The Hepatitis_withoutdupl_05_v07 dataset is another subset of Hepatitis data also used especially for outlier detection. It also contains 70 data points, of which 3 are outliers. The experimental results of the proposed algorithm (denoted by PESA) and three best cases of the rest methods are presented in Fig. 6.11 for $k = 10, 20, 30$. It is clear from the figure that our algorithm performs better than all other methods for all k's even though it only significantly outperforms others for $k = 30$. The precision of our PESA algorithm is less than 0.2 when k is set to 10 and 20. Unfortunately, all other methods perform unsatisfactorily on this data with precisions below 0.1.

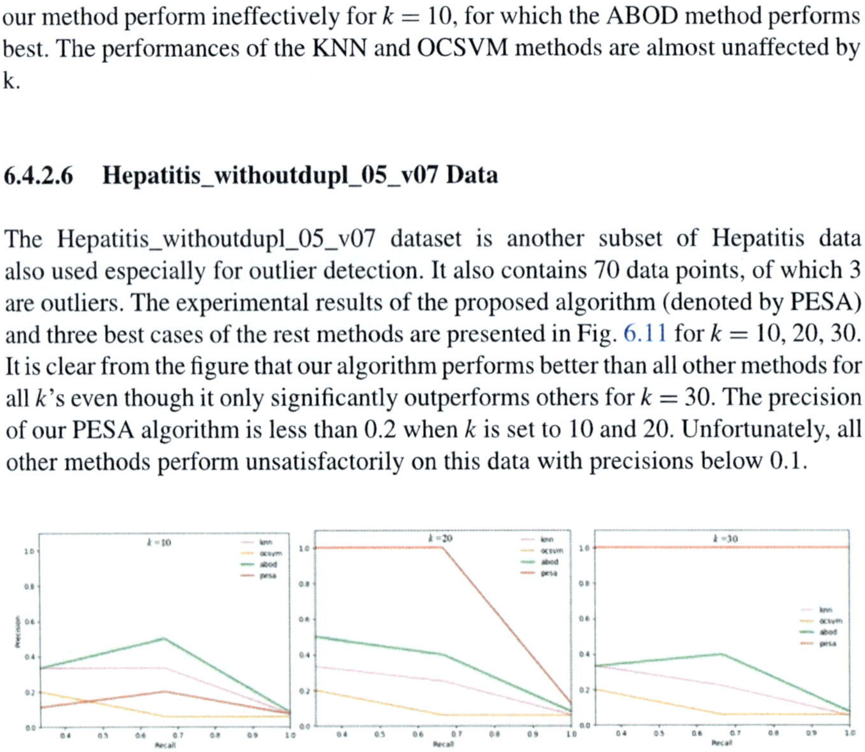

Fig. 6.10 Experimental results by the precision-recall plot for Hepatitis_withoutdupl_05_v06 data

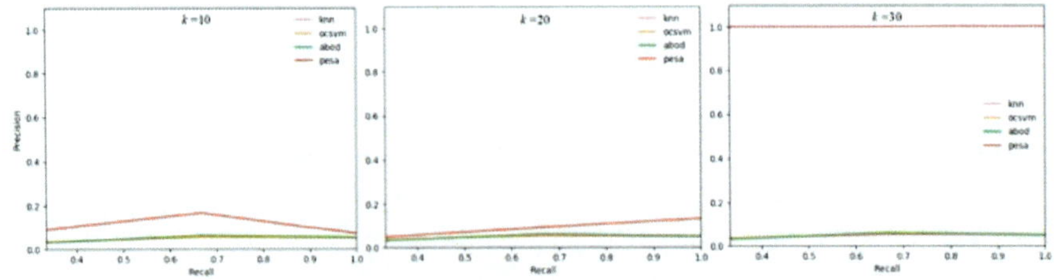

Fig. 6.11 Experimental results by the precision-recall plot for Hepatitis_withoutdupl_05_v07 data

6.4.2.7 Hepatitis_withoutdupl_norm_10_v08 Data

The Hepatitis_withoutdupl_norm_10_v08 dataset is another subset of hepatitis data set designed for outlier detection purpose as well and contains 74 data points, of which 7 are outliers. The experimental results of the proposed algorithm (denoted by PESA) and three best cases of the rest methods are presented in Fig. 6.12 for k = 10, 20, 30. Apparently, no algorithms perform very well on this dataset since the precisions of these 4 methods are all around 0.2. Overall, the KNN method performs better than others, however, its performance degrades with the increasing k's. Our method performs worse than the others for different k's one way or the other.

6.4.2.8 Parkinson's Data

The Parkinson's dataset is composed of a range of biomedical voice measurements from 31 people, 23 with Parkinson's disease. Each column in the table is a particular voice measure, and each row corresponds one of 195 voice recordings from these individuals of which 48 individuals have the disease. The main aim of the data is to discriminate healthy people from those with Parkinson's disease. The experimental results of the proposed algorithm (denoted by PESA) and three best cases of the rest methods are presented in Fig. 6.13. Three different k values, that is, 10, 20, and 30, are used to evaluate these methods. MOGAAL performs well at the beginning only when k is equal to 10 and worse when k is set to 20 and 30. Therefore, ABOD

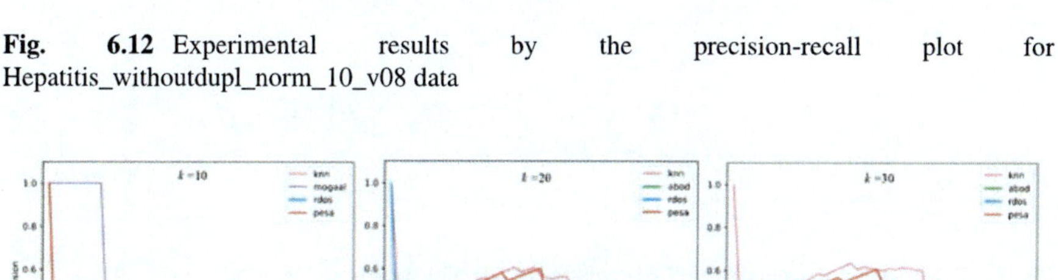

Fig. 6.12 Experimental results by the precision-recall plot for Hepatitis_withoutdupl_norm_10_v08 data

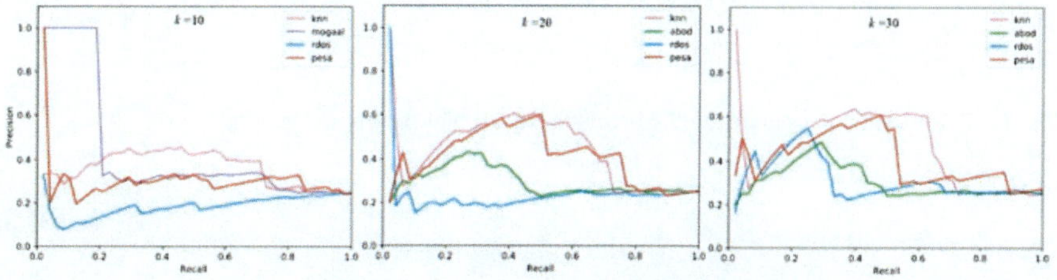

Fig. 6.13 Experimental results by the precision-recall plot for Parkinson's data

is chosen for the rest comparisons. The precision of our methods can achieve 1 at beginning when k is equal to 10. For $k = 20$ and 30, our method is among top 2 best detectors and KNN performs a little better than ours.

6.4.2.9 Phoneme Data

The Phoneme dataset consists of 2 clusters. Class '1' and class '2' contain 3818 and 1586 patterns, respectively. We select the points in the second class as outliers. The experimental results of the proposed algorithm (denoted by PESA) and three best cases of the rest methods are presented in Fig. 6.14 for 3 different k values, 10, 20, and 30, to evaluate these methods. From these figures, it can be seen that the precisions of all the algorithms are around 0.4. The final precisions of these 4 methods are the same. ABOD and KNN perform better than others most of time. Further, the performances of these methods are not very sensitive to the parameter k. Our method performs well at the beginning and can achieve a precision of 1 for $k = 10$ and 30. KNN can perform better than ABOD at the end.

6.4.2.10 Seismic-Bumps Data

The Seismic-bumps dataset describes the problem of high energy seismic bumps forecasting in a coal mine and consists of 2584 data points of which 170 are regarded as outliers. The experimental results of the proposed algorithm (denoted by PESA) and three best cases of the rest methods are presented in Fig. 6.15 for 3 different k values, 10, 20, and 30. The performances of KNN and ABOD are similar with each other and with our method, but better than our method on the last 50% outliers. KDEOS performs worst on this data set. The precision of our method can achieve 1 when k is set to 10. These methods are all less sensitive to the parameter k.

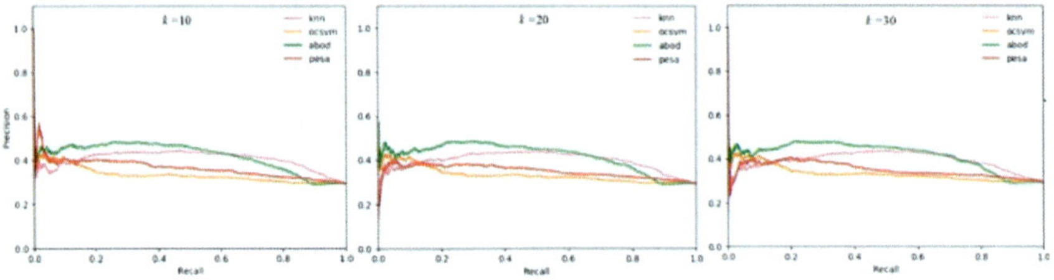

Fig. 6.14 Experimental results by the precision-recall plot for Phoneme data

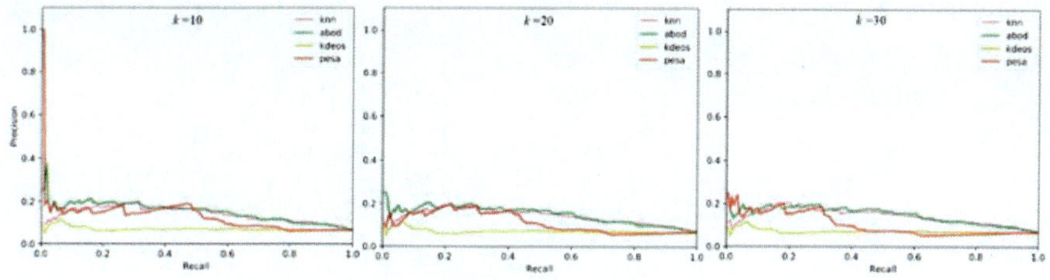

Fig. 6.15 Experimental results by the precision-recall plot for Seismic-bumps data

6.4.2.11 Statlog-Australian-Credit Data

The Statlog-asustralian-credit dataset concerns credit card applications and consists of 2 clusters. Class '1' and class '2' contain 383 and 307 patterns, respectively. We select the points in the second class as outliers. The experimental results of the proposed algorithm (denoted by PESA) and three best cases of the rest methods are presented in Fig. 6.16 for $k = 10$, 20, and 30. It is evident that, for the first 10% outliers, our method performs better than all other methods while performs the worst for the rest outliers. *K*NN and ABOD perform better than LOF when k is equal to 10. However, the results of the three competing methods are similar for $k = 20$ and 30.

6.4.2.12 Statlog-German-Credit Data

The Statlog-german-credit dataset concerns credit card applications and consists of 2 clusters. Class '1' and class '2' contain 700 and 300 patterns, respectively. We select the points in the second class as outliers. The experimental results of the proposed algorithm (denoted by PESA) and three best cases of the rest methods are presented in Fig. 6.17 for $k = 10$, 20, and 30. OCSVM performs the worst at the beginning on this data set. The results of the other three algorithms are very similar for all k's while our method performs better to some extent. The precision of our algorithm can achieve 1 at beginning and subsequently down to around 0.3. By contrast, OCSVM can outperform others at the end.

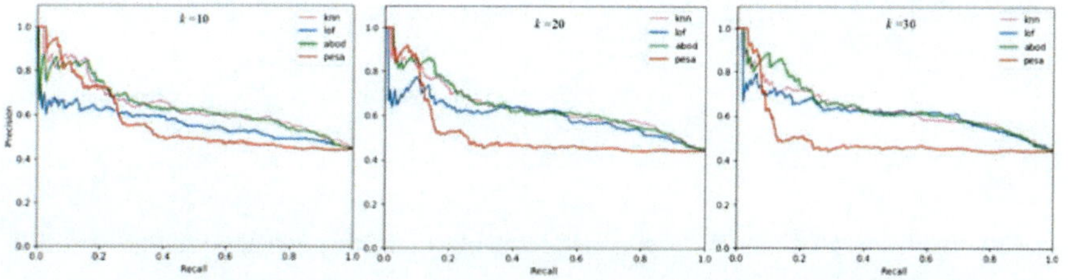

Fig. 6.16 Experimental results by the precision-recall plot for Statlog-australian-credit data

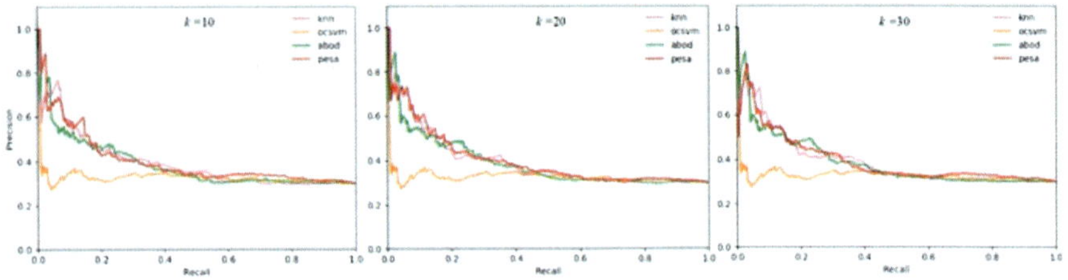

Fig. 6.17 Experimental results by the precision-recall plot for Statlog-german-credit data

6.4.2.13 Statlog-German-Credit-Numeric Data

The Statlog-german-credit-numeric dataset concerns credit card applications and consists of 2 clusters. Class '1' and class '2' contain 700 and 300 patterns, respectively. We select the points in the second class as outliers. The experimental results of the proposed algorithm (denoted by PESA) and three best cases of the rest methods are presented in Fig. 6.18 for $k = 10$, 20, and 30. The performances of these four algorithms are very similar with KNN performing better on some outliers than other methods, indicating a small influence of k's on this data set. The precision of the proposed algorithm can achieve 1 at beginning and subsequently down to around 0.3.

6.4.2.14 Wholesale-Region Data

The Wholesale-region dataset is the set of clients of a wholesale distributor in business area and consists of 3 clusters, class '1', class '2', and class '3' with 77, 47, and 316 patterns, respectively. We select the points in the second class as outliers. The experimental results of the proposed algorithm (denoted by PESA) and three best cases of the rest methods are presented in Fig. 6.19 for $k = 10$, 20, and 30. For the final precision, our method is the best one among these algorithms. The results of the other three methods are similar. It can be seen that k has little influence on the performances

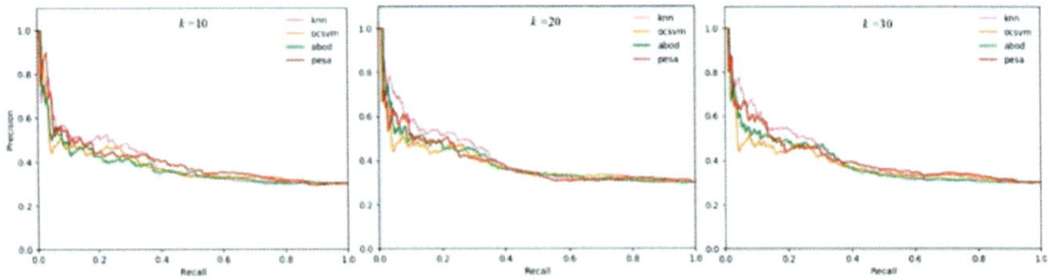

Fig. 6.18 Experimental results by the precision-recall plot for Statlog-german-credit-numeric data

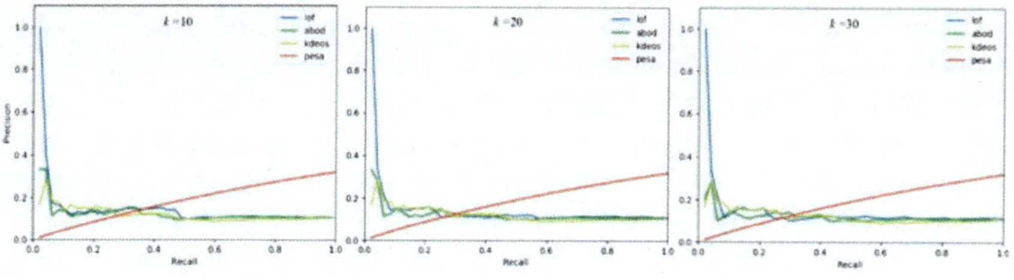

Fig. 6.19 Experimental results by the precision-recall plot for Wholesale-region data

of outlier detection methods on this dataset. Our method always performs better than others with regard to the final result. The precision of LOF can achieve to 1 at beginning. However, it degrades rapidly subsequently. The performances of ABOD and KDEOS are similar.

6.5 Discussion

Distance-based and density-based outlier detection methods are good outlier detectors. However, they are very sensitive to parameter k and a small change in k can lead to changes in the scores and, correspondingly, the ranking. From Sect. 6.3, we know that our method is based on a low frequency of the second eigenvalue in the set of eigenvalues of the spectral clustering for each point's $k + 1$ nearest neighbor set. k is a very critical parameter of our proposed method. For a suitable value to be chosen for k, our goal is to promote eigenvalues other than the second smallest one to appear as frequently as possible. In other words, a desired k should be large enough for each point's $k + 1$ nearest neighbor set to include normal points so much that these normal points dominate the neighborhood. From the experiments, we see that, for syn_Data1 and the real data, where outlying groups have outliers more than one data point, the optimal k's for our methods are much larger than those for the other methods, while for syn_Data2, where there is only one data point for each outlying group, the optimal k's for our methods is the same as those for the other methods.

6.6 Conclusions

In this chapter, we have proposed an effective spectral clustering-based outlier detection method that can detect both global and local outliers. Traditional spectral clustering-based outlier detection algorithms have a quadratic running time complexity with data sizes and are very time- and space-consuming for modern large datasets. To partially circumvent this problem, we apply the spectral clustering process upon N mini-datasets, each consisting a data point and its k-nearest neighbors. To identify potential outliers, our algorithm first locates those data points in the eigenspace whose second smallest eigenvalue appears least frequently as outlier candidates that deviate from the main patterns. Candidate outliers are then ranked based on the notion of distance-based outlier scores that are assigned to each data point. To demonstrate the utility of our proposed outlier detection mechanism, we have performed a detailed comparison of its performance with state-of-the-art distance-based and density-based outlier detection methods. Experimental results show the ability of our algorithm to rank the best candidates for being an outlier with high recall. Our study also manifests that, in reality, it is usually difficult to detect all the outliers that fit user's intuitions by a single outlier detection method. Thus, it is more meaningful to incorporate our proposed outlier detection method as a component into current outlier detection framework.

Acknowledgements This chapter was modified from the paper published by our group in "machine learning and data mining in pattern recognition" [35]. The related contents are reused with permission.

References

1. Hodge, V. J., & Austin, J. (2004). A survey of outlier detection methodologies. *Artificial Intelligence Review, 22*(2), 85–126.
2. Chandola, V., Banerjee, A., & Kumar, V. (2009). Anomaly detection: A survey. *ACM Computing Surveys, 41*(3), 15.1–15.58.
3. Malik, J., Belongie, S., Leung, T., et al. (2001). Contour and texture analysis for image segmentation. *International Journal of Computer Vision, 43*(1), 7–27.
4. Bach, F. R., & Jordan, M. I. (2004). Blind one-microphone speech separation: A spectral learning approach. In *Proceedings of the 18th Annual Conference on Neural Information Processing Systems* (NIPS'04), Vancouver, BC, Canada, pp. 65–72.
5. Ding, C., He, X., Zha, H., et al. (2001). A min-max cut algorithm for graph partitioning and data clustering. In *Proceedings of the IEEE International Conference on Data Mining* (ICDM'01), California, USA, pp. 107–114.
6. Shi, J., & Malik, J. (2000). Normalized cuts and image segmentation. *IEEE Transactions on Pattern Analysis and Machine Intelligence, 22*(8), 888–905.
7. Stoer, M., & Wagner, F. (1997). A simple min-cut algorithm. *Journal of the ACM, 44*(4), 585–591.
8. Hagen, L., & Kahng, A. (1992). New spectral methods for ratio cut partitioning and clustering. *IEEE Transactions on Computer-Aided Design, 11*(9), 1074–1085.
9. Zelnik-Manor, L., & Perona, P. (2004). Self-tuning spectral clustering. In *Proceedings of the 18th Annual Conference on Neural Information Processing Systems* (NIPS'04), Vancouver, BC, Canada, pp. 1601–1608.
10. Bojchevski, A., Matkovic, Y., & G¨unnemann, S. (2017). Robust spectral clustering for noisy data. In *Proceedings of 23rd ACM SIGKDD International Conference on Knowledge Discovery and Data Mining* (KDD'17), Halifax, NS, Canada, pp. 737–746.
11. Wu, L., Chen, P.-Y., Yen, I.E.-H., Xu, F., Xia, Y., & Aggarwal, C. (2018). Scalable spectral clustering using random binning features. In *Proceedings of the 24th ACM SIGKDD International Conference on Knowledge Discovery and Data Mining* (KDD'18), London, United Kingdom, pp. 2506–2515.

12. Tan, M., Zhang, S., & Wu, L. (2018). Mutual kNN based spectral clustering. *Neural Computing and Applications*. https://doi.org/10.1007/s00521-018-3836-z
13. Tong, T., Zhu, X., & Du, T. (2019). Connected graph decomposition for spectral clustering. *Multimedia Tools and Applications, 78*(23), 33247–33259.
14. Yang, X., Yu, W., Wang, R., Zhang, G., & Nie, F. (2020). Fast spectral clustering learning with hierarchical bipartite graph for large-scale data. *Pattern Recognition Letters, 130,* 345–352.
15. Pang, Y., Xie, J., Nie, F., & Li, X. (2020). Spectral clustering by joint spectral embedding and spectral rotation. *IEEE Transactions on Cybernetics, 50*(1), 247–258.
16. Jiang, M. F., Tseng, S. S., & Su, C. M. (2001). Two-phase clustering process for outlier detection. *Pattern Recognition Letters, 22*(6–7), 691–700.
17. Yu, D., Sheikholeslami, G., & Zang, . (2002). FindOut: Finding outliers in very large datasets. *Knowledge and Information Systems, 4,* 387–412.
18. Wang, C. H. (2008). Recognition of semiconductor defect patterns using spatial filtering and spectral clustering. *Expert Systems with Applications, 34*(3), 1914–1923.
19. Xiang, T., & Gong, S. (2008). Spectral clustering with eigenvector selection. *Pattern Recognition, 41*(3), 1012–1029.
20. Filippone, M., Camastra, F., & Masulli, F. (2008). A survey of kernel and spectral methods for clustering. *Pattern Recognition, 41*(1), 176–190.
21. Luxburg, U. V. (2007). A tutorial on spectral clustering. *Statistics and Computing, 17*(4), 395–416.
22. Yang, P., & Huang, B. (2008). An outlier detection algorithm based on spectral clustering. In *Proceedings of 2008 Pacific-Asia Workshop on Computational Intelligence and Industrial Application* (PACIIA 2008), Wuhan, China, pp. 507–510.
23. He, Z. Y., Xu, X. F., & Deng, S. C. (2003). Discovering cluster based local outliers. *Pattern Recognition Letters, 24*(9–10), 1641–1650.
24. Yang, P., & Huang, B. (2008). A spectral clustering algorithm for outlier detection. In *Proceedings of the 2008 International Seminar on Future Information Technology and Management Engineering* (FITME'08), Leicestershire, United kingdom, pp. 33–36.
25. Lin, H., & Zhu, Q. (2012). A spectral clustering-based dataset structure analysis and outlier detection progress. *Journal of Computational Information Systems, 8*(1), 115–124.
26. Tyuryukanov, I., van der Meijden, M. A. M. M., Terzija, V., & Popov, M. (2018). Spectral MST-based graph outlier detection with application to clustering of power networks. In *Proceedings of the 20th Power Systems Computation Conference* (PSCC'18), Dublin, Ireland.
27. Aggarwal, C. C. (2013). Outlier Analysis. Springer.
28. Sun, H., Huang, J., Han, J., Deng, H., Zhao, P., & Feng, B. (2010). gSkeletonClu: Density-based network clustering via structure-connected tree division or agglomeration. In *Proceedings of the 10th IEEE International Conference on Data Mining* (ICDM'10), pp. 481–490.
29. Xiong, L., Chen, X., & Schneider, J. (2011). Direct robust matrix factorization for anomaly detection. In *Proceedings of the 11th IEEE International Conference on Data Mining* (ICDM'11), Vancouver, BC, Canada, pp. 844–853.
30. Kriegel, H.-P., Schubert, M., & Zimek, A. (2008). Angle-based outlier detection in high-dimensional data. In *Proceedings of the 14th ACM SIGKDD International Conference on Knowledge Discovery and Data Mining* (KDD'08), Las Vegas, Nevada, USA, pp. 444–452.
31. Knorr, E. M., Ng, R. T., & Tucakov, V. (2000). Distance-based outliers: Algorithms and applications. *The VLDB Journal, 8*(3–4), 237–253.
32. Ramaswamy, S., Rastogi, R., & Shim, K. (2000). Efficient algorithms for mining outliers from large data sets. In *Proceedings of the ACM International Conference on Management of Data* (SIGMOD'00), Dallas, pp.427–438.
33. Breuning, M. M., Kriegel, H. P., Ng, R. T., & Sander, J. (2000). LOF: Identifying density-based local outliers. In *Proceedings of the 2000 ACM SIGMOD International Conference on Management of Data* (SIGMOD'00), Dallas, TX, United states, pp.93–104.
34. Angiulli, F. and Pizzuti, C. (2002). Fast outlier detection in high dimensional spaces. In *Proceedings of the 6th European Conference on Principles of Data Mining and Knowledge Discovery* (PKDD'02), Helsinki, pp.15–26.

35. Wang, Y., Wang, X., & Wang, X. L. (2016). A spectral clustering based outlier detection technique. In *Proceedings of 12th International Conference on Machine Learning and Data Mining*, New York, USA, pp.15–27.
36. UCI: The UCIKDD Archive, University of California, Irvine, CA. https://kdd.ics.uci.edu/.
37. Aggarwal, C., & Yu, P. (2001). Outlier detection for high-dimensional data. In Proceedings of *the 2001 ACM International Conference on Management of Data* (SIGMOD'01), Santa Barbara, CA, USA, pp. 37–46.
38. Janssens, J., Huszar, F., Postma, E., & van den Herik, H. (2012). Stochastic outlier selection.
39. Liu, Y., Li, Z., Zhou, C., Jiang, Y., Sun, J., Wang, M., He, X. (2019). Generative adversarial active learning for unsupervised outlier detection. *IEEE Transactions on Knowledge and Data Engineering*, https://arxiv.org/abs/1809.10816.
40. Ru, X., Liu, Z., Huang, Z., et al. (2016). Normalized residual-based constant false-alarm rate outlier detection. *Pattern Recognition Letters, 69,* 1–7.
41. Tang, B., & He, H. (2017). A local density-based approach for outlier detection. *Neurocomputing, 241,* 171–180.
42. Erfani, S. M., Rajasegarar, S., Karunasekera, S., & Leckie, C. (2016). High-dimensional and large-scale anomaly detection using a linear one-class SVM with deep learning. *Pattern Recognition, 58,* 121–134.

Chapter 7
Enhancing Outlier Detection by Filtering Out Core Points and Border Points

Abstract Outlier detection is an important task in data mining and has high practical value in numerous applications such as astronomical observation, text detection, fraud detection, and so on. At present, a large number of popular outlier detection algorithms are available, including distribution-based, distance-based, density-based, and clustering-based approaches. However, traditional outlier detection algorithms face some challenges. For one example, most distance-based and density-based outlier detection methods are based on k-nearest neighbors. Therefore, even though the outlier data occupy a relatively small amount in the dataset, the existing approaches need to perform local outlier factor calculation on all data during the outlier detection, which greatly reduces the efficiency of the algorithms. For another example, some methods can only detect the global outliers, but fail to detect the local outliers. Last but not the least, most outlier detection algorithms do not accurately distinguish between boundary points and outliers. To partially solve these problems, it is realized that the outlier detection problem is related to the clustering problem by complementarity. According to density-based clustering, there are three kinds of data points, namely core points, border points, and outliers. If indicators can be extracted from the data that make outliers have much larger deviation values than the other two kinds of data points, outlier detection problems can be fulfilled. Therefore, in this chapter, we propose to augment some boundary indicators to classical outlier detection algorithms. Experiments performed on both synthetic and real data sets demonstrate the efficacy of enhanced outlier detection algorithms.

Keywords Outlier detection · Distance-based outlier detection · Density-based outlier detection · Density-based clustering · Boundary detection · K-nearest neighbors

7.1 Introduction

With the rapid development of information technology, a large amount of information has been produced from the real word. How to find import and useful information from these massive and multi-dimensional data has become an urgent problem. Therefore, data mining and database technologies come into being consequently.

X. Wang et al., *New Developments in Unsupervised Outlier Detection*,
https://doi.org/10.1007/978-981-15-9519-6_7

In practice, data often come from different information individuals, departments, enterprises, and countries. These complex data may contain a small portion of the data objects which differ significantly from other data objects in behavior or model. These data objects are referred to as outliers. In many fields, outliers are more important than the normal data objects, as they imply some useful information. Being a special subset of a dataset that are significantly distinct from the other data objects and have a very small percentage in number, outliers often contain real and unimaginable knowledge. Outlier detection is devoted to discovering anomalies from a large amount of informational data. Being an important part of data mining now, outlier detection not only has attracted the attention of researchers in many disciplines including data mining, statistics, and information theory [1, 2], but also is considered as a critical task in many practical applications, such as network intrusion detection [3–7], fraud detection [8–10], industrial damage detection [11–14], and healthcare monitoring [15–17].

At present, the study on outlier detection becomes very active. Many outlier detection algorithms have been proposed. Outlier detection methods can be divided into distribution-based methods, depth-based methods, distance-based methods, density-based methods, and clustering-based methods. However, none of them have been proved to be completely applicable to every situation. Each type of outlier detection algorithms has its advantages and disadvantages. In distribution-based methods, an object is considered as an outlier if it deviates from a standard distribution (e.g. normal, Poisson, etc.) too much [18]. However, the underlying distribution is usually unknown and there are many practical applications which do not follow a standard distribution. As a result, distribution-based methods have a limited number of applications. Being an improvement, depth-based methods assign a depth value to each data object and map it to the corresponding layer, where data objects on the shallow layers are more likely to be outliers than those on the deeper ones. Unfortunately, these methods suffer high computational complexity for more than three dimensions of data. Distance-based methods, also known as adjacency-based methods, believe that data objects are outliers if they are far away from the majority of data points and address more globally oriented outliers in databases [19]. Density-based methods usually assign to each data object a measure of outlier degree as the classic LOF algorithm does and then regard those data objects which possess the largest outlier degrees as outliers [20]. In comparison with distance-based methods, these methods address more locally oriented outliers. Finally, clustering-based methods obtain outliers as a by-product and regard those data items that reside in the smallest clusters as outliers [21].

However, traditional outlier detection algorithms face some challenges. For one example, most distance-based and density-based outlier detection methods are based on k-nearest neighbors and, therefore, are very sensitive to the value of k. For another example, some methods can only detect global outliers, but fail to detect local outliers. Last but not the least, in many existing outlier detection algorithms, the boundary points are mistakenly classified to be outliers. To partially solve these problems, in this chapter, we propose to augment classical outlier detection algorithms with some boundary indicators so as to enhance traditional methods for outlier detection.

When compared with some classic outlier detection algorithms on sample datasets, the enhanced methods are more accurate while being less sensitive to the parameter k.

The rest of the chapter is organized as follows. In Sect. 7.2, we review some existing work on classic DBSCAN clustering algorithms and density-based clustering for outlier detection. We then present our proposed enhancer in Sect. 7.3. In Sect. 7.4, a performance evaluation is conducted and the results are analyzed. Finally, conclusions are made in Sect. 7.5.

7.2 Related Work

K-means algorithms are designed to find spherical-shaped clusters and have difficulty finding clusters of arbitrary shapes. Given such data, they would likely to inaccurately identify non-convex regions, particularly when noise or outliers are included in the clusters.

Being a very popular branch of techniques in data clustering, density-based clustering algorithms rely on a local density criterion to form clusters and can detect clusters of different shapes, sizes, and densities. In density-based clustering, clusters are defined as areas of higher density than the remainder of the data set [22]. Objects in these sparse areas—that are required to separate clusters—are usually considered to be noise (i.e. outliers) and border points. Therefore, density-based clustering algorithms are well known for identifying clusters of very different local densities and existing in different regions of data space.

An important branch of density-based clustering begins with DBSCAN, which was proposed by Ester et al. in 1996 [22] and allows users to determine two parameters, the neighborhood of a given radius (Eps, or ε) and a minimum number of points (MinPts, or τ) in the neighborhood, to complete the clustering procedure. More specifically, the algorithm decides for each point of a cluster that its neighborhood of a given radius (ε) has to contain at least a minimum number of points (τ) where ε and τ are input parameters.

7.2.1 Density-Based Clustering with DBSCAN

The DBSCAN approach works by using the density characteristics of data points to merge them into clusters. Therefore, the individual data points in dense regions are used as building blocks after classifying them on the basis of their density. The density of a data point is defined by the number of points that lie within a radius ε of that point (including the point itself). The densities of these spherical regions are used to classify the data points into core points, border points, or noise points (i.e. outliers). These notions are defined as follows:

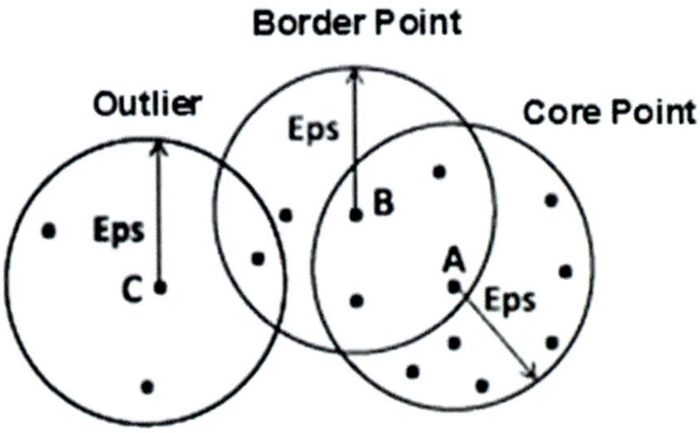

Fig. 7.1 Examples of core points, border points, and outliers

- Core point: A data point is defined as a core point, if it contains at least τ (i.e. MinPts) data points.
- Border point: A data point is defined as a border point, if it contains less than (i.e. MinPts) points, but it also contains at least one core point within a radius ε (i.e. Eps).
- Noise point: A data point that is neither a core point nor a border point is defined as a noise point.

Examples of core points, border points, and noise points (i.e. outliers) are illustrated in Fig. 7.1 for $\tau = 10$. The data point A is a core point because it contains 10 data points within the illustrated radius Eps. On the other hand, data point B contains only 6 points within a radius of Eps, but it contains the core point A. Therefore, it is a border point. The data point C is a noise point (i.e. an outlier) because it contains only 4 points within a radius of Eps, and it does not contain any core point.

After the core points, border points, and noise points (i.e., outliers) have been determined, the DBSCAN clustering algorithm proceeds as follows. Initially, all objects in a given data set D are marked as 'unvisited.' DBSCAN randomly selects an unvisited object p, marks p as 'visited,' and checks whether the ε-neighborhood of p contains at least MinPts objects. If yes, a new cluster C is created for p, and all the objects in the ε-neighborhood of p are added to a candidate set, *S*. DBSCAN iteratively adds to *C* those objects in *S* that do not belong to any cluster. In this process, for an object q in *S* that carries the label 'unvisited,' DBSCAN marks it as 'visited' and checks its ε-neighborhood. If the ε-neighborhood of q has at least MinPts objects, those objects in the ε-neighborhood of q are added to *S*. DBSCAN continues adding objects to *C* until *C* can no longer be expanded, that is, *S* is empty. At this time, cluster *C* is complete and thus is output. To find the next cluster, DBSCAN randomly selects an unvisited object from the remaining data objects. The clustering process continues until all objects are visited. With appropriate settings of the user-defined parameters, ε and τ, the algorithm is effective in finding arbitrary-shaped clusters in spite of the noise in the data.

The parameters τ (i.e. MinPts) and ε (i.e. Eps) are related to one another in an intuitive way, which is useful for parameter setting. In particular, after the value of τ has been set by the user, the value of ε can be determined in a data-driven way.

The idea is to use a value of ε that can capture most of the data points in clusters as core points. This can be achieved as follows. For each data point, its τ-nearest neighbor distance is determined. Typically, the vast majority of the data points inside clusters will have a small value of the τ-nearest neighbor distance. However, the value of the τ-nearest neighbor distance often increases suddenly for a small number of noisy points (or points at the fringes of clusters). Therefore, the key is to identify the tail of the distribution of τ-nearest neighbor distances. Statistical tests, such as the Z-value test, can be used in order to determine the value of Eps at which the τ-nearest neighbor distance starts increasing abruptly. This value of the τ-nearest neighbor distance at this cutoff point provides a suitable value of Eps.

The major time complexity of DBSCAN is in finding the neighbors of the different data points within a distance of Eps. For a database of size N, the time complexity can be $O(N^2)$ in the worst case. However, for some special cases, the use of a spatial index for finding the nearest neighbors can reduce this to approximately O(NlogN) distance computations. The O(logN) query performance is realized only for low-dimensional data, in which nearest neighbor indices work well.

Then, density-based spatial clustering of applications with noise (DBSCAN) groups points that are closely packed together, expanding clusters in any direction where there are nearby points. The DBSCAN method therefore can discover clusters of arbitrary shapes and does not require the number of clusters as an input parameter. However, DBSCAN assumes clusters of similar density and may have problems separating nearby clusters. Furthermore, the parameters required by most popular density-based clustering algorithms, such as DBSCAN, are hard to determine while having significant impacts on the clustering results. The disadvantage of DBSCAN is that it requires two parameters to be carefully selected to cluster a dataset. The two parameters relate to the minimum cluster size and the neighborhood radius. The challenge is that the two parameters require carefully tuning for different datasets. However, modern large real-world data sets often have a much skewed distribution that is difficult to be revealed by using only one global parameter setting. For example, in the data set depicted in Fig. 7.2, it is impossible to detect clusters A, B, C_1, C_2, and C_3 simultaneously using one global density parameter.

In 1999, Ankerst et al. proposed OPTICS to solve this problem [23]. OPTICS is a DBSCAN variant that handles different densities much better. Basically, it creates an augmented ordering of the dataset representing its density-based clustering structure, which contains information equivalent to the density-based clustering corresponding to a broad range of parameter settings. However, it only creates an augmented ordering of the dataset representing its density-based clustering structure instead of producing clusters of a dataset explicitly. In addition, it is short at separating the overlapping clusters appropriately. The key drawback of DBSCAN and OPTICS is that they expect some kind of density drop to detect cluster borders. On data sets with, for example, overlapping Gaussian distributions, the cluster borders produced by these algorithms will often look arbitrary, because the cluster density decreases continuously. Other more sophisticated density-based methods, such as DENCLUE [24], use gradient ascent on the kernel-density estimates to create the building blocks.

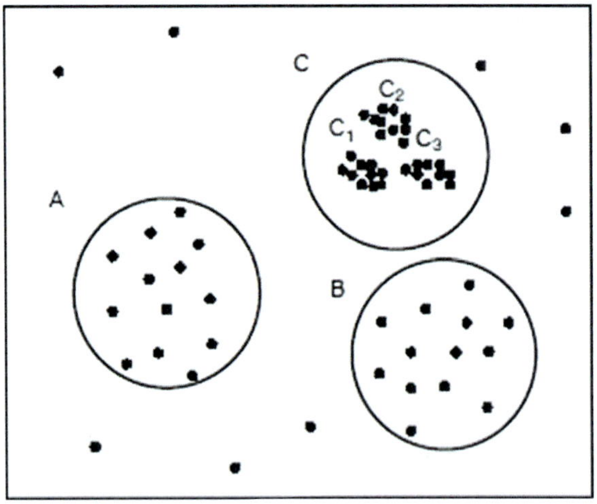

Fig. 7.2 Clusters with respect to different densities

In 2007, Duan et al. proposed a local-density-based spatial clustering algorithm relying on a local-density-based notion of clusters to discover clusters with arbitrary shape and density as well as to be insensitive to noise data [25]. It takes advantage of the LOF (Local Outlier Factor) to detect the noises and outliers in comparison with other density-based clustering algorithms. It can discover clusters of different densities and existing in different regions of data space. But it needs a user to input four parameters which have a significant influence on the clustering result but are hard to determine.

In 2011, Chen et al. proposed the APScan algorithm which combines the affinity propagation (AP) clustering approach with DBSCAN [26]. More specifically, their technique first uses AP to partition the dataset into patches and calculates the density of each patch to output a normalized density list to be used for the next algorithm stage based on a modified DBSCAN variation. Although the AP algorithm does not require the number of clusters to be pre-specified, the algorithm still requires a 'preference' parameter to be used [27].

In 2014, Rodriguez and Laio propose an approach based on the idea that cluster centers are characterized by a higher density than their neighbors and by a relatively large distance from other points with higher densities. This idea forms the basis of a clustering procedure in which the number of clusters arises intuitively, outliers are automatically spotted and excluded from the analysis, and clusters are recognized regardless of their shape and of the dimensionality of the space in which they are embedded. [28].

In 2016, Hou et al. proposed the DSets-DBSCAN algorithm which is actually an image enhancement technique to be applied onto the pairwise similarity matrix of the input data using histogram equalization transformations to make the clustering results invariant [29]. A closer study of their algorithm shows an implicit parameter (*MinPts*) is used (as in DBSCAN) to denote the minimum number of points in the neighborhood. The authors proposed a set of heuristic rules based on their investigation of several datasets and fixed this parameter to a value of 3, and showed that their algorithm was less influenced by this parameter compared to DBSCAN.

In the same year, Qi and Wang proposed a density-based clustering algorithm for high-dimensional data based on feature selection (FS-DBSCAN) [30]. They assume that there exist a number of optimal features which contribute to clustering while a number of features are redundant and irrelevant in such datasets. In order to cluster the data objects efficiently and get a higher classification accuracy, the proposed algorithm aims to eliminate the redundant and irrelevant features, select the optimal features and weight them simultaneously.

Also in the same year, Zhu et al. proposed a density-ratio-based method that can be implemented in two approaches [31]. One approach is to modify a density-based clustering algorithm to do density-ratio-based clustering by using its density estimator to compute density-ratio. The other approach involves rescaling the given dataset only. An existing density-based clustering algorithm, when applied to the rescaled dataset, can find all clusters with varying densities that would otherwise be impossible had the same algorithm been applied to the unscaled dataset.

In 2019, Messaoud et al. proposed INFS-DBSCAN (Infinite Feature Selection DBSCAN) [32] which combines the qualities of DBSCAN for outlier detection and Inf-FS [33] for feature selection. INFS-DBSCAN is a two-step algorithm. In the first step, a machine learning method is used to automatically select most relevant features (attributes) to a given predictive model. Simply, unneeded, irrelevant, and redundant features are identified or removed from the dataset as such attributes do not contribute to the efficiency of the model. In the second step, DBSCAN is implemented on the preprocessed dataset to detect outlying points. Infinite feature selection is a recent filter-based algorithm proposed by Roffo et al. [33], also known as Inf-FS. The main idea of Inf-FS is to assign a ranking score to each feature followed by a cross-validation step to select most relevant features. Unlike most filter methods, the algorithm evaluates the importance of a given feature while considering all the possible subsets of features.

7.2.2 *Density-Based Clustering for Outlier Detection*

In 2018, Rahman et al. proposed parameter independent density-based clustering (PIDC) algorithms by utilizing two novel concepts for neighborhood functions which they termed as unique closest neighbor (UCN) and unique neighborhood set (UNS) [34]. In their paper, they proposed density-based clustering algorithms which are independent of the parameter settings. They utilized two novel concepts for neighborhood functions to develop their parameter independent clustering algorithms, which was termed as unique closest neighbor and unique neighborhood set. The *UNS* for a particular sample point is unique and relies only on the distribution of the records of the dataset. They discussed two derivatives of the proposed parameter independent density-based clustering (PIDC) algorithms, termed PIDC-WO and PIDC-O. PIDC-WO has been designed for datasets that do not contain explicit outliers, whereas PIDC-O provides very good performance even on datasets with the presence of outliers. PIDC-O uses a two-stage processing where the first stage identifies and removes outliers before passing the records to the second stage to perform the density-based clustering. Their algorithms are computationally efficient in the sense that they do not use the random sampling aspect and multiple repetitions to find many candidate solutions, and then narrow to a 'best' clustering solution. They showed that their algorithms, while remaining parameter-free, perform effectively both for the clustering and outlier detection objectives.

In 2019, Su et al. proposed a safe nonoutlier- object elimination approach named as rough clustering based on multi-level queries (RCMLQ) to preprocess the datasets to eliminate the nonoutlier objects to the utmost [35]. They first redefined a local outlier factor called local deviation coefficient (LDC) by taking full advantage of the distribution of the object and its neighbors. Then, an efficient local outlier detection approach named as efficient density-based local outlier detection for scattered data (EDLOS) was proposed based on the LDC and RCMLQ. The RCMLQ greatly reduces the amount of data that needs to be quantified for local outlier factor while the LDC is more sensitive to the degree of anomaly of the scattered datasets. Therefore, the E2DLOS improves the existing local outlier detection approaches in both time efficiency and detection accuracy.

In the same year, Wang et al. proposed an outlier detection method based on OPTICS, known as OD-OPTICS, which adds a preprocessing process and modifies the core distance *(CD)* computing method [36]. First, the radius filtration strategy, which provides the key radii, is performed, and the crucial distances of the point set are reflected. Then, for the purpose of filtrating invalid radii and selecting the most appropriate one, the covering space model was proposed. With the influence of the three rules that they established, the basic distance between two neighboring points can be calculated by the covering space. Moreover, the *CD* calculation was optimized such that it could magnify the difference value between normal points and outliers.

Also in 2019, Messaoud et al. proposed a new outlier detection technique based on a feature selection strategy to avoid the curse of dimensionality, named infinite feature selection DBSCAN (INFS-DBSCAN) [33]. The main purpose of their proposed method is to reduce the dimensions of a high-dimensional data set in order to efficiently identify outliers using clustering techniques. Their approach combines the qualities of DBSCAN [22] for outlier detection and Inf-FS [34] for feature selection. Thus, INFS-DBSCAN is a two-step algorithm. In the feature selection step, a data mining method is used to automatically select most relevant features (attributes) to a given predictive model. Simply, unneeded, irrelevant, and redundant features are identified or removed from the dataset as such attributes do not contribute to the efficiency of the model. In the cluster-based outlier detection step, DBSCAN is implemented on the preprocessed dataset to detect outlying points.

Also in 2019, Nagamani and Chittineni proposed a density-based outlier detection (DBOD) method which divides the network into subnetworks and identifies outliers on them [37]. The dataset is divided into multiple subsets and checked for exceptions which make the task of outlier detection easy. The exceptions are then consolidated from various subsets. In this way, the neighborhood density-based anomalies can be figured effectively.

7.3 The Proposed Enhancer for Outlier Mining

7.3.1 A Simple Idea

Distance-based (DB) outlier detection methods are good at identifying global outliers. To identify the relatively small number of DB outliers, *k*-nearest neighbors (*k*NN) for each data point are first computed, together with the corresponding distances between each data to their *k*-nearest neighbors. The average of these distances or the *k*-th distance is used as an outlier score in distance-based outlier detection method and top-n ones are returned as outliers. However, there is no reason to assume that this must be the case for some outliers due to the existence of boundary data points. To meet this challenge, an outlier indicator can be an aid. More specifically, a problem with distance-based outlier detection is that these methods do not take the outlying degrees of a data point's *k*-nearest neighbors into consideration in the detection process. As a result, false positives can happen. For example, for the sample dataset in Fig. 7.3, although DB or DB-Max outlier scores can be calculated for each data point and four boundary data points of cluster C_2 can be mistakenly detected as DB or DB-Max outliers, there are no outstanding outliers. To prevent the false positives from happening, there must be some ways to differentiate between boundary points and outliers existing in a dataset in the first place. To do so, as a first degree approximation, the ratios of the distances of each data point and its *k*NN to their first nearest neighbors within a cluster can be assumed to follow a uniform

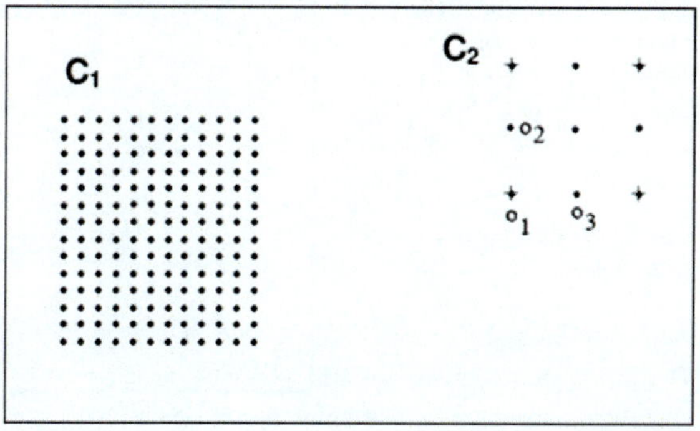

Fig. 7.3 An illustration of outlier indicator, reprinted from Ref. [44], copyright 2018, with permission from Springer

distribution and the corresponding mean and standard deviation can thus be calculated. The ratio of the standard deviation of the ratios over the mean of the ratios can be used to judge to some degree whether outliers exist or not. For the sample dataset shown in Fig. 7.3, if k is set to be 2 (i.e. 2NN), the ratio of the distance of data point o_1 to its first nearest neighbor (say, o_2) over the distance of data point o_2 to its first nearest neighbor is the same as that of data point o_2 to its first nearest neighbor and that of data point o_3 to its first nearest neighbor. The corresponding ratio of the standard deviation over the mean is 0, indicating there are not outstanding global outliers.

The state-of-the-art outlier definitions and some classical outlier mining algorithms are able to identify some outliers, either global outliers, or local outliers. However, for distance-based outlier detection methods, boundary points could be misclassified as outliers, while, for density-based outlier detection methods, global outliers may have low outlier scores and therefore be missed. Taking the dataset in Fig. 7.4 as an example, data point A is farthest away from its six closest neighbors and therefore should be identified as a global outlier. However, for $k = 6$, the LOF-based outlier detection method assigns a higher outlier score to data point B than to A and, therefore, fails to identify A as the most significant global outlier. However, if we find a way to filter the boundary points like point B in advance, the outlier ranking of point A will be higher, and it will be easier to identify point A as an outlier in the final experimental results.

Through the above analysis, a method is in need to distinguish the boundary points and outliers, filter out the boundary points in advance, reduce the impact of boundary points on outlier detection in the calculation process, and further improve the outlier ranking so as to improve the accuracy of outlier detection results. Based on these observations, we propose a new outlier detection algorithm, which enhances current state-of-the-art outlier mining methods by filtering out boundary points from outlier candidates, and formularize it in the following subsections.

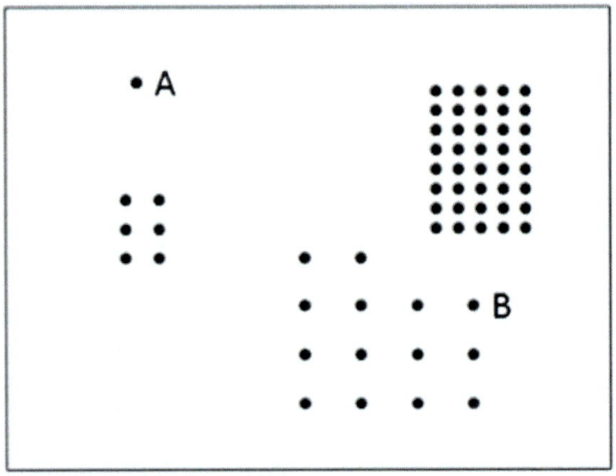

Fig. 7.4 An illustration of a difference between global outlier and local outlier definitions, reprinted from Ref. [44], copyright 2018, with permission from Springer

7.3.2 Some Definitions

In this chapter, we propose a ratio-based outlier indicator to separate border points in clusters of different densities from being mixed with the outliers.

Definition 1 (*k-Distance of an object p*) For any positive integer *k*, the *k*-Distance of object *p*, denoted as *k*-Distance(*p*), is defined as the *distance*(*p*,*o*), or simply, *d*(*p*,*o*), between *p* and an object $o \epsilon D$ such that:

(1) for at least *k* objects $o' \epsilon D \backslash \{p\}, d(p,o') <= d(p,o)$;
(2) for at most *k*-1 objects $o' \epsilon D \backslash \{p\}, d(p,o') < d(p,o)$.

Definition 2 (*k-nearest neighbors of an object p*) For any positive integer *k*, given *k*-Distance(*p*), *k*-nearest neighbors of *p* contain the first *k* closest objects whose distance from *p* is not greater than *k*-Distance(*p*), denoted as $kNN_{k\text{-}Distance(p)}(p)$, for which, *kNN*(*p*) is used as shorthand.

For outlier detection, we are more interested in those data points whose distance to its first nearest neighbor is significantly larger than the average values of the distances of the point's kNN to their first nearest neighbor. To quantify the significance of a data point's positioning outside some cluster, the uniform distribution is used as a first degree approximation for the distances associated with a data point and its kNN's to their first nearest neighbors. To distinguish between boundary points and outliers, we therefore formulate an outlier indicator using the distances associated with the first nearest neighbor of the data point and its kNN as in the following to focus our attention on the small number of outstanding global and local outliers.

To characterize boundary points, we observe that, for the boundary points in a dataset, the ratio of the distance between the boundary point and its nearest neighbor to the distance between the k-th nearest neighbor of the boundary point and its nearest neighbor (i.e., the nearest neighbor of the k-th nearest neighbor of the boundary point) is close to the value 1, which is actually equal to 1 in the ideal case illustrated in Fig. 7.4 when *k* is not set to a too large value, say $k = 2$. Although these points are distributed in clusters of different densities, for the boundary points in the dataset, the ratio of the distance to their nearest neighbor over the distance from their nearest neighbor to the nearest neighbor of theirs is equal to 1, while for outlier A, the ratio is much larger than 1.

Based on this observation and analysis, we propose a new concept of boundary factor to filter out the boundary points, reduce the interference of border points to outlier detection, and enhance outlier detection algorithms.

Definition 3 (*Distance ratio measurement of an object p*) Given *k*-nearest neighbors of an object p, in the following, l_1 of an object p denotes the distance to its nearest neighbor (say, *q*), and l_2 denotes the distance of such nearest neighbor (i.e., *q*) to its nearest neighbor, the proposed distance ratio measurement of an object *p* is defined as,

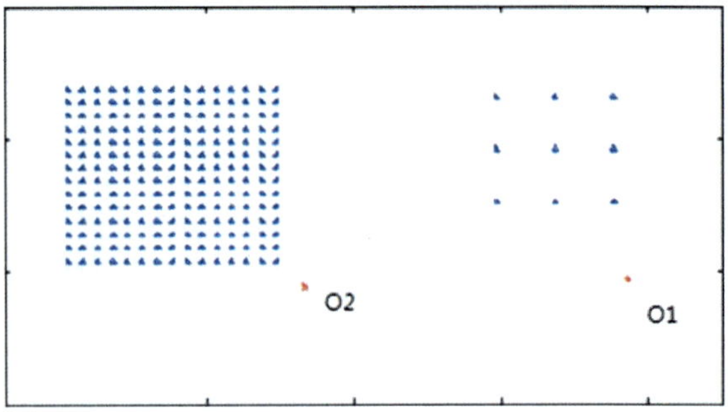

Fig. 7.5 An illustration of the effect of outlier indicator, reprinted from Ref. [44], copyright 2018, with permission from Springer

$$\text{ratio}(p) = \frac{l_1}{l_2} \tag{7.1}$$

Definition 4 (*Outlier indicator of an object p*) Given k-nearest neighbors of an object p, there exists a set of $k + 1$ ratios associated with it, in the following, ratio[0] denotes the distance ratio measurement of object p and ratio[i] denotes the distance ratio measurement of the i-th nearest neighbor of an object p, the proposed outlier indicator, Indicator, is defined based on these ratios in the following,

$$\text{Mean}_{\text{ratio}}(p) = \frac{1}{k+1}\sum_{i=0}^{k} \text{ratio}[i] \tag{7.2}$$

$$\text{Std}_{\text{ratio}}(p) = \sqrt{\frac{1}{k+1}\sum_{i=0}^{k}(\text{ratio}[i] - \text{Mean}_{\text{ratio}}(p))^2} \tag{7.3}$$

$$\text{Indicator}(p) = \frac{\text{Std}_{\text{ratio}}(p)}{\text{Mean}_{\text{ratio}}(p)} \tag{7.4}$$

To manifest the effectiveness of the outlier indicator, for the test sample dataset shown in Fig. 7.5, we calculate the outlier indicators for all the data points and use the sum of their mean and standard deviation as a threshold to highlight the potential outliers and the results shown in Fig. 7.5 demonstrate that outliers O_1 and O_2, denoted by red color, are well-identified because their indicators are much larger than the others, which are actually zero for this idea situation.

7.3.3 *Our Proposed Outlier Detection Algorithm*

To find global outliers, we follow the notion of kNN-based distance outlier definition and calculate the distance-based outlier factors (i.e. DB-Max) for all the data points, sort them in a nonincreasing order, and searching for the largest factor values. If their corresponding outlier indicators are significantly larger than a threshold value, *SOM*, top-*n* data points can be regarded as the global outliers. For local outliers, we follow the notion of kNN-based density-based outlier definition and calculate the local outlier factors (i.e., LOF) for all the data points, sort them in a nonincreasing order, and searching for the largest factor values. We combine three proposed factors to create our kNN-based outlier detection algorithm. To improve the readability, our proposed outlier detection algorithm is presented in a pseudo-code format in Table 7.1.

To determine a threshold for indicators in a data set, let the outlier indicators of all points in the dataset be computed, based on which the average of the indicators, mean, and the corresponding standard deviation, std, are calculated. The threshold of indicators, *SOM*, for finding potential outliers is defined as,

$$\text{SOM} = \text{mean(Indicator)} + f \times \text{std(Indicator)} \tag{7.5}$$

To summarize, the numerical parameters the algorithm needs from the user include the data set, *S*, the loosely estimated number of outliers (i.e. the percentage of outlier candidates in the original data set), *n*, and the number of nearest neighbors, *k*.

Table 7.1 A combined outlier detection algorithm

Input:	*S*: a set of *N* data objects;
	k: the number of nearest neighbors;
	som: threshold
	n: the required number of top outliers.
Output:	*Index*: the indices of top-*n* outliers
Begin:	
	Compute *k*-nearest neighbors for each data point;
	Compute the global and local outlier scores, DB-MAX and LOF, and outlier indicators;
	Compute the *mean* and *std* of the outlier indicators, and the threshold value, *som*;
	Sort the outlier scores, DB-MAX and LOF, in a nonincreasing order;
	While(*Index*.size < n)
	{
	if(DB-MAX.next > *som*) *Index*.push_back(DB-MAX.next.index);
	if(LOF.next > *som*) *Index*.push_back(LOF.next.index);
	}
	Return *Index*.
End	

7.3.4 *The Complexity Analysis*

From the description in the previous subsections, it can be seen that our algorithm mainly consists of two steps, an initial *k*-nearest neighbors' search for each of N data points, and the calculation of three scores and top outliers' mining. Therefore, most of the calculation time is spent on the first step, that is, the *k*-nearest neighbors' search. As a result, the total time complexity for the average case can be $O(N^2)$ or O(NlogN) if some index structure can be used. Since the number of outliers is small, the second step takes a nearly linear time on average. Physically, the resource consumed by our algorithm includes the space for hold the whole data set in memory, space to store their *k* nearest neighbors and three scores, and some space for the index structure if used.

7.4 Experiments and Results

In this section, we compare the effectiveness of the proposed outlier detection method with several state-of-the-art outlier detection methods, including the DB [38], DB-Max [39], LOF [20], the INFLO [40], and RBDA [41] methods. In the first set of experiments, two two-dimensional synthetic data sets are used to show that our proposed outlier detection method augmented with the outlier indicator can outperform classic outlier detection algorithms in classification accuracy. Further, it is important for an outlier detection method to work effectively on real-world datasets. Therefore, in the second set of experiments, a real multi-dimensional dataset obtained from the UCI machine learning repository [42] is used to check the technical soundness of this study and to illustrate the effectiveness of our method in real-world situation. All the data sets are briefly summarized in Table 7.2.

We implement all the algorithms in Java and perform all the experiments on a computer with AMD A6-4400 M Processor 2.70 GHz CPU and 4.00 G RAM. The operating system running on this computer is Windows 7. In our evaluation, we focus on the outlier detection accuracy rate of these outlier detection algorithms on different data sets. The results show that, overall, our proposed outlier detection algorithm is superior over other state-of-the-art outlier detection algorithms.

Table 7.2 Description of all datasets

Data name	Data size	Dimension	#of outliers
syn_Data1	82	2	10
syn_Data2	473	2	6
LYMPHOGRAPHY	148	18	6

7.4.1 Performance of Our Algorithm on Synthetic Data

In this subsection, we use two synthetic datasets to show that the proposed outlier detection method performs better than the traditional outlier detection methods. The two synthetic datasets, syn_Data1 and syn_Data2 are shown in the first plot of Fig. 7.6 and Fig. 7.7, respectively. For this set of experiments, the parameter k's is set to be 3 for all the methods and the results for syn_Data1 and syn_Data2 are plotted in Fig. 7.6 and Fig. 7.7, respectively.

The first synthetic dataset, syn_Data1, consists of 82 instances, including six single outliers (A, B, C, D, E and F, and four clusters of different densities consisting of 36, 8, 12, and 16 uniformly distributed instances, respectively. From the results depicted in Fig. 7.6, we can see that DB and DB-Max neither can mine the two local outliers, that is, B and C, nor have the same ranks for A, D, E and F. RBDA, INFLO, LOF and our outlier detection method detect all six outliers correctly. The plot at bottom left corner shows the corresponding *Indicator* values (which are actually 0 for boundary and inner points) thresholded by Eq. (7.5), which correctly identifies the six outliers.

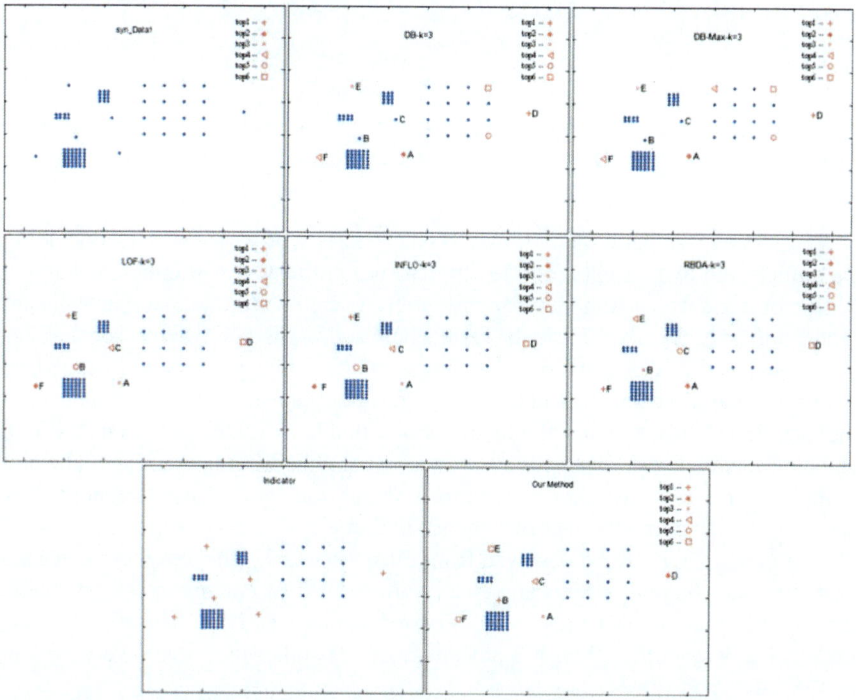

Fig. 7.6 Outlier detecting results on syn_Data1 for $k = 3$, reprinted from Ref. [44], copyright 2018, with permission from Springer

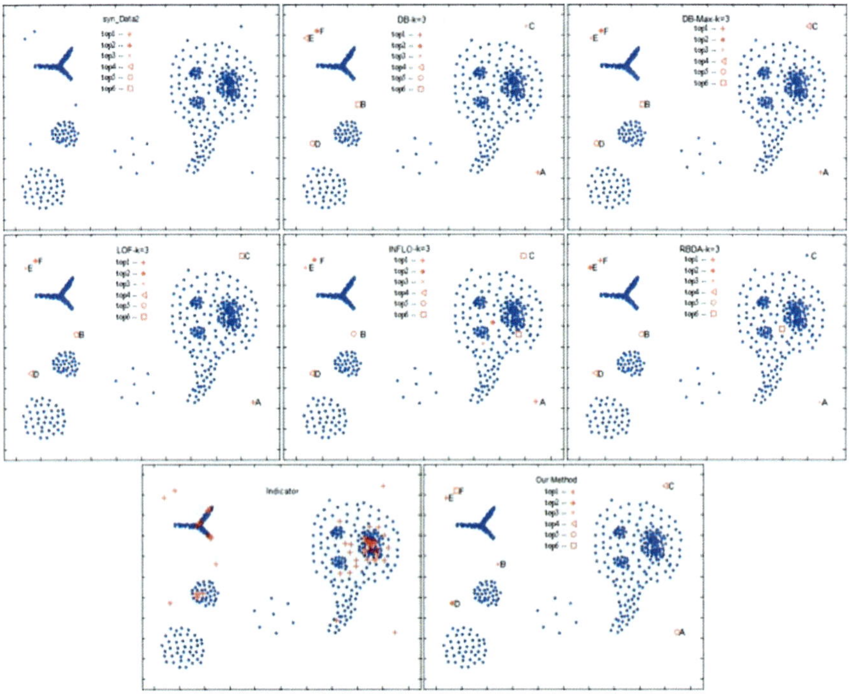

Fig. 7.7 Outlier detecting results on syn_Data2 for k = 3, reprinted from Ref. [44], copyright 2018, with permission from Springer

The second synthetic dataset, syn_Data2, consists of 473 instances, including six outliers and five clusters of different densities clusters. A particular challenging feature of this data set is that three denser clusters are buried into one sparse cluster on the upper-right corner. From the results depicted in Fig. 7.7, we can see that this is a global outlier detection situation while the detection process is disturbed by the immediate connection of clusters with different densities. For detecting top 6 outliers, RBDA misses C but all other methods, that is, DB, DB-Max, LOF, INFLO, and our method detect all six outliers correctly but with different rankings. The plot at the bottom left corner shows the corresponding Indicator values thresholded by Eq. (7.5), which correctly identifies the six outliers.

To summarize, it can be observed from Figs. 7.6 and 7.7 that our method has no problems detecting all outliers and clearly offers the best ranking in two synthetic datasets while all other methods do not perform competently with detecting all the outliers one way or the other. The advantage of our outlier indicator is very evident on these 2-dimensional data sets.

7.4.2 Performance of Our Algorithm on Real Data

It has been pointed out by Aggarwal and Yu that one way to test how well an outlier detection algorithm works is to run the method on a dataset and test the percentage of points which belongs to the rare classes [43]. In order to test how well our outlier indicator works on real dataset, we compare its ability in finding outliers with other methods in a real dataset, LYMPHOGRAPHY, which is downloaded from UCI [42]. This dataset has 148 instances with 18 attributes and contains a total of 4 classes. Classes 2 and 3 have 81 and 61 instances, respectively. The remaining two classes have totally 6 instances (2 and 4, respectively) and are regarded as outliers (i.e. rare classes) for they are small in size.

To quantitatively measure the performance of an outlier detection method, a popular metric, called recall, is used. Assuming that a dataset $D = D_o \cup D_n$ where D_o denotes the set of all outliers and D_n denotes the set of all normal data. Given any integer $m \geq 1$, if O_m denotes the set of outliers among objects in the top m positions returned by an outlier detection scheme, recall is defined as,

$$\text{recall} = \frac{|O_m|}{|D_O|} \tag{7.6}$$

In Eq. (7.6), recall shows the percentage of detected outliers in all outliers.

Table 7.3 shows the experimental results of the proposed outlier detection method in comparison with five other methods, DB, DB-Max, LOF, INFLO, RBDA, respectively, for four values of k's (i.e., 7, 10, 20, 30) and six values of m's (6, 7, 8, 9, 10, 15). In the table, n denotes the correct number of outliers among returned m ones, and r denotes the corresponding recall.

From the experimental results shown in Table 7.3, it can be seen that the proposed method mines all the outliers correctly for all m's and all k's and thus performs the best. RBDA method and LOF method perform next since they mine outliers as well as our method for $k = 20$ and $k = 30$ but does not do well in cases for $k = 7$ and $k = 10$. Overall, with increasing k's, RBDA, LOF, and INFLO methods work better and better while DB and DB-Max work worse and worse.

7.5 Conclusions

In this chapter, we have proposed a combined outlier detection method that can detect both global and local outliers. Traditional distance-based and density-based outlier detection algorithms can effectively detect two different kinds of outliers separately but not both at the same time. Further, classic distance-based outlier detection algorithms do not differentiate global outliers from boundary data points. To partially circumvent these problems, we propose a novel outlier detection approach

Table 7.3 Experimental results for LYMPHOGRAPHY data

	DB		DB-Max		LOF		INFLO		RBDA		OUR	
m	n	r	n	r	n	r	n	r	n	r	n	r
k = 7												
6	5	0.83	5	0.83	5	0.83	4	0.67	5	0.83	**6**	**1.00**
7	**6**	**1.00**	5	0.83	5	0.83	5	0.83	5	0.83	**6**	**1.00**
8	**6**	**1.00**	**6**	**1.00**	**6**	**1.00**	5	0.83	**6**	**1.00**	**6**	**1.00**
9	**6**	**1.00**	**6**	**1.00**	**6**	**1.00**	**6**	**1.00**	**6**	**1.00**	**6**	**1.00**
10	**6**	**1.00**	**6**	**1.00**	**6**	**1.00**	**6**	**1.00**	**6**	**1.00**	**6**	**1.00**
15	**6**	**1.00**	**6**	**1.00**	**6**	**1.00**	**6**	**1.00**	**6**	**1.00**	**6**	**1.00**
k = 10												
6	5	0.83	5	0.83	5	0.83	4	0.67	5	0.83	**6**	**1.00**
7	**6**	**1.00**	5	0.83	5	0.83	5	0.83	**6**	**1.00**	**6**	**1.00**
8	**6**	**1.00**	5	0.83	**6**	**1.00**	5	0.83	**6**	**1.00**	**6**	**1.00**
9	**6**	**1.00**	**6**	**1.00**	**6**	**1.00**	**6**	**1.00**	**6**	**1.00**	**6**	**1.00**
10	**6**	**1.00**	**6**	**1.00**	**6**	**1.00**	**6**	**1.00**	**6**	**1.00**	**6**	**1.00**
15	**6**	**1.00**	**6**	**1.00**	**6**	**1.00**	**6**	**1.00**	**6**	**1.00**	**6**	**1.00**
k = 20												
6	5	0.83	5	0.83	**6**	**1.00**	5	0.83	**6**	**1.00**	**6**	**1.00**
7	5	0.83	5	0.83	**6**	**1.00**	5	0.83	**6**	**1.00**	**6**	**1.00**
8	**6**	**1.00**	**6**	**1.00**	**6**	**1.00**	**6**	**1.00**	**6**	**1.00**	**6**	**1.00**
9	**6**	**1.00**	**6**	**1.00**	**6**	**1.00**	**6**	**1.00**	**6**	**1.00**	**6**	**1.00**
10	**6**	**1.00**	**6**	**1.00**	**6**	**1.00**	**6**	**1.00**	**6**	**1.00**	**6**	**1.00**
15	**6**	**1.00**	**6**	**1.00**	**6**	**1.00**	**6**	**1.00**	**6**	**1.00**	**6**	**1.00**
k = 30												
6	5	0.83	5	0.83	**6**	**1.00**	5	0.83	**6**	**1.00**	**6**	**1.00**
7	5	0.83	5	0.83	**6**	**1.00**	**6**	**1.00**	**6**	**1.00**	**6**	**1.00**
8	5	0.83	5	0.83	**6**	**1.00**	**6**	**1.00**	**6**	**1.00**	**6**	**1.00**
9	**6**	**1.00**	5	0.83	**6**	**1.00**	**6**	**1.00**	**6**	**1.00**	**6**	**1.00**
10	**6**	**1.00**	**6**	**1.00**	**6**	**1.00**	**6**	**1.00**	**6**	**1.00**	**6**	**1.00**
15	**6**	**1.00**	**6**	**1.00**	**6**	**1.00**	**6**	**1.00**	**6**	**1.00**	**6**	**1.00**

Reprinted from Ref. [44], copyright 2018, with permission from Springer

which combines the global outlier detection and local outlier detection in a separate and simultaneous way and, when augmented with an outlier indicator, can outperform traditional outlier detection approaches. To demonstrate the utility of our proposed outlier detection mechanism, we have performed a detailed comparison of its performance with state-of-the-art distance-based and density-based outlier detection methods. Experimental results show the ability of our algorithm to rank the best candidates for being an outlier with high recall.

Acknowledgements This chapter was modified from the paper published by our group in machine learning and data mining in pattern recognition [44]. The related contents are reused with permission.

References

1. Knorr, E.M., & Ng, R.T. (1999). A unified notion of outliers: Properties and computation. In: *Proceedings of the 3rd International Conference on Knowledge Discovery and Data Mining* (KDD'97) (pp. 219-222). CA, USA: Newport Beach.
2. Knorr, E. M., Ng, R. T., & Tucakov, V. (2000). Distance-based outliers: Algorithms and applications. *The VLDB Journal, 8*(3), 237–253.
3. Mehnaz, S., & Bertino, E. (2017). Ghostbuster: A fine-grained approach for anomaly detection in file system accesses. In *Proceedings of the ACM Conference on Data and Application Security and Privacy* (CODASPY'17) (pp. 3–14). Scottsdale, AZ, United states.
4. Iturbe, M., Garitano, I., Zurutuza, U. and Uribeetxeberria, R. (2017). Towards large-scale, heterogeneous anomaly detection systems in industrial networks: A survey of current trends. *Security and Communication Networks, 2017*(6), Art. no. 9150965.
5. Wang, Y., Wu, Z., Zhu, Y., & Zhang, P. (2018). Research on anomaly detection algorithm based on generalization latency of telecommunication network. *Future Generation Computer Systems, 85,* 9–18.
6. Gogoi, P., Bhattacharyya, D. K., Borah, B., & Kalita, J. K. (2011). A survey of outlier detection methods in network anomaly identification. *Computer Journal, 54*(4), 570–588.
7. Bhuyan, M. H., Bhattacharyya, D. K., & Kalita, J. K. (2012). Survey on incremental approaches for network anomaly detection. *International Journal of Communication Networks and Information Security, 3*(3), 226–239.
8. Agarwal, D. (2005) An empirical Bayes approach to detect anomalies in dynamic multidimensional arrays. In *Proceedings of the 5th IEEE International Conference on Data Mining* (ICDM'05) (pp. 26–33). Houston, TX, United states.
9. Pimentel, M. A. F., Clifton, D. A., Clifton, L., & Tarassenko, L. (2014). A review of novelty detection. *Signal Processing, 99,* 215–249.
10. Avdiienko, V., Kuznetsov, K., Rommelfanger, I., Rau, A., Gorla, A., & Zeller, A. (2017). Detecting behavior anomalies in graphical user interfaces. In *Proceedings of the International Conference on Software Engineering Companion* (ICSE-C'17) (pp. 201–203). Buenos Aires, Argentina.
11. Keogh, E., Lin, J., Lee, S.-H., & van Herle, H. (2010). Finding the most unusual time series subsequence: Algorithms and applications. *Knowledge and Information Systems, 11*(1), 1–27.
12. Cai, L., Thornhill, N., Kuenzel, S., & Pal, B. C. (2017). Real-time detection of power system disturbances based on k-nearest neighbor analysis. *IEEE Access, 5,* 5631–5639.
13. Mccarren, A., Mccarthy, S., Sullivan, C.O., & Roantree, M. (2017). Anomaly detection in agri warehouse construction. In *Proceedings of 2017 Australasian Computer Science Week Multiconference* (ACSW'17) (pp. 1–10). Geelong, VIC, Australia.
14. Stojanovic, N., Dinic, M. and Stojanovic, L. (2018). A data-driven approach for multivariate contextualized anomaly detection: Industry use case. In *Proceedings of the 5th IEEE International Conference on Big Data* (Big Data'17) (99. 1560–1569). Boston, MA, United states.
15. Vidmar, G., & Blagus, R. (2014). Outlier detection for healthcare quality monitoring: A comparison of four approaches to over-dispersed proportions. *Quality and Reliability Engineering International, 30*(3), 347–362.
16. Yan, K., You, X., Ji, X., Yin, G., & Yang, F. (2016). A hybrid outlier detection method for health care big data. In *Proceedings of the 6th IEEE International Conference on Big Data and Cloud Computing* (BDCloud'16) (pp. 157–162). Atlanta, GA, United states.

17. Gu, F., Niu, J., Das, S. K., He, Z., & Jin, X. (2017). Detecting breathing frequency and maintaining a proper running rhythm. *Pervasive and Mobile Computing, 42,* 498–512.
18. Barnett, V., & Lewis, T. (1994). *Outliers in statistical data.* New York: Wiley.
19. Knorr, E.M., & Ng, R.T. (1998). Algorithms for mining distance-based outliers in large datasets. In *Proceedings of the International Conference on Very Large Data Bases* (VLDB'98) (pp. 392–403), New York.
20. Breuning, M.M., Kriegel, H.P., Ng, R.T., Sander, J. (2000). LOF: Identifying density-based local outliers. In *Proceedings of the 2000 ACM SIGMOD International Conference on Management of Data* (SIGMOD'00) (pp. 93–104). Dallas, TX, United states.
21. Jiang, M. F., Tseng, S. S., & Su, C. M. (2001). Two-Phase Clustering Process for Outliers Detection. *Pattern Recognition Letters, 22*(6–7), 691–700.
22. Ester, M., Kriegel, H.-P., Sander, J., Xu, X. (1996). A density-based algorithm for discovering clusters in large spatial databases with noise. In *Proceedings of the 2nd ACM SIGKDD International Conference on Knowledge Discovery and Data Mining* (KDD'96) (pp. 226–231). Portland, Oregon, USA.
23. Ankerst, M., Breunig, M. M., Kriegel, H. P., et al. (1999). OPTICS: Ordering points to identify the clustering structure. In *Proceedings of the 1999 ACM SIGMOD International Conference on Management of Data* (SIGMOD'99), *28*(2), 49–60.
24. Hinneburg A., & Keim D.A. (1998). An efficient approach to clustering in large multimedia databases with noise. In *Proceedings of the 4th International Conference on Knowledge Discovery and Data Mining* (KDD'98) (pp. 58–65). New York, NY, USA.
25. Duan, L., Xu, L., Liu, Y., & Lee, J. Cluster-based outlier detection. *Annals of Operations Research, 168*(1), 151–168.
26. Chen, X., Liu, W., Qiu, H., & Lai, J. (2011). APSCAN: A parameter free algorithm for clustering. *Pattern Recognition Letters, 32*(7), 973–986.
27. Chen, Y.Q., Wang, X., Xu, R., Bai, X., & Meng, X. (2010). An adaptive affinity propagation document clustering. In *Proceedings of the 2010 7th International Conference on Informatics and Systems* (INFOS'10) (pp. 1–7). Cairo, Egypt.
28. Rodriguez, A., & Laio, A. (2014). Clustering by fast search and find of density peaks. *Science, 344*(6191), 1492–1496.
29. Hou, J., Gao, H., & Li, X. (2016). DSets-DBSCAN: A parameter-free clustering algorithm. *IEEE Transactions on Image Processing, 25*(7), 3182–3193.
30. Qi, X., & Wang, P. (2016). A density-based clustering algorithm for high-dimensional data with feature selection. In *Proceedings of the 2016 International Conference on Industrial Informatics-Computing Technology, Intelligent Technology, Industrial Information Integration* (ICIICII'16) (pp. 114–118). Wuhan, Hubei, China.
31. Zhu, Y., Ting, K. M., & Carman, M. J. (2016). Density-ratio based clustering for discovering clusters with varying densities. *Pattern Recognition, 60,* 983–997.
32. Messaoud, T.A., Smiti, A. and Louati, A. (2019). A novel density-based clustering approach for outlier detection in high-dimensional data. In *Proceedings of the 14th International Conference on Hybrid Artificial Intelligence Systems* (HAIS'19) (pp. 322–331). León, Spain.
33. Roffo, G., Melzi, S. and Cristani, M. (2015). Infinite feature selection. In *Proceedings of the 15th IEEE International Conference on Computer Vision* (ICCV'15) (pp. 4202–4210). Santiago, Chile.
34. Rahman, M. A., Ang, K. L.-M., & Seng, K. P. (2018). Unique neighborhood set parameter independent density-based clustering with outlier detection. *IEEE Access, 6,* 44707–44717.
35. Su, S., Xiao, L., Ruan, L., Gu, F., Li, S., Wang, Z., et al. (2019). An efficient density-based local outlier detection approach for scattered data. *IEEE Access, 7,* 1006–1020.
36. Wang, Y. F., Yu, J., Su, G. P., & Qian, Y. R. (2019). A new outlier detection method based on OPTICS. *Sustainable Cities and Society, 45,* 197–212.
37. Nagamani, C., & Chittineni, S. (2019). Efficient neighborhood density based outlier detection inside a sub network with high dimensional data. *Ingenierie des Systemes d'Information, 24*(1), 107–111.

38. Angiulli, F., & Pizzuti, C. (2002). Fast outlier detection in high dimensional spaces. In *Proceedings of the 6th European Conference on Principles of Data Mining and Knowledge Discovery* (PKDD'02) (pp. 15–26). Helsinki.
39. Ramaswamy, S., Rastogi, R., & Shim, K. (2000). Efficient algorithms for mining outliers from large data sets. In *Proceedings of the ACM International Conference on Management of Data* (SIGMOD'00) (pp. 427–438). Dallas.
40. Jin, W., Tung, A.K.H., Han, J., & Wang, W. (2006). Ranking outliers using symmetric neighborhood relationship. In *Proceedings of the 10th Pacific-Asia Conference on Knowledge Discovery and Data Mining* (PAKDD'06) (pp. 577–593). Singapore.
41. Huang, H., Mehrotra, K., & Mohan, C. K. (2013). Rank-based outlier detection. *Journal of Statistical Computation and Simulation, 83*(3), 518–531.
42. UCI: The UCI KDD Archive, University of California, Irvine, CA. http://kdd.ics.uci.edu/.
43. Aggarwal, C., & Yu, P. (2001). Outlier detection for high-dimensional data. In *Proceedings of the 2001 ACM International Conference on Management of Data* (SIGMOD'01) (pp. 37–46). Santa Barbara, CA, USA.
44. Li, X., Wang, X., & Wang, X.L. (2018). Enhancing outlier detection by an outlier indicator. In *Proceedings of the 14th International Conference on Machine Learning and Data Mining* (pp. 393–405). New York, USA.

Part III
Applications

Chapter 8
An Effective Boundary Point Detection Algorithm Via *k*-Nearest Neighbors-Based Centroid

Abstract Outlier detection techniques and clustering techniques are important areas of data mining. Clustering is all about finding groups of data points, whereas outlier analysis is all about finding data points that are far away from these clusters. Clustering and outlier detection, therefore, share a well-known complementary relationship. A simplistic view is that every data point is either a member of a cluster or an outlier. Data points on the boundary regions of a cluster may also be considered weak outliers. However, the study of boundary points is sometimes more meaningful than clusters and outliers. There has been many research work done on boundary point detection. However, with the data obtained becoming more and more complex, existing boundary point detection algorithms have problems such as low precision, parameter dependence, and difficulty in separating outliers. In this chapter, we propose a boundary point detection algorithm, CENTROID-B, based on the concept of kNN-based centroid which has low dependence on parameters, high precision and can detect outliers at the same time. The experimental results on different types of data sets show that the proposed boundary point detection algorithm is effective and manifests high accuracy. Euclidean minimum spanning tree algorithms run typically with quadratic computational complexity, which is not practical for large-scale multi-dimensional datasets. In this chapter, we propose a new two-level approximate Euclidean minimum spanning tree algorithm for large-scale multi-dimensional datasets. In the first level, we perform the proposed outlier and boundary point detection for a given data set to identify a small amount of boundary points. In the second level, we run standard Prim's algorithm on the reduced dataset to complete an approximate Euclidean minimum spanning tree. Experiments on sample data sets demonstrate the efficiency of the proposed method, while keeping high approximate precision.

Keywords Boundary point detection · *k*-nearest neighbor-based centroid · Minimum spanning tree · Euclidean minimum spanning tree · Approximate Euclidean minimum spanning tree

X. Wang et al., *New Developments in Unsupervised Outlier Detection*,
https://doi.org/10.1007/978-981-15-9519-6_8

8.1 Introduction

With the development of information technology, more and more data has been accumulated in various fields. Clustering, boundary point detection, and outlier detection techniques are designed to analyze and interpret these data. Outlier detection techniques are intended to detect anomalous patterns that differ from normal behavior patterns, while boundary point detection techniques are used to find patterns that should behave abnormally but not. Unlike outliers which are also called isolated points, abnormal points, and so on, and are observations that deviate significantly from other observations so much that people suspect that they are produced by a different mechanism [1], a boundary point is a point at the border of a cluster. Being between the clusters and the outliers, the boundary points may have the characteristics of clusters and outliers at the same time. It was Ester et al. who first proposed the concept of cluster boundaries in 1996; that is, boundary points are noncores that are reachable from a certain core point [2]. Since then, many boundary point detection algorithms have been developed which can be roughly divided into: (1) density-based methods [3], (2) grid-based methods [4], (3) angle-based methods [5, 6]. The density-based boundary point detection algorithms mainly use the distribution characteristics of the cluster boundary points, that is, the density mutation occurs near the boundary. This type of algorithms separates the boundary points from the outliers, especially on a uniform dataset. The grid-based boundary point detection algorithms usually divide the data space into grids and then calculate the boundary degree of each grid. If the boundary degree meets certain conditions, the grid is the boundary grid, and the data objects in the grid are boundary points. The problem with this type of algorithms is that the size of the grid is difficult to determine. If the grid is too large, the accuracy of the boundary point detection is not satisfactory enough. If the grid is too small, the time complexity may be too high. Moreover, such algorithms are often difficult to separate cluster points from the boundary points. The angle-based algorithms mainly use the angular relationship between the space vectors formed by the object and other objects in the neighborhood. For the boundary point, the angle between the vector and other vectors in the neighborhood is less than 180 degrees. Angle-based algorithms often need parameters to obtain certain neighborhoods. And the choice of neighborhood radius often varies from dataset to dataset. In addition, the time complexity of such algorithms tends to be relatively high.

In general, existing boundary point detection algorithms have the problems of highly relying on input parameters, high time complexity, and being difficult to accurately separate them from outliers. In order to solve these problems, in this chapter, based on the degree of deviation of a data point from the normal data patterns, we propose a boundary point detection algorithm which uses a *k*-nearest neighbors (kNN) based centroid to separate the small number of outlying and dangling data points from the normal data patterns. The concept of a centroid is to define an imaginary point around which the neighbors in the data space aggregate. The proposed boundary point detection algorithm first detects the outliers and boundary points at the same time based on the idea that outliers and boundary points are always far from a local cluster centroid. Then based on the fact that the neighborhoods of the outliers and boundary points have different characteristics, we then propose a method for calculating the degree of boundary to evaluate the extent to which data points are outliers or boundary points. Finally, according to the different features of the boundary points and the outliers, the outliers are distinguished from the candidate boundary points.

Further, as an important application of the presented boundary detection algorithm, we will propose a fast approximate Euclidean minimum spanning tree (EMST) based on the set of boundary data points detected. Finding a minimum spanning tree (MST) for a given connected graph is a fundamental problem with diverse application domains in data mining, and many efficient MST algorithms have been developed. In today's MST tasks, usually, a set of N d-dimensional data points is given and the problem is commonly solved in the Euclidean setting, giving rise to the so-called Euclidean minimum spanning tree (EMST) problem. In this case, there are V = N vertices and E = N(N-1)/2 edges in the complete graph, and standard EMST algorithms, such as Kruskal's [7] and Prim's [8], have a time complexity roughly equal to $O(dN^2)$. For large-scale high-dimensional datasets, standard EMST algorithms will lose its time performance. Fortunately, in many practical applications, an exact MST can be generally replaced by an approximate one without degrading the quality of the final application.

Being a compact representation of a given data set, EMST has been extensively used in image segmentation [9, 10], cluster analysis [11–13], classification [8–14], and manifold learning [9–15]. In particular, we are interested in EMST construction for cases where data consist of well-separated clusters and the number of clusters is significantly smaller than the number of data points [9, 16, 17]. In these situations, in-cluster edges have weights that are significantly smaller than those longest edges corresponding to the cluster breakers and can be found quickly by fast *k*-nearest neighbor search algorithms. However, the location of the longest edges that correspond to the cluster breakers in such EMSTs cannot be determined quickly and is often the bottleneck for fast EMST algorithms.

In this chapter, we propose a new clustering-inspired EMST algorithm that is both computationally efficient and competent with the state-of-the-art EMST algorithms. Basically, our novel data-dependent clustering-inspired EMST algorithm tries to identify the relatively small number of longest edges in an EMST by finding and retaining the relatively small number of boundary points and outliers before the complete EMST is constructed. Being composed of two-level structures, the working principle behind our novel data-dependent EMST algorithm is illustrated in Fig. 8.1.

During the first level, an index-based fast search algorithm to find *k*-nearest neighbors (kNN) for each data point is performed to form a few number of cluster-wise EMSTs and a relatively small group of the boundary points (including outliers) are identified using the proposed boundary detection algorithm and retained. During the second level, Prim's EMST algorithm runs upon the boundary points (including outliers) to locate the small number of cluster-breaking longest edges. To be as general as possible, our proposed algorithm has no specific requirements on the dimensionality of the datasets and the format of the distance measure, although Euclidean

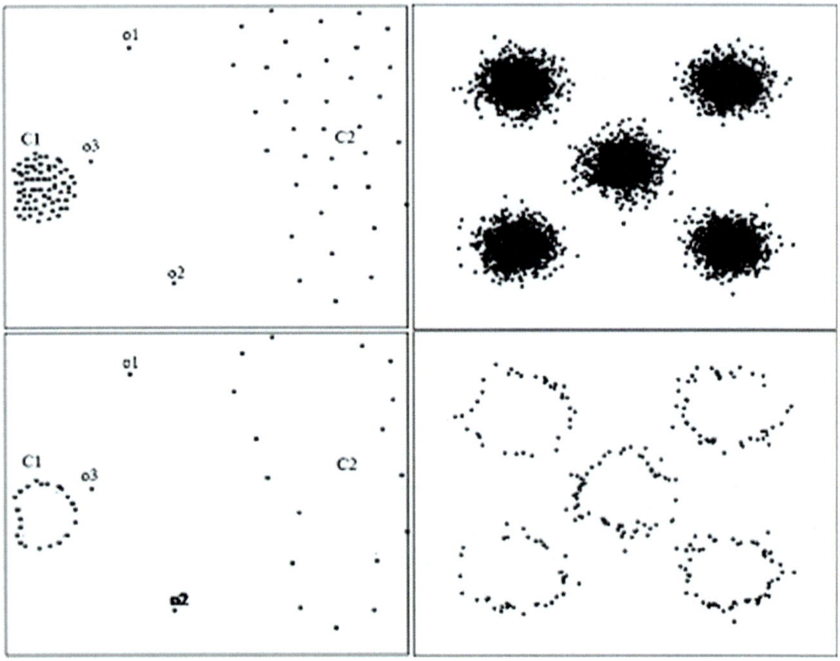

Fig. 8.1 An illustration of the simple idea, reprinted from Ref. [50]. Copyright 2018, with permission from Springer

distance is used as the edge weight in our experiments. Experiments conducted on sample datasets show that our proposed algorithm can obviously improve the run time performance of EMST while keeping high precision.

The rest of this chapter is organized as follows. Section 8.2 gives a review on some related work. Section 8.3 presents the proposed boundary point detection method. Section 8.4 presents the proposed boundary point-based fast approximate EMST algorithm. Section 8.5 gives the experimental evaluations. In Sect. 8.6, conclusions are made.

8.2 Related Work

Work related to the methods presented in this chapter falls into two main categories: density-based outlier and boundary point detection methods and EMST algorithms.

8.2.1 *Outlier and Boundary Point Detection*

Being an important branch of data clustering techniques, density-based clustering produces clusters that consist of a set of data objects spread in the data space over a contiguous region of high density of objects and that are separated from each other by contiguous regions of low density of objects. Data objects located in low-density regions are typically considered as noise (i.e. outliers) and border points. The most popular density-based clustering method is DBSCAN [2]. The key drawback of DBSCAN is that they expect some kind of density drop to detect cluster borders. For our purpose, we need to detect outliers and border points which are objects in the sparse areas and are required to separate clusters.

The existence of outliers can make data modeling more difficult and imprecise. Many efforts have been devoted to detecting them. It has been generally agreed that distance-based outlier definitions can detect more globally orientated outliers and regard being an outlier as a binary property while density-based outlier definitions can detect more locally distributed outliers. Rather than manipulating distances of a data object to its k-nearest neighbors, density-based approaches such as the local outlier factor (LOF) of an object proposed by Breunig et al. in 2000 consider the outlying status of an object by using the ratio between the local density around the object and the average of the local densities around its neighboring objects. For an object, LOF captures the degree to which it is an outlier by looking at the densities of its neighbors. The higher the difference between the density in the local neighborhood of an object and those around its k-nearest neighbors, the higher its LOF value and the more distinctly the object is considered to be an outlier. It has been proved that a LOF value of approximately 1 indicates that the corresponding object is located within a region of homogeneous density (i.e. the object is deep inside a cluster). Following the notion of local outlier factor, several extensions and refinements to the basic LOF model have been proposed, such as a connectivity-based outlier factor (COF) [18] proposed in 2002 by Tang et al., a local outlier integral (LOCI) based on the concept of a multi-granularity deviation factor (MDEF) [19] proposed by Gibbons et al. in 2003, a spatial local outlier measure named SLOM [20] proposed by Sun and Chawla in 2004, INFLO [21] proposed by Jin et al. in 2006, a resolution-based outlier factor (ROF) [22] proposed by Fan et al. in the same year, a local distance-based outlier factor (LDOF) [23] proposed by Zhang et al. in 2009, a rank-based outlier factor (RBDF) [24] proposed by Huang et al. in 2013, a kernel density estimation-based outlier score (KDEOS) [25] proposed by Schubert et al. in 2014, a normalized residual-based outlier factor (NROF) [26] proposed by Ru et al. in 2016, and a local outlier factor named RDOS and proposed by Tang et al. in 2017 [27].

Most of these outlier factors are based on k-nearest neighbors. However, in high-dimensional data space, these methods are bound to deteriorate due to the notorious 'curse of dimensionality' where concepts like proximity, distance, or nearest neighbors become less meaningful with increasing dimensionality. The issue exists whether the notion of these outliers is still meaningful for high-dimensional data. To circumvent this problem, an angle-based outlier detection (ABOD) method designed especially for high-dimensional data was proposed by Kriegel in 2008 [28] and plays an important role in identifying outliers in high-dimensional spaces. Unfortunately, the ABOD algorithm runs in cubic time (with a quadratic time heuristic). To improve, a novel random projection-based technique was proposed in 2012 that is able to estimate the angle-based outlier factor for all data points in time near linear in the size of the data [29].

Boundary points are generally located around the margin of densely distributed data such as a cluster. Although outliers and boundary points are different by definition, they are generally located around the margin of the region with high density, such as that of a cluster. Thus, this chapter focuses on simultaneously identifying outliers and boundary points of clusters based on local geometrical measures and is based on the idea that boundary points and outliers are characterized by a lower local density, that is, larger deviation from normal data points in their *k*-nearest neighborhood.

8.2.2 EMST Algorithms

For a given connected and weighted graph G = (E, V), Boruvka's algorithm begins with each vertex of a graph being a tree, and for each consecutive iteration, it selects the shortest edge from a tree to another tree and combines them. This process continues until all the trees are combined into one tree [30]. Proposed independently by Jarník [31], Prim [8], and Dijkstra [32] in 1930, 1957 and 1959, respectively, the famous Prim's algorithm first arbitrarily selects a vertex as a tree and then repeatedly adds the shortest edge that connects a new vertex to the tree, until all the vertices are included. Proposed in 1956, Kruskal's algorithm starts with sorting all the edges by their weights in a nondecreasing order, treats each vertex as a tree, and iteratively combines the trees by adding edges in the sorted order excluding those leading to a cycle until all the trees are combined into one tree [7]. The time complexity of these classic MST algorithms is O(ElogV).

To construct an MST in the Euclidean setting, standard Prim's algorithm requires a quadratic running time. To be more efficient, in 1978, Bentley and Friedman [33] proposed to use a Kd-tree in the Prim's algorithm to enhance the search for the next edge to add to the tree, which can reach an O(NlogN) running time for most data distributions. In 1985, Preparata and Shamos [34] gave a lower bound for the EMST problem of (NlogN), which has been the tightest known lower bound. In 1993, Callahan and Kosaraju proposed the well-separated pair decomposition (WSPD) [35] which forms the basis of most recent EMST algorithms. The WSPD partitions data points into a set of pairs of tree nodes such that the nodes in any pair are farther apart than the diameter of either node. It can be shown that the WSPD has O(N) pairs of nodes, and that the MST is a subset of the edges formed between the closest pair of points in each pair of nodes. In 2000, Narasimhan and Zachariasen applied WSPD to compute neighbors of components for Boruvka's algorithm to find edges of the MST [36]. However, the constant in the O(N) size of the WSPD grows exponentially with the data dimension and is often very large in practice. In 2010, March et al. presented a new dual-tree algorithm for efficiently computing the EMST [37], which is superficially similar to the method in [36] except that the WSPD is replaced by the new dual-tree data structure. They used adaptive algorithm analysis to prove the tightest (and possibly optimal) runtime bound for the EMST problem to-date.

In addition to the exact EMST problem, approximate EMST algorithms have been also proposed. In 1988, Vaidya [38] employed a group of grids to partition a data set into cubical boxes of identical size. For each box, a representative point is determined. Within a cubical box, points are connected to the representative. Any two representatives of two cubical boxes are connected if the corresponding edge length is between two specific thresholds. In 1993, Callahan and Kosaraju [35] proposed to utilize WSPD of a data set to extract a sparse graph from the complete graph and then apply an exact MST algorithm to it. More recently, in 2009, Wang et al. [39] employed a divide-and-conquer scheme to construct an approximate EMST, which is superficially similar to WSPD method. Their goal was to detect longest edges in an EMST at an early stage for clustering. At the same year, Lai et al. proposed a two-stage Hilbert curve-based approximate EMST algorithm for clustering [40]. In 2014, Wang et al. proposed a fast EMST algorithm [41], which is superficially similar to the method in [33] except that the *K*d-tree is replaced by the iDistance indexing structure for fast kNN search in high-dimensional datasets. The authors argued that the algorithm has an expected O(NlogN) running time, but did not prove this rigorously. In 2015, Zhong et, al. proposed a fast two-stage approximate Euclidean minimum spanning tree (FAEMST) algorithm which employs a divide-and-conquer scheme to produce an approximate EMST with theoretical time complexity of $O(N^{1.5})$ [42]. In the first stage, K-means is employed to partition a dataset into $N^{1/2}$ clusters. Then an exact EMST algorithm is applied to each cluster and the produced $N^{1/2}$ EMSTs are connected to form an approximate EMST. In the second stage, the clusters produced in the first stage form $N^{1/2}-1$ neighboring pairs, and the dataset is repartitioned so that the neighboring boundaries of a neighboring pair are put into a cluster. With these $N^{1/2}-1$ clusters, another approximate EMST is constructed. Finally, the two approximate EMSTs are combined into a graph and a more accurate approximate EMST is generated from it.

8.3 Boundary Point Detection Based on *k*NN Centroid

Usually, outliers and boundary points occupy a small portion of the data sets. The opposite of them is the normal data. One of the differences between the outliers and the boundary points is that the boundary points usually gather around the normal data while the outliers are far away from the clusters.

8.3.1 Definitions

In this subsection, we provide the underlying definitions for the proposed boundary point detection algorithm.

Definition 1 (*k-Distance of an object p*). For any positive integer k, the k-Distance of object p, denoted as k-distance(p), is defined as the distance(p,o), or simply, dist(p,o) or $d(p,o)$, between p and an object $o \in D$ such that:

(1) for at least k-objects $o' \in D \backslash \{p\}$, $d(p,o') \leq d(p,o)$;
(2) for at most k-1 objects $o' \in D \backslash \{p\}$, $d(p,o') < d(p,o)$.

Definition 2 (*k-Distance neighborhood of an object p*). Given the k-distance of p, the k-distance neighborhood of p contains every object whose distance from p is not greater than the k-distance(p), i.e. $N_{k\text{-distance}(p)}(p) = \{q \in D \backslash \{p\} | d(p,q) \leq k\text{-distance}(p)\}$. These objects are called the k-nearest neighbors of p. For simplicity, $N_k(p)$ is used as a shorthand for $N_{k\text{-distance}(p)}(p)$.

Definition 3 (*k-nearest neighborhood centroid*). Given the k-nearest neighborhood $N_k(p)$ of object p, the kNN-based centroid for p is defined as,

$$\text{CENTROID}_{N_k(p)} = \frac{\sum_{q \in N_k(p)} \vec{q}}{k} \tag{8.1}$$

In other words, the kNN-based centroid of p is the vector mean of the k-nearest neighbors of the object p.

Even though Definition 3 provides the calculation method for the centroid, in order to reduce the interference of the complex density distribution on the outlier detection results, it is necessary to determine a corresponding neighborhood size for each data object.

Definition 4 (*The size of local neighborhood for kNN based centroid*). In order to minimize the influence of density changes on the calculation of the centroid, we calculate the corresponding number of local nearest neighbors for each data object. And that number should make the distance between the data object p and its corresponding centroid CENTROID as small as possible, such number is denoted as *mpt* and can be determined by,

$$\min_{mpt}\left(dist\left(\text{CENTROID}_{N_{mpt}(p)}, p\right)\right), 2 < mpt < 4d \tag{8.2}$$

where d is the dimension of the data set.

In the process of finding the optimized centroid by minimizing the distance between the centroid and the data point given in Eq. (8.2) with varying *mpt*, the number of nearest neighbors *mpt* can be obtained that best reflects the local distribution of the objects around data point *p*. It should be noted that the minimum value of *mpt* is also data-specific. When searching for the best centroid on the multi-dimensional data set, the minimum value of *mpt* can be appropriately incremental.

Definition 5 (*Centroid-based local outlier factor of an object p*) Let a centroid be computed upon a data point *p*'s *k*-nearest neighbors, *dist*[1] denote the first nearest neighbor distance associated with *p*'s first nearest neighbor, *k*NN-centroid-based local outlier factor is defined as,

$$\mathrm{CLOF(p)} = \max\left(\frac{dist(p, \mathrm{CENTROID}_{N_{mpt}(p)})}{dist[1]}, \frac{dist[1]}{dist(p, \mathrm{CENTROID}_{N_{mpt}(p)})}\right) \tag{8.3}$$

Centroid-based local outlier factor of an object *p* captures the degree *p* is deviated from its nearest neighboring cluster. It is not difficult to see that the larger the *CLOF* of *p* is, the further the point *p* is away from the boundaries of its nearest cluster. For the centroid-based boundary point detection algorithm, named CENTROID-B, the key to determine whether the data object is a boundary point is the centroid-based local outlier factor of the data object. The detection of the boundary points and outliers is related to the distance between the data object and the centroid. Most *d*-dimensional normal data points have a small deviation distance from their centroid. For the algorithm proposed in this chapter, no input parameters are required, and the only needed one is the threshold corresponding to the boundary points and the outlier as,

$$\mathrm{Threshold} = \mathrm{mean(CLOF)} + \mathrm{std(CLOF)} \tag{8.4}$$

Any data points having a CLOF larger than the threshold can be regarded as the boundary point candidates. Outliers and boundary points are different in distribution. Outliers are usually far away from other outliers, while border points tend to have other boundary points in their neighbors. Accordingly, the different characteristics of the boundary points and the outliers in the neighborhood distribution can be utilized to differentiate outliers in the boundary point candidates from the boundary points. However, the threshold value is sometimes affected by the distribution of data in the dataset. Optimizing for specific data sets can result in better boundary point detection results. According to the different distribution characteristics of the outliers and the boundary points, a portion of the outliers can be distinguished among the candidate boundary points, although it is difficult to distinguish all the outliers in the presence of the outliers. To do so, the following outlier indicator was proposed.

Definition 6 (*Outlier indicator*) Given *mpt* and $N_{mpt}(p)$ obtained for data object *p*, for each point in $N_{mpt}(p)$, such as q_i, calculate the distance dist[i] between q_i and its nearest neighbor, and the set of dist[i] ($1 \leq i \leq mpt$) is denoted as list[$N_{mpt}(p)$]. Then the outlier indicator is defined as,

$$\text{indicator(p)} = \frac{std(list[N_{mpt}(p)])}{mean(list[N_{mpt}(p)])} \tag{8.5}$$

where $std()$ represents the standard deviation and $mean$ () represents the mean of k-distances.

To distinguish the small number of outliers from the boundary point candidates, the following threshold based on the outlier indicators is defined as,

$$\text{Threshold} = \text{mean(indicators)} + \text{std(indicators)} \tag{8.6}$$

Any boundary point candidates which have an indicator value larger than the threshold can be regarded as the outliers.

8.3.2 The Proposed Boundary Point and Outlier Detection Algorithm

The boundary point detection algorithm proposed in this chapter uses the centroid-based local outlier factor to find candidate boundary points, including outliers. The outliers and boundary points are then separated from each other according to the outlier indicators and the differences in the neighborhood distributions of the boundary points and the outliers. The boundary point detection algorithm is presented in a pseudo-code format in Table 8.1. The basic steps of the algorithm are as follows:

(1) Read in all data points from the data set;
(2) Calculate *mpt* corresponding to each data point, so that the data point is the smallest distance from its centroid, and calculate the centroid corresponding to the *mpt*;
(3) Calculate the CLOF factor and outlier indicator corresponding to each data object, their mean's and standard deviation's, and determine the boundary point candidate set B by Eq. (8.4);
(4) For each point p in B, if its outlier indicator is larger than the threshold obtained from Eq. (8.6), exclude it from the boundary point candidate list;
(5) For the remaining data points in B, declare them to be the boundary points and return them as the output (Table 8.1).

8.3.3 The Complexity Analysis

From the description in the previous subsections, it can be seen that our algorithm mainly consists of three steps, an initial k-nearest neighbor search for each of N

Table 8.1 Boundary point detection algorithm

Input:	*data*: a set of N data points			
Output:	a set of boundary points and a set of outliers			
Begin:				
1:	for each point p in data {			
2:		compute k = mpt(p) and corresponding kNN and C_dist;		
3:		kNNMap.put(p,kNN); BMap.put(p, C_dist);}		
4:	thresholddb = average(BMap.values) + std(BMap.values);			
5:	for each point p in data {			
6:		if (BMap.get(p) > thresholddb)		
			CandidateSet.add(p);	
7:	for each point p in CandidateSet {			
8:		Compute $N_B(p)$;		
9:		if ($N_B(p) < 1$)		
			CandidateSet.remove(p);	
		NMap.put(p, $N_B(p)$);}		
10:	for each point p in data {			
11:		kNN = kNNMap.get(p);		
12:		for each point q in kNN {		
13:			distq = distance(q.kNNMap.get(q).get(0);	
14:			list.add(distq);	}
15:		indicator = std(list)/average(list);		
16:		iMap.put(p,indicator);}		

(continued)

Table 8.1 (continued)

17:	threshold = average(iMap.values) + std(iMap.values);			
18:	for point p in CandidateSet{			
19:		if (iMap.get(p) > threshold && NMap.get(p) < 2d-2)		
			Oset.put(p);	
20:		else		
			Bset.put(p);	
21	return Bset, Oset;			
End				

data points, the calculation of the centroid-based local outlier factors and the outlier indicators, and finally top outliers' and boundary points' mining. Therefore, most of the calculation time is spent on the first two steps, *k*-nearest neighbor based *mpt* search and the calculation of the CLOF's and the outlier indicators. As a result, the total time complexity for the average case can be $O(N^2)$ or $O(N\log N)$ if some index structure can be used. Since the number of outliers is small, the third step takes a nearly linear time on average. Physically, the resource consumed by our algorithm includes the space for holding the whole data set in memory, space to store their *k*-nearest neighbors, and some space for the CLOF's and the outlier indicators and some index structure if used.

8.4 The Proposed Fast Approximate EMST Algorithm

EMST-based applications often require the construction of an EMST in the first place. To speed up, in this section, we describe a new hybrid scheme that integrates kNN searching, the detection of a small number of longest edges, and EMST construction into one process, to enhance EMST algorithm for modern large databases. Basically, the design of this more efficient scheme is motivated by the following two observations. Firstly, for some application problems, such as clustering, an EMST consists in a major amount of in-cluster edges which have weights that are significantly smaller than those of edges corresponding to cluster breakers and can be found quickly by fast kNN search algorithms. This is because, according to Boruvka's algorithm, in the Euclidean setting, at least half of data points' edge to its nearest neighbor must be in an EMST. Secondly, for clustering-based applications, EMST algorithms can be more efficient if a small number of boundary points (including outliers) can be correctly identified and quickly retained for the efficient localization of a few longest edges which correspond to the cluster breakers.

8.4.1 Our Clustering-Inspired EMST Algorithm

Standard EMST-based clustering algorithms usually take a quadratic running time to ensure the properties of MSTs to be satisfied. From our point of view, this is neither efficient nor necessary. This observation is based on the following facts.

First, a vertex can have a limited number of nearest neighbors usually in an EMST. At any point of EMST construction, data points can be classified into two different kinds of tree nodes, those already in the tree and those not in it yet. The nodes in the tree have a limited number of edges to the nodes not in it. For example, at the beginning, the number of data points added to the tree is relatively small (compared to the whole database) and the next edge to add to the tree must be the nearest neighbor of one of the data points in the tree from the data points not in it. With the help of some index structure, the nearest neighbor of each node in the tree from the nodes not in it can be calculated efficiently.

Second, data points in a given dataset for density-based clustering purpose can be categorized into three different categories, that is, core points (i.e. inliers), boundary (or border) points, and outliers. Most of the time, the number of the boundary points and outliers is significantly less than the number of inner points. If these small amount of boundary points (including outliers) can be separated from inliers quickly, it will be more efficient on the whole to perform standard Prim's algorithm on this much size-reduced dataset to fully identify the small number of longest edges. The other edges in the EMST can be found through k-nearest neighbors search. The assumption underlying the proposed method is that boundary points and outliers are located around the margin of densely distributed data region such as a cluster. Boundary points are generally characterized by a lower density and smaller angle variance. If data clusters are connected, that is, all the clusters overlap, there is no need to perform Prim's algorithm on the identified boundary points.

Based on these findings, our clustering-inspired approximate EMST algorithm can be summarized in the following,

(1) To establish the search structure, a random data point not marked done yet is taken, and its distances to the rest data points which are not marked done either are calculated. The data point with the largest distance is selected as the reference point and the tree root node and marked done (using a first Boolean array). The distances between the reference point and the rest of dataset not marked done yet are calculated as the index. Next, the data points are sorted according to their distances to the reference point in a nondecreasing order and their original sequential positions in the dataset are remembered in the sorted order.
(2) Initialize the search radii of kNNs of a tree node and those of these kNNs' kNN's, find all the needed kNNs (the search starts from the data point and proceeds bidirectionally until its kNN are identified) and mark it done (using the first Boolean array), and calculate the corresponding kNN-centroid-based local outlier score of this tree node.
(3) Add the edge corresponding to the nearest neighbor of the tree node and the data point at the other end of the edge to the tree; if all the nearest neighbors of the data points currently in tree are added to the tree, we can declare the end of a detected cluster and the start of another cluster, and goto (1), else if some data points are not marked done yet, search through the distance array for the edge with the smallest distance to the tree and remember the vertex at the other end of this edge, select this vertex as the tree node candidate, and goto (2).
(4) If all the data points are marked done (using the first Boolean array), calculate the threshold given in Eq. (8.4), apply standard Prim's algorithm to those data points whose scores are larger than the threshold to locate longest edges.
(5) Finally return an approximate MST represented by the parent array to remember the tree edges and the distance array to remember the edge weights.

Physically, the resource consumed by our algorithm includes the space for the whole data set to reside in memory, some temporary space to store the nearest neighbors of some data items' for calculating the short edges in the EMST, the space for distance and parent arrays, and the space for the indexing structure. Unlike traditional iDistance [43], the selection of reference points in our approach is dynamically adapted to the data being analyzed [44]. This data dependency hopes to produce improved performance over traditional iDistance transforms. To summarize, the numerical parameters our algorithm needs from the user include the data set, while the output includes an approximate MST represented by the parent array and the distance array.

8.4.2 *Time Complexity Analysis*

From the description in the previous subsections, our final clustering-inspired EMST algorithm includes a running time to estimate the centroid-based local outlier factors for all data points, an O(NlogN) for calculating short edges in cluster-wise EMST and an $O(M^2)$ (M«N) Prim's algorithm where M is the number of data points whose centroid-based local outlier scores are larger than the threshold at the end of the first stage. We expect our proposed efficient approximate EMST algorithm to run with a $O(LQ\log Q + M^2)$ time complexity, where L denotes the number of clusters and Q denotes the number of data points in the largest cluster detected at the first stage, where an O(QlogQ) is the index-based search for calculating intra-cluster EMST edge for each data point in the largest cluster since an approximate Q number of data points involved in the search structure. The worst-case time complexity of our proposed final approximate EMST algorithm could degenerate to $O(N^2)$, particularly when a data set contains quite a large number of well-separated clusters.

8.5 Experiments and Results

8.5.1 *Performance Evaluation of the Proposed Boundary Point Detection Algorithm*

In this subsection, we prove the effectiveness of the proposed algorithm for various boundary point detection tasks with different features through experimental results and analysis. First, several two-dimensional datasets are employed to visually prove that the proposed algorithm is effective. Further, based on the assumption that a better boundary point detection algorithm can make clustering algorithms perform better, more experiments are performed to K-means cluster several multi-dimensional datasets and the results are compared with several state-of-the-art boundary point

Table 8.2 Summary of data sets

Dataset Name	# of Points	Dimension	# of Clusters
FLAME	240	2	2
GAUSSIAN	2000	2	2
XOR	2000	2	4
t4	8000	2	6
t5	8000	2	6
t7	10,000	2	9
t8	8000	2	8
Wine	178	13	3
Abalone	4177	9	3
WDBC	569	32	2

detection algorithms. An overview of the set of data used in this set of experiments is given in Table 8.2.

All the algorithms are implemented in Java and run on a computer with Intel i5-6300HQ CPU and 2 GB RAM. The operating system running on this computer is Windows 10. The results show that, overall, our proposed *k*NN-centroid-based boundary point detection algorithm is competent with state-of-the-art boundary detection algorithms in classification accuracy.

8.5.1.1 Performance on Two-Dimensional Data

In this subsection, several two-dimensional synthetic data sets are used to show that the proposed method can effectively identify boundary points and outliers in various scenarios. A particular challenging feature of these datasets is that clusters are of different sizes and have complex shapes and different densities. We exclusively use two-dimensional data sets due to their convenience for visual inspection. The relative effects of our proposed boundary point detection algorithm are studied in comparison with BPDAD [45], IDD [46], and BEPS [47] methods.

The FLAME dataset is a two-dimensional dataset containing 240 data points. The two points A and B are outliers and treated as special boundary points in most boundary point detection algorithms. The results of several boundary point detection algorithms are plotted in Fig. 8.2, with the blue dots representing the in-cluster points, the red dots representing the boundary points, and the green dots representing the outliers. Figure 8.2a is the result of the proposed CENTROID-B algorithm, and (b–d) are the results of the BPDAD, IDD, and BEPS algorithms, respectively. It can be clearly seen from the figure that the CENTROID-B algorithm achieves the best results, not only to identify the boundary points relatively more accurately, but also to separate outliers A and B from the boundary points. For the rest of competing algorithms, BPDAD performs better than the other two methods and clearly identifies

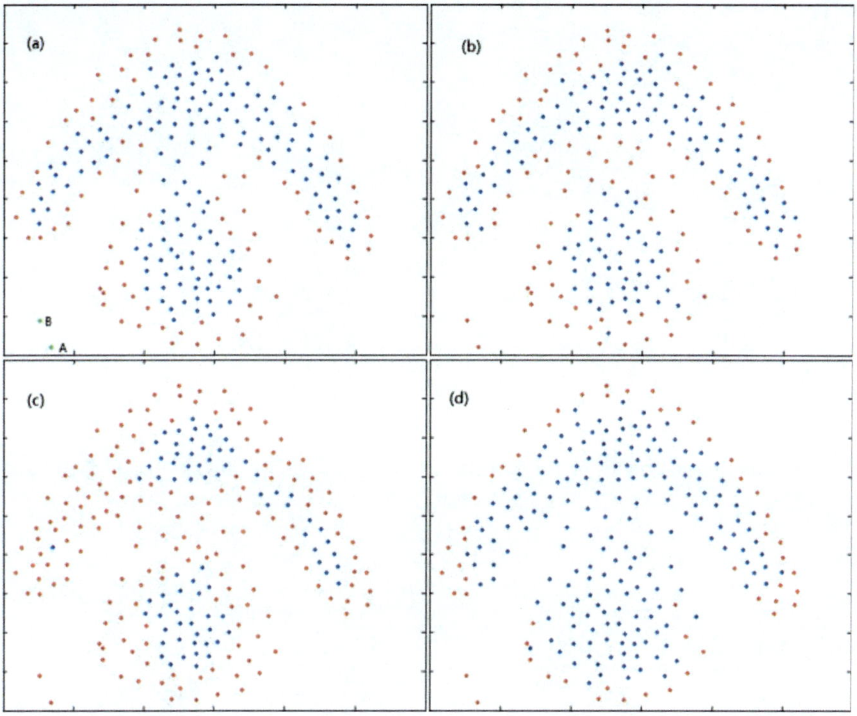

Fig. 8.2 Experimental results on the FLAME data set by the **(a)** CENTROID-B, **(b)** BPDAD, **(c)** IDD, **(d)** BEPS algorithms

most boundary points. However, both BPDAD and IDD misclassify many in-cluster points. On the contrary, BEPS misses many boundary points.

The GAUSSIAN dataset is a two-dimensional dataset containing 2000 data objects and consists of two clusters that follow a Gaussian distribution [45]. The relative effects of our proposed boundary point detection algorithm are studied in comparison with the BPDAD, IDD, and BEPS methods. The results from the two best performing algorithms are presented in Fig. 8.3 in which the blue dots represent the in-cluster points, the red dots represent the boundary points, and the green dots represent the outliers. From the figure, it can be seen that, in comparison to the BPDAD algorithm, the CENTROID-B algorithm is not only less likely to misclassify in-cluster points but also can simultaneously distinguish between outliers (the green dots) and boundary points (the red dots).

The XOR GAUSSIAN dataset is a two-dimensional dataset containing 2000 data objects and consists of four data clusters that are Gaussian-distributed [45]. The results from the two best performing algorithms are presented in Fig. 8.4 in which the blue dots represents the in-cluster points, the red dots represents the boundary points, and the green dots represents the outliers. From the figure, it can be seen that, in terms of the performance comparison, the CENTROID-B algorithm is almost the same

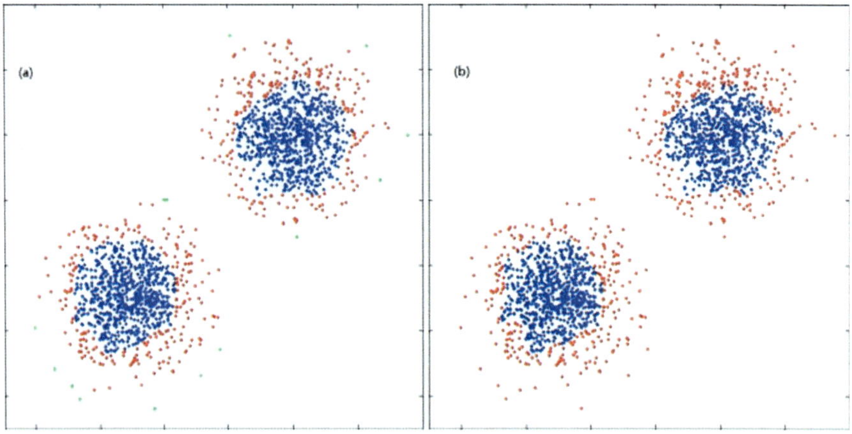

Fig. 8.3 Experimental results on the GAUSSIAN dataset by the **(a)** CENTROID-B **(b)** BPDAD algorithms

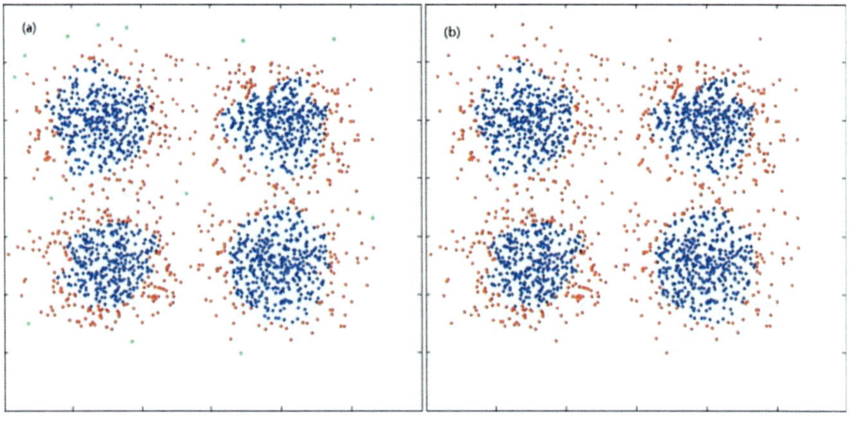

Fig. 8.4 Experimental results on the XOR data set by the (**a**) CENTROID-B (**b**) IDD algorithms

as the IDD algorithm when detecting the boundary points, but CENTROID-B can distinguish the outlier points from the boundary points while performing the detection (Fig. 8.4).

To further test the performances of the proposed boundary detection algorithm, the detection results of the CENTRIOD-B algorithm on four two-dimensional datasets, T4, T5, T7, and T8, are shown in Fig. 8.5 in which the blue dots represent the in-cluster points, the red dots represent the boundary points, and the green dots represent the outliers. As shown in Fig. 8.5, the T4 dataset consists of six clusters with different shapes and sizes. The T5 dataset contains 8000 two-dimensional data objects, which has six clusters of different shapes and sizes surrounded by a lot of noise. T7 is a data set containing 10,000 two-dimensional data objects which constitutes nine clusters

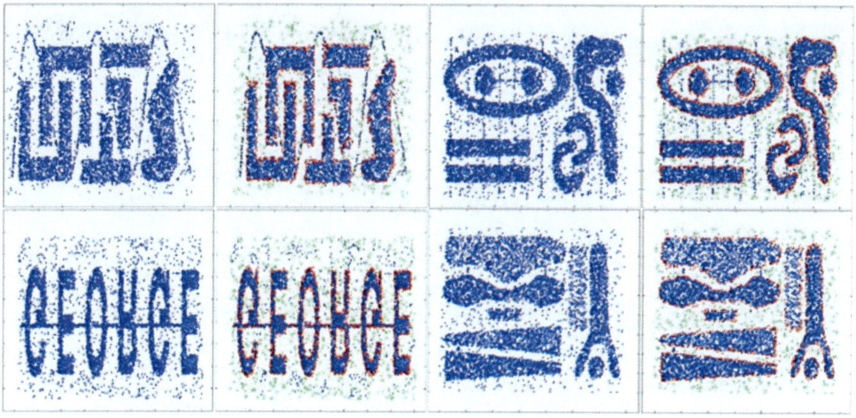

Fig. 8.5 Detection results of CENTROID-B on (upper left) T4, (lower left) T5, (upper right) T7 and (lower right) T8

of different shapes and distributions and also contains a large amount of noise. T8 is a data set containing 8000 two-dimensional data objects, which constitute 8 clusters of different shapes and sizes surrounded by a lot of noise. There also exist interference lines composed of noise across the clusters. These datasets are extensively used in the study of boundary point detection. From the figure, it can be seen that the CENTROID-B algorithm can not only eliminate the influence of the interference line, but also distinguish outliers from the clearly identified cluster boundaries. Although there are some misclassifications, the CENTROID-B algorithm has achieved the best detection results compared with other algorithms.

8.5.1.2 Clustering Performance

In this subsection, the effectiveness of the proposed algorithm is demonstrated by experiments on multi-dimensional data sets. As pointed out by Li et al. one way to test how well a boundary point detection algorithm works is to run the method on the dataset and test the clustering performance by removing the detected boundary points and outliers [6]. Therefore, our experiment on multi-dimensional data is to compare the clustering effect obtained by the proposed algorithm and state-of-the-art boundary point detection algorithms. The effect of clustering is evaluated by the precision which is defined to be TP/(TP + FP), where TP is true positive and FP is false positive.

The Wine dataset, downloaded from the UCI machine learning repository, is a 13-dimensional dataset containing 178 data objects. As can be seen from Table 8.3, the precision of the direct result of the K-means clustering algorithm on the Wine dataset is 70.22%. When the BEPS algorithm is used to remove the boundary points, the clustering precision is reduced to 68.89%. By using the BPDAD and

Table 8.3 Clustering precision achieved using each method on Wine data

Method	# of boundary points removed	Cluster	Precision (%)
/	/	K-means	70.22
BEPS	43	K-means	68.89
BPDAD	98	K-means	73.49
CENTROID-B	97	K-means	74.07

Table 8.4 Clustering precision achieved using each method on Abalone data

Method	# of boundary points removed	Cluster	Precision (%)
/	/	K-means	81.95
BEPS	1539	K-means	42.24
BPDAD	245	K-means	50.09
CENTROID-B	1524	K-means	53.37

CENTROID-B algorithms to remove the boundary points, the clustering performances are improved, reaching 73.49% and 74.07%, respectively. Therefore, it can be said that the CENTROID-B algorithm achieves the best boundary point detection results on the Wine dataset (Table 8.3).

The Abalone dataset is a nine-dimensional dataset containing 4177 data objects. There is a large overlap in the data points of this dataset. This dataset is divided into three categories based on the attribute of gender. Table 8.4 shows the precision of K-means clustering after removing boundary points detected by different boundary point detection algorithms. As can be seen from Table 8.4, the precision of *K*-means algorithm on the Abalone dataset is 81.95%. But the precision of clustering is reduced after the boundary points are removed using the boundary point detection algorithms. Nevertheless, the precision of clustering after removing the boundary points using the CENTOID-B algorithm is better than BEPS and BPDAD. Therefore, the CENTROID-B algorithm achieves the best boundary point detection results on the Abalone data set.

The Wisconsin diagnostic breast cancer (WDBC) data set is a 32-dimensional data set containing 569 data objects of two categories depending on whether the tumor is malignant or benign. The properties of the data objects are calculated digitally based on the images extracted from the fine needles of the breast mass. As can be seen from Table 8.5, the precision of the *K*-means clustering algorithm on the WDBC data set is 62.57%. When the boundary points are removed using the BEPS algorithm, the clustering precision is reduced to 55.56%. After using the BPDAD and CENTROID-B algorithms to remove the boundary points, the clustering precision has been improved, reaching 66.14% and 66.27%, respectively. The result of CENTROID-B is only slightly better than that of BPDAD, but it can be said that the CENTROID-B algorithm has achieved the best boundary point detection results on the WDBC dataset.

Table 8.5 Clustering precision achieved using each method on WDBC data

Method	Removed boundary points	Cluster	Precision (%)
		K-means	62.57
BEPS	182	K-means	55.56
BPDAD	253	K-means	66.14
CENTROID-B	231	K-means	66.27

8.5.2 Performance Evaluation of the Fast Approximate EMST Algorithm

In this subsection, two sets of experiments are conducted to evaluate the proposed approximate EMST algorithm. In the first set of experiments, we evaluate the performance of our proposed EMST algorithm on three two-dimensional synthetic datasets to investigate the influences of kNN-centroid-based local outlier scores on the detection of boundary points and outliers. The performance is also compared with those of two other outlier definitions, INFLO and RBDA, to show the possible impact of different outlier definitions on the detection of boundary points. With some sensible knowledge of the impact of kNN-centroid-based local outlier scores on the detection of boundary and outlying points, in the second set of experiments, we evaluate our AEMST-inspired algorithm on two real multi-dimensional data sets without any assumptions on the data distribution, and compare the overall performance with the state-of-the-art approximate EMST algorithms to check the technical robustness of this study. All the data sets used in this set of experiments are briefly summarized in Table 8.6. The results show that, overall, our clustering-inspired AEMST algorithm is superior to other state-of-the-art AEMST algorithms.

We implemented all the algorithms in C++. All the experiments were performed on a computer with Intel Core 2 Duo Processor E6550 2.33 GHz CPU and 2 GB RAM. The operating system running on this computer was Windows XP. We used the timer utilities defined in the C standard library to report the CPU time.

Table 8.6 Sets of data

Data set	Data size (N)	Dimensionality	# of clusters
Data1	6,142	2	6
Data2	7,634	2	9
Data3	6,566	2	8
UJIndoorLocT	19,937	529	/
SliceLocalization	53,500	386	/

8.5.2.1 Performance on Small Two-Dimensional Data

To find an EMST of a data set using our algorithm, we need to specify the threshold value of kNN-centroid-based local outlier scores for the detection of boundary points (including outliers). Therefore, we first focus our study on the behavior of the kNN-centroid-based local outlier scores on the performance of our algorithm. The performance of the proposed method is evaluated on three CHAMELEON datasets obtained from [48] and consisting of clusters of different sizes, shapes, and orientations. The geometric shapes of three datasets are displayed in the plots on the first row of Fig. 8.6 together with the corresponding EMSTs.

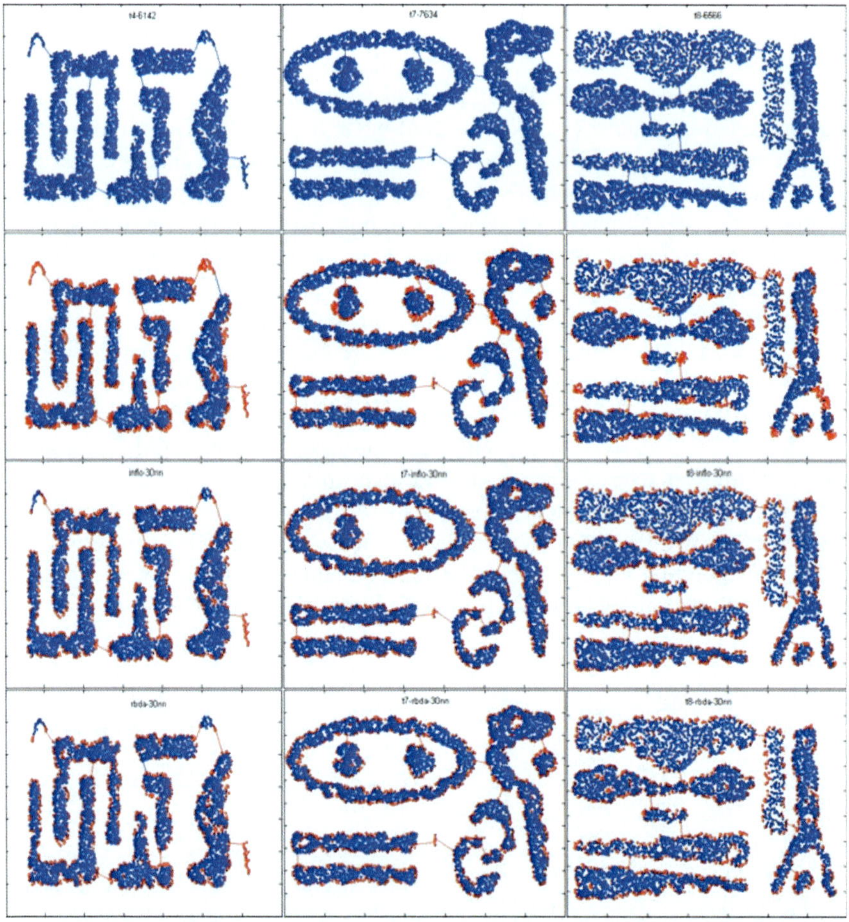

Fig. 8.6 (first row) Three original datasets, (second row) the thresholded CLOF values, (third row) the thresholded INFLO values, (fourth row) the thresholded RBDA values, reprinted from ref. [50], copyright 2018, with permission from Springer

Table 8.7 Running time performance in seconds

Data Name	MST(BF)	FAEMST	Ours
UJIndoorLocT	888	462	171
SliceLocalization	3,907	864	421

Shown in the last three rows of Fig. 8.6 are plots of values of CLOF, INFLO, and RBDA scores for each point when the number of nearest neighbors, k, is set to be 30, respectively. The red dots denote those data points whose outlier score is larger than the threshold calculate from Eq. (8.4). The blue ones denote those data points whose outlier score is less than or equal to the calculated threshold. It can be seen from the figures that overall suitable score values from our method can identify boundary points (including outliers) more accurately than two other methods. To summarize, using CLOF scores can very easily identify the boundary points (including outliers). For the datasets in this experiment, we observe that an estimated lower bound number of data points to retain to identify the longest edges is a fifth of the original data size.

8.5.2.2 Performance on Multi-Dimensional Real Data

In this subsection, we evaluate the efficiency of our algorithm versus three other EMST algorithms on two real high-dimensional data sets, UJIndoorLocT and Slice-Localization, which are downloaded from the UCI KDD Archive [49]. The UJIndoor-LocT dataset contains 19,937 instances with 529 attributes. The SliceLocalization dataset contains 53,500 instances with 386 attributes.

The running time performances of our algorithm (i.e. Ours), FAEMST algorithm, and the Prim's algorithm (i.e. MST(BF)) are summarized in Table 8.7. From the table, we can see that our algorithm outperforms the other three algorithms. It is faster than Prim's algorithm (i.e. the MST(BF) algorithm in the table) by a factor of 5 in running time performance, and faster than the FAEMST algorithm by a factor of 2.

8.6 Conclusions

In this chapter, a local kNN-centroid-based boundary point detection algorithm is proposed to solve the problems that the existing boundary point detection algorithms rely on input parameters, cannot separate outliers and so on. The algorithm calculates the centroid of each data point's kNN, calculates the distance between the point and the centroid, and takes those points which are far from the centroid as the candidate boundary points. Then the outlier indicators are used to find the outliers contained in the boundary points. Experiments on two-dimensional and multi-dimensional data sets prove that the algorithm can detect outliers while performing boundary point detection. It also obtains high boundary point detection precision on datasets with arbitrary shapes, densities, and interferences from noise.

EMST algorithms are of growing importance in many application domains. A central problem in such algorithms for modern large multi-dimensional real datasets is the classic quadratic time complexity on the construction of an EMST. In this chapter, we present a more efficient method that can quickly identify the longest edges in an EMST so as to save some computations. We conduct experiments to evaluate our algorithm against the Prim's algorithm and one other state-of-the-art EMST algorithm on two UCI data sets. The experimental results show that our proposed EMST algorithm is very effective and stable.

Acknowledgements This chapter was modified from the paper published by our group in "machine learning and data mining" [50]. The related contents are reused with permission.

References

1. Hawkins, D. M. (1980). *Identification of outliers*. London: Chapman and Hall.
2. Ester, M., Kriegel, H.-P., Sander, J. and Xu, X. (1996). A density-based algorithm for discovering clusters in large spatial databases with noise. In *Proceedings of the 2nd ACM SIGKDD International Conference on Knowledge Discovery and Data Mining* (KDD'96), Portland, Oregon, USA, pp. 226–231.
3. Breuning, M. M., Kriegel, H. P., Ng, R. T., & Sander, J. (2000). LOF: Identifying density-based local outliers. In *Proceedings of the 2000 ACM SIGMOD International Conference on Management of Data* (SIGMOD'00), Dallas, TX, United states, pp. 93–104.
4. Qiu, B. Z., Yue, F., & Shen, J. Y. (2007). BRIM: an efficient boundary points detecting algorithm. In *Proceedings of the 11th Pacific-Asia Conference on Knowledge Discovery and Data Mining* (PAKDD'07), Nanjing, China, pp. 761–768.
5. He, Y. Z., Wang, C. H., & Qiu, B. Z. (2013). Clustering boundary points detection algorithm based on gradient binarization. *Applied Mechanics and Materials*, 263–266(n PART 1): 2358–2363.
6. Li, X., Wu, X., Lv, J., He, J., Guo, J., & Li, M. (2018). Automatic detection of boundary points based on local geometrical measures. *Soft Computing, 22*(11), 3663–3674.
7. Kruskal, J. B. (1956). On the shortest spanning subtree of a graph and the traveling salesman problem. *Proceedings of the American Mathematical Society, 7*(1), 48–50.
8. Prim, R. C. (1957). Shortest connection networks and some generalizations. *Bell System Technical Journal, 36*(6), 1389–1401.
9. An, L., Xiang, Q. S., & Chavez, S. (2000). A fast implementation of the minimum spanning tree method for phase unwrapping. *IEEE Transactions on Medical Imaging, 19*(8), 805–808.
10. Xu, Y., & Uberbacher, E. C. (1997). 2D image segmentation using minimum spanning trees. *Image and Vision Computing, 15*(1), 47–57.
11. Zahn, C. T. (1971). Graph-theoretical methods for detecting and describing gestalt clusters. *IEEE Transactions on Computers, C-20*(1), 68–86.
12. Xu, Y., Olman, V., & Xu, D. (2002). Clustering gene expression data using a graph-theoretic approach: An application of minimum spanning trees. *Bioinformatics, 18*(4), 536–545.
13. Zhong, C., Miao, D., & Wang, R. (2010). A graph-theoretical clustering method based on two rounds of minimum spanning trees. *Pattern Recognition, 43*(3), 752–766.
14. Juszczak, P., Tax, D. M. J., Pekalska, E., & Duin, R. P. W. (2009). Minimum spanning tree based one-class classifier. *Neurocomputing, 72,* 1859–1869.

15. Yang, C. L. (2005). Building k edge-disjoint spanning trees of minimum total length for isometric data embedding. *IEEE Transactions Pattern Analysis and Machine Intelligence, 27*(10), 1680–1683.
16. Gower, J. C., & Ross, G. J. S. (1969). Minimum spanning trees and single linkage cluster analysis. *Applied Statistics, 18*(1), 54–64.
17. Balcan, M-F., Blum, A., & Vempala, S. (2008). A discriminative framework for clustering via similarity functions. In *Proceedings of the 40th Annual ACM Symposium on Theory of Computing* (STOC'08), Victoria, BC, Canada, pp. 671–680.
18. Tang, J., Chen, Z., Fu, A. W. C., & Cheung, D. W. (2002). Enhancing effectiveness of outlier detections for low density patterns. In *Proceedings of the 6th Pacific-Asia Conference on Knowledge Discovery and Data Mining (PAKDD'02)*, Taipei, Taiwan, pp. 535–548.
19. Gibbons, P.B., Papadimitriou, S., Kitagawa, H. and Faloutsos, C. (2003). LOCI: Fast outlier detection using the local correlation integral. In *Proceedings of the IEEE 19th International Conference on Data Engineering (ICDE'03)*, Bangalore, India, pp.315–326.
20. Sun, P. and Chawla, S. (2004). On local spatial outliers. In *Proceedings of the 4th IEEE International Conference on Data Mining (ICDM'04)*, Brighton, UK, pp. 209–216.
21. Jin, W., Tung, A. K. H., Han, J., & Wang, W. (2006). Ranking outliers using symmetric neighborhood relationship. In *Proceedings of the 10th Pacific-Asia Conference on Knowledge Discovery and Data Mining (PAKDD'06)*, Singapore, pp. 577–593.
22. Fan, H., Zaiane, O. R., Foss, A., & Wu, J. (2006). A nonparametric outlier detection for efficiently discovering top-N outliers from engineering data. In *Proceedings of the 10th Pacific-Asia Conference on Advances in Knowledge Discovery and Data Mining* (PAKDD'06), Singapore, Singapore, pp. 557–566.
23. Zhang, K., Hutter, M., & Jin, H. (2009). A new local distance-based outlier detection approach for scattered real-world data. In *Proceedings of the 13th Pacific-Asia Conference on Knowledge Discovery and Data Mining (PAKDD'09)*, pp. 813–822.
24. Huang, H., Mehrotra, K., & Mohan, C. K. (2013). Rank-based outlier detection. *Journal of Statistical Computation and Simulation, 83*(3), 518–531.
25. Schubert, E., Zimek, A., & Kriegel, H. P. (2014). Generalized outlier detection with flexible kernel density estimates. In *Proceedings of the 14th Siam International Conference on Data Mining* (SDM'14), Philadelphia, pp. 542–550.
26. Ru, X., Liu, Z., Huang, Z., et al. (2016). Normalized residual-based constant false-alarm rate outlier detection. *Pattern Recognition Letters, 69,* 1–7.
27. Tang, B., & He, H. (2017). A local density-based approach for outlier detection. *Neurocomputing, 241,* 171–180.
28. Kriegel, H.-P., Schubert, M., & Zimek, A. (2008). Angle-based outlier detection in high-dimensional data. In *Proceedings of the 14th ACM SIGKDD International Conference on Knowledge Discovery and Data Mining* (KDD'08), Las Vegas, Nevada, USA, pp. 444–452.
29. Pham, N., & Pagh, R. (2012). A near-linear time approximation algorithm for angle-based outlier detection in high-dimensional data. In *Proceedings of the 18th ACM SIGKDD Conference on Knowledge Discovery and Data Mining* (SIGKDD'12), Beijing, China, pp. 877–885.
30. Borůvka, O. (1926). O jistém problému minimálním (About a Certain Minimal Problem). *Práce moravské přírodovědecké společnosti v Brně,* III 37–58 (in Czech with German summary).
31. Jarník, V. (1930). O jistém problému minimálním (About a Certain Minimal Problem). *Práce moravské přírodovědecké společnosti v Brně,* VI 57–63 (in Czech).
32. Dijkstra, E. W. (1959). A note on two problems in connexion with graphs. *Numerische Mathematik, 1*(1), 269–271.
33. Bentley, J., & Friedman, J. (1978). Fast algorithms for constructing minimal spanning trees in coordinate spaces. *IEEE Transactions on Computers, C-27*(2), 97–105.
34. Preparata, F. P., & Shamos, M. I. (1985). *Computational Geometry*. New York: Springer.
35. Callahan, P., & Kosaraju, S. (1993). Faster algorithms for some geometric graph problems in higher dimensions. In *Proceedings of the 4th Annual ACM-SIAM Symposium on Discrete Algorithms,* Austin, TX, USA, pp. 291–300.

36. Narasimhan, G., Zachariasen, M., & Zhu, J. (2000). Experiments with computing geometric minimum spanning trees. In *Proceedings of the 2nd Workshop on Algorithm Engineering and Experimentation* (ALENEX'00), pp. 183–196.
37. March, W. B., Ram, P., & Gray, A. G. (2010). Fast Euclidean minimum spanning tree: algorithm, analysis, and applications. In *Proceedings of 16th ACM SIGKDD International Conference on Knowledge Discovery and Data Mining* (KDD'10), Washington, USA, pp. 603–612.
38. Vaidya, P. M. (1988). Minimum spanning trees in k-dimensional space. *SIAM Journal on Computing, 17*(3), 572–582.
39. Wang, X., Wang, X., & Wilkes, D. M. (2009). A divide-and-conquer approach for minimum spanning tree-based clustering. *IEEE Transactions Knowledge and Data Engineering, 21*(7), 945–958.
40. Lai, C., Rafa, T., & Nelson, D. E. (2009). Approximate minimum spanning tree clustering in high-dimensional space. *Intelligent Data Analysis, 13*(4), 575–597.
41. Wang, X., Wang, X. L., & Zhu, J. (2014). A new fast minimum spanning tree based clustering technique. In *Proceedings of the 2014 IEEE International Workshop on Scalable Data Analytics (ICDMW'14)*, Shenzhen, China, pp. 1053–1060.
42. Zhong, C., Malinen, M., Miao, D., & Fränti, P. (2015). A fast minimum spanning tree algorithm based on K-means. *Information Sciences, 295,* 1–17.
43. Yu, C., Ooi, B. C., Tan, K. L., & Jagadish, H. V. (2005). iDistance: An adaptive B+-tree based indexing method for nearest neighbor search. *ACM Transactions on Data Base Systems (TODS), 30*(2), 364–397.
44. Wang, X., Wang, X. L., & Wilkes, D. M. (2012). Modifying iDistance for a fast CHAMELEON with application to patch based image segmentation. In *Proceedings of the 9th IASTED International Conference on Signal Processing, Pattern Recognition and Applications*, Crete, Greece, pp. 107–114.
45. Li, Y., & Maguire, L. (2011). Selecting critical patterns based on local geometrical and statistical information. *IEEE Transactions on Pattern Analysis and Machine Intelligence, 33*(6), 1189–1201.
46. Li, X., Lv, J., & Yi, Z. (2018). An efficient representation-based method for boundary point and outlier detection. *IEEE Transactions on Neural Networks and Learning Systems, 29*(1), 51–62.
47. Rodriguez, A., & Laio, A. (2014). Clustering by fast search and find of density peaks. *Science, 344*(6191), 1492.
48. Karypis, G., Han, E. H., & Kumar, V. (1999). CHAMELEON: A hierarchical clustering algorithm using dynamic modeling. *IEEE Transactions on Computers, 32*(8), 68–75.
49. https://archive.ics.uci.edu/ml/datasets.html.
50. Wang, X. L., Wang, X., & Li, X. (2018). A fast two-level approximate Euclidean minimum spanning tree algorithm for high-dimensional data. In *Proceedings of the 14th International Conference on Machine Learning and Data Mining*, New York, USA, pp. 273–287.

Chapter 9
A Nearest Neighbor Classifier-Based Automated On-Line Novel Visual Percept Detection Method

Abstract The *k*-nearest neighbor classifier is a very effective and simple nonparametric technique in pattern classification. However, it only can classify new data items into existing categories, but not the data items coming from any new categories that have not been identified beforehand. Also, its classification performance is highly influenced by the efficiency of the *k*-nearest neighbors search structure employed. In this chapter, we present a fast approximate nearest neighbor search tree-based novelty filter for the multiple percept detection and incremental learning tasks in image sequences. We begin with an introduction to the concept of novelty detection in general and an image patch-based perceptual learning system as a basis for visual novelty detection in specific. The proposed on-line novel visual percept detection method is next presented. Finally, the performance of the proposed filter is compared with that of the well-known Grow-When-Required neural network approach for a novelty detection task in an indoor environment and with that of efficient support vector data description method for a novelty detection task in an outdoor environment.

Keywords *k*-nearest neighbors classifier · Approximate nearest neighbor search · Quantization tree · Image segmentation · Perceptual grouping · Novelty detection

9.1 Introduction

k-nearest neighbor algorithm is a traditional nonparametric and most famous technique among machine learning algorithms [1–4]. An instance-based k-nearest neighbors (*k*NN) classifier works by first locating the *k*-nearest neighbors of an instance in an instance space and then using a majority voting strategy to label the unknown instance with the located nearest neighbors. Due to its simple concept and implementation and its excellent asymptotic classification performance in Bayes sense with minimal error rates [5, 6], *k*NN-based classifiers have a wide range of applications in the field of pattern recognition [7, 8], medical imaging [9, 10], electronic intelligence systems [11], classification and identification of radar signal [12], fractal analysis [13], SAR technologies [14], image processing [15], remote sensing [16, 17], and biomedical research [18–21]. Although *k*NN-based classification has several

X. Wang et al., *New Developments in Unsupervised Outlier Detection*,
https://doi.org/10.1007/978-981-15-9519-6_9

benefits, there are still some issues to be resolved. For one example, the instance-based k-nearest neighbor classifier can only classify new data items into existing categories, but not the data items coming from the new categories that have not been identified by the system beforehand. For another example, its classification performance is highly influenced by the efficiency of the k-nearest neighbor search structure employed. Fortunately, novelty detection techniques based on an instance-based single nearest neighbor classifier can be employed for incremental learning in such tasks. Novelty detection can be defined as the task of classifying test data that differ in some respect from the data that are available during training. Being one of the fundamental requirements of a good classification or identification system, novelty detection techniques can identify new or unknown data or signals that a machine learning system is not aware of at the time of training the model. Novelty detection has gained much research attention in many application domains, including the detection of mass-like structures in mammograms [22] and other medical diagnostic problems [23, 24], faults and failure detection in complex industrial systems [25], structural damage [26], intrusions in electronic security systems, credit card or mobile phone fraud detection [27, 28], video surveillance [29, 30], mobile robotics [31, 32], sensor networks [33], astronomy catalogs [34, 35], and text mining [36]. In mobile robotic and video surveillance applications, novelty detection has been extensively employed to recognizing novel objects in images and video streams. It is deemed a very important task and gaining attention because of the availability of large amounts of video data and because of the lack of automated methods for extracting important details from such media.

In the novelty detection problems, a set of data indicative of normal system behaviors is given and the goal is to learn a description of normality by constructing a model of it, $M(\theta)$ (where θ represents the free parameters of the model), from the data under the assumption that the data used to train the learning system constitute the basis to build a model of normality. A relatively few number of previously unseen patterns x are then compared with the model of normality and each often assigned a novelty score $z(x)$, which may or may not be probabilistic. Large novelty scores $z(x)$ should correspond to increased 'abnormality' with respect to the model of normality. Deviations from normality are then detected according to a decision boundary that is usually referred to as the novelty threshold $z(x) = T$. By comparing the scores to the novelty threshold, those test data for which the threshold is exceeded are then deemed to be 'abnormal'. In other words, a novelty threshold $z(x) = T$ is defined such that x is classified as 'normal' if $z(x) < T$, or 'abnormal' otherwise. Different types of models for constructing M, methods for setting their parameters θ, and methods for determining novelty thresholds T have been proposed in the literature. Markou and Singh distinguished between two main categories of novelty detection techniques: statistical approaches [37] and neural network-based approaches [38]. In statistical approaches, the statistical properties of data are used to estimate whether a new test data point comes from the same distribution as the training data points or not. Generally, there are two main statistical novelty detection techniques, parametric modeling, which assumes that the data come from a family of known distributions, and nonparametric modeling, in which no assumption on the form of the data distribution is made. The former is somewhat efficient in practice, but may often be limited by the assumptions about the distributions, while the latter is more flexible but often more computationally expensive, especially for large data sets. Neural network-based approaches come from a wide range of flexible nonlinear regression and classification models, data reduction models, and nonlinear dynamical models that make no a priori assumptions on the properties of data, demand a relatively small number of parameters for optimization, and therefore, have been extensively used for novelty detection [39, 40]. However, neural networks have increased computational complexity in training and typically cannot be retrained to learn new units with computational efficiency in both time and space comparable to that of statistical models, particularly when the dimensionality of the data is high.

Early work in neural network-based on-line novelty detection for mobile robots was conducted by Marsland et al. through a Grow-When-Required (GWR) neural network [41]. The GWR network is a self-organizing learning mechanism which is able to determine whether an input is novel or not through the use of a model of habituation. Capable of continuous learning, this unsupervised approach was used in mobile robots having sonar readings [42]. In 2004, this GWR network was successfully applied to visual novelty detection based on local color histograms [43, 44]. To scale to high-dimensional visual data (such as visual inputs acquired from a moving platform), more recently in 2007, Neto and Nehmzow proposed to use an incremental principle component analysis to reduce the dimensionality of input to the novelty filter while trying to preserve discriminability between different classes of features as much as possible [45]. Their approach combined both unsupervised continuous learning and color vision with no explicit image segmentation or restrictions in field of view. Being another popular neural network-based approach, support vector data description (SVDD) [46] uses a minimum volume sphere to enclose all or most of the data, either in the feature space directly or by replacing the dot products between patterns with the corresponding kernel functions. The novelty detection is performed by using the obtained boundary as the decision boundary to determine whether a test sample is novel or not. Although it is specially designed for handling large datasets, the computation cost scales linearly with the number of support vectors. To improve the prediction speed, a fast SVDD (F-SVDD) method was proposed in 2010 [47] to adopt a novel pre-image method to approximate the sphere center of SVDD as a single vector instead of a linear combination of support vectors. Noticing that the decision hypersphere of F-SVDD becomes a sphere in the input space using the given Gaussian kernel, which is not suitable for many real-world cases, Peng and Xu proposed an efficient SVDD (E-SVDD) [48]. Based on the observation that most real datasets contain at most a few clusters, E-SVDD assigns each cluster a unique pre-image and then uses these pre-images to represent the sphere center of SVDD, improving the prediction performance of SVDD while maintaining the computation efficiency.

Different metrics are used to evaluate the effectiveness and efficiency of novelty detection methods. The effectiveness of novelty detection techniques can be evaluated according to how many novel data points are correctly identified as well as according to how many normal data are incorrectly classified as novel data. The latter is also known as the false alarm rate. Novelty detection techniques should aim to have a high detection rate while keeping the false alarm rate low. Generally speaking, it is in this sense that the novelty threshold is regarded as one of the most critical parameters in novelty detection tasks and directly affects the false negative rate. The efficiency of novelty detection approaches is evaluated according to both time and space complexity. In addition, depending on the specific novelty detection task, the amount of memory required to implement the technique is typically considered to be an important performance evaluation metric. Efficient novelty detection techniques should be scalable to large and high-dimensional datasets.

In this chapter, we present a fast approximate nearest neighbor search tree-based novelty filter for mobile robotic and video surveillance applications. First, it identifies novel samples using statistically obtained thresholds. Next, these samples are accumulated and evaluated whether they represent truly novel patterns and form dense clusters in both the feature space and the image space. By this way, the novelty detection problem becomes the identification of novel percepts in the segmented images. Finally, the proposed on-line novelty detection mechanism completes the learning process by inserting the verified novel samples into the search tree of image vocabulary percepts. One important contribution of the proposed method is its significant efficiency in the tree construction over vocabulary tree (VT) [49] for image patch-based highly sparse high-dimensional data while maintaining reasonably good search capability. Another important contribution of the proposed method is the automated strategy for setting the novelty threshold. Finally, the proposed novelty detection strategy incorporates a capability for automatically learning multiple newly discovered objects by using spatial information to cluster the detected novel samples into different categories and by inserting them into the tree on-line to allow for fast incremental updating of the knowledge base. To demonstrate the soundness of this manuscript, the performance of the proposed filter is compared with that of the well-known GWR neural network approach for a novelty detection task in an indoor environment and with that of the E-SVDD method for a novelty detection task in an outdoor environment. The ultimate goal of the research presented here is to relieve human operators by equipping autonomous robots with such automatic inspection capability.

9.2 A Percept Learning System

Novel object detection is an important application of novelty detection in both robotics and computer vision. Due to the lack of examples from the novel targets, model selection in such scenarios is a very challenging research topic [50] and requires a synthesis of techniques from computer vision, machine learning, and pattern recognition. In the following, an image patch-based perceptual learning system is first described.

A primary goal of the human perceptual system is to recognize and locate objects and to keep track of them as new images are captured. Perceptual learning is the ability to construct compact representations of sensory events based on their statistical properties in the perceptual level as opposed to the behavior or cognitive level [51–53]. However, not having human's sophisticated visual system, the robot perceives the world differently, and there is no guarantee that the robot vision system can easily detect or recognize objects defined by human designers. As a result, to operate in realistic environments and to recognize objects in a given image, a robot must have the ability to form percepts on its own by natural association among features of sensory information, and, based on that, to segment the image into nonoverlapping but meaningful regions, denoting reliable objects [54]. With high spatial frequencies being filtered out, image segmentation reduces the amount of storage and, when done well, makes behavior resistant to some loss of information. The approach proposed in this chapter is based on a computer vision system inspired loosely by neurobiological evidences [55], a simplified description of which is given in the following.

9.2.1 Feature Generation

Color is an identifying feature that is local and largely independent of view and resolution. However, the color of a single pixel has little information by itself and may even contain only noise. If the color pixels in a simple $N \times N$ (pixels) patch of an image are considered, the histogram of these N^2 colors in the patch can describe a color pattern which may not only contain considerable more information but also potentially be less sensitive to noise. Also, this color histogram description remains relatively independent of translation and rotation and is also somewhat independent of scale such as the change of distance to the object [56]. For texture-based measure, Gabor filters have been frequently used in image processing due to their match with the receptive fields of primate's V1 simple cells. As a result, the combination of color histogram and Gabor texture measure is a good appearance feature to use for segmenting images and identifying objects.

To obtain feature vectors from small patches of an image in this research, as shown in Fig. 9.1, a moving window of size $N \times N$ (pixels) is used which is shifted by M

Fig. 9.1 Image segmentation model

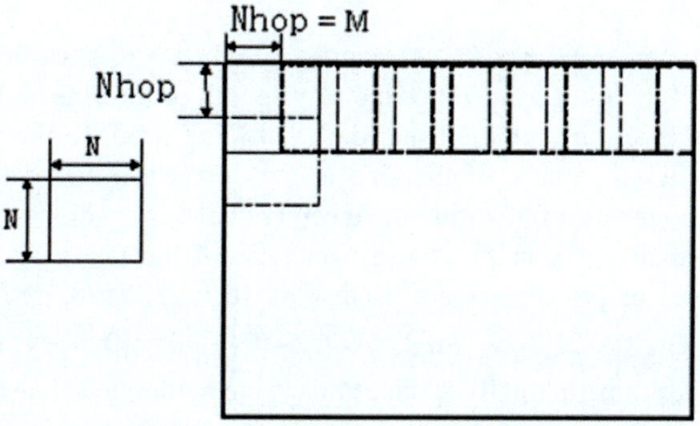

pixels in the row and column directions. The window shift controls the resolution of the result and should not exceed the border of the image.

The images and videos taken from a camcorder are stored in RGB color space. To be more robust to lighting conditions, the RGB color images are converted to the HSV (Hue, Saturation, Value) color space for use [57]. The conversion formula is given by

$$H = \begin{cases} \left(0 + \frac{G-B}{\text{MAX}-\text{MIN}}\right) \times 60 \text{ if } R = \text{MAX} \\ \left(2 + \frac{B-R}{\text{MAX}-\text{MIN}}\right) \times 60 \text{ if } G = \text{MAX} \\ \left(4 + \frac{R-G}{\text{MAX}-\text{MIN}}\right) \times 60 \text{ if } B = \text{MAX} \end{cases} \quad S = \frac{\text{MAX}-\text{MIN}}{\text{MAX}} \quad V = \text{MAX} \tag{9.1}$$

where MAX is the maximum value of (*R*, *G*, *B*), and MIN is the minimum. Hue describes each color by a normalized number in the range [0.0 … 1.0], starting at red and cycling through yellow, green, cyan, blue, magenta, and back to red. Saturation represents the purity of a color such as the 'redness' of red. Value describes the brightness of the color. To construct color features, a histogram of color measurements in the HSV space for each moving window (i.e., each image patch) in an image is computed as follows. The hues are broken evenly into 100 bins. The saturations and values are evenly distributed into ten bins each ranging from 0.0 to 1.0. Each color can be represented by combining the three bins, one hue bin, one saturation bin, and one value bin. All possibilities of the combinations equal 10,000 different color bins for the histogram. The total number of pixels in the selected region divides these 10,000 numbers, resulting in a highly sparse feature vector. The conjunction of the very high-dimensional feature space and small image patches allows for the image patches to be efficiently represented as sparse vectors. And the use of a sparse histogram representation for the color space makes the calculations largely independent of the feature space dimension while taking advantage of the high-dimensional properties.

First introduced by Gabor [58], complex Gabor filters are complex exponentials with a Gaussian envelope and provide a mathematical approximation of the spatial receptive field of a simple cell in primate's visual area V1 as

$$G(x, y, \theta, \lambda, \varphi, \sigma, \gamma) = e^{-\frac{\tilde{x}^2+\gamma^2\tilde{y}^2}{2\sigma^2}} e^{j\left(2\pi\frac{\tilde{x}}{\lambda}+\varphi\right)} \tag{9.2}$$

where $\tilde{x} = x\cos\theta + y\sin\theta$, $\tilde{y} = y\cos\theta - x\sin\theta$, θ specifies the orientation of the filter, λ specifies the wavelength of the sine and cosine wave and determines the spacing of light and dark bars that produce the maximum response, φ specifies the phase of the sine and cosine wave and determines where the ON-OFF boundaries fall within the receptive field, σ specifies the radius of the Gaussian, and finally, γ specifies the aspect ratio of the Gaussian. In this work, the following set of parameters are used: $\theta \in \{0, \pi/8, 2\pi/8, 3\pi/8, 4\pi/8, 5\pi/8, 6\pi/8, 7\pi/8\}$; $\lambda \in \left\{\sqrt{2}, 2\sqrt{2}, 3\sqrt{2}, 4\sqrt{2}, 5\sqrt{2}\right\}$, $\varphi \in \{0, \pi/2\}$, $\sigma = \lambda$, and, finally, $\lambda = 1$, resulting in totally 80 sine and cosine modulated Gaussian filters. Given each image, the convolutions of itself with each pair of Gabor filters are calculated as preparation for the generation of the orientation features. Then the magnitude of each complex Gabor filter is calculated and the convolution results within each $N \times N$ image patch, as used in color histogram generation, are averaged to obtain 40 texture measures, which are appended to the color histogram to complete the formation of the feature vector. The use of local color histogram and Gabor filters as descriptors is quite standard, and the values of some of the parameters used here have been justified [59].

9.2.2 Similarity Measure

With the set of features being defined, the next step is to construct a multidimensional space of perceptual similarities and to ascribe to each physical stimulus a particular position within this feature space. In this process, two major issues are to set up an appropriate similarity measure and to design an algorithm to cluster the input patterns based on the adopted similarity measure. It is desired that the similarity measure is given in numerical form. In a d-dimensional space, the most widely used mathematical models for similarities are L_1-norm and L_2-norm, which are obtained by setting $r = 1$ or 2, respectively, in the following Minkowski power formula,

$$d_{ij} = \left(\sum_{l=1}^{d} \left| x_{il} - x_{jl} \right|^r \right)^{1/r} \tag{9.3}$$

In the multidimensional scaling literature, these two similarity measures are also called the City-block distance and the Euclidean distance and are believed to represent two types of processing. As indicated by [60–62], for unitary or holistic stimuli, such as the hue, saturation, and brightness of colors, the closest approximation to an invariant relation between data and distances has uniformly been achieved in a space endowed with the familiar Euclidean distance; on the other hand, for analyzable or separable stimuli, such as the size and brightness difference (i.e., orientation), the closest approach to invariance between data and distances has generally been achieved with the City-block distance. To implement these suggestions, the Euclidean distance is used for the n-dimensional HSV color histogram component of the feature vector, and the City-block distance is used for the m-dimensional texture measure component of the feature vector. That is

$$d_{ij} = \sqrt{ \sum_{c=1}^{n} \left(x_{ic} - x_{jc} \right)^2 + \sum_{t=n+1}^{n+m} \left| x_{it} - x_{jt} \right| } \tag{9.4}$$

9.2.3 Percept Formation

Each feature vector appears as a point in the feature space and patterns pertaining to different classes will fall into different regions of the feature space. The learning process is to partition the feature space into sensible clusters based on some properties in common. In the proposed research, a minimum spanning tree (MST)-based clustering method is used. Being a graphical analysis of the feature vectors, a minimum spanning tree is a connected and weighted graph with no closed loops. If the distance associated with each edge of an MST denotes the weight, every edge in the MST is the shortest distance between two subtrees connected by that edge. Therefore, removal of the largest edges produces separate clusters. To begin this method, a minimum spanning tree is constructed. Next the edges of the tree are sorted in a nonincreasing manner and removed until the desired number of clusters is formed. Each cluster is assigned a name describing its visual property, the so-called visual word, and a closest color for displaying purposes. A detailed description of the MST-based clustering algorithm used here can be found in [63] and will not be dwelt on here further.

9.2.4 A Fast Approximate Nearest Neighbor Classifier

With the set of feature vectors being labeled, the next step is to classify new observations. Real-time processing of the information coming from video demands for efficient search algorithms. Design of searching methods that scale well with the size of the database and the dimensionality of the data is a challenging task. Trees present an efficient way to index local image regions, making the nearest neighbor search more efficient by pruning. Scalable vocabulary tree (SVT) is a two-stage recognition scheme that provides a more efficient training of the tree which is built by hierarchically K-means clustering on a large set of representative descriptor vectors [49]. Instead of K defining the final number of clusters or quantization cells, K defines the branch factor (i.e., the number of children of each node) of the tree. More specifically, an initial K-means process is run on the training data, defining K cluster centers. The training data are then partitioned into K groups, where each group consists of the descriptor vectors closest to a particular cluster center. The same process is then applied to each newly generated group of descriptor vectors, recursively defining quantization cells by splitting each quantization cell into K new parts. The tree is determined level by level, up to some maximum number of levels, and each division into K parts is only defined by the distribution of the descriptor vectors that belong to its parent quantization cell.

For very large databases consisting of high-dimensional highly sparse feature vectors, we choose to follow the notion of vocabulary tree but with important variations to generate a K-way approximate nearest neighbor search tree. As shown in Fig. 9.2, given the obtained database, at the first level, there is only one node, the tree root. At the second level, a set of K (which is 3 in Fig. 9.2) representative patterns are

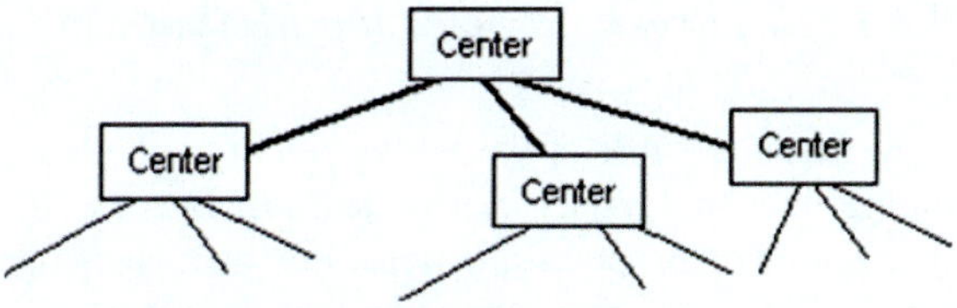

Fig. 9.2 Three-way approximate nearest neighbor search tree, reprinted from Ref. [65], copyright 2012, with permission from Elsevier

randomly selected from the whole database as the cluster centers. Then the whole database is clustered into K subsets by assigning each feature vector to its closest center according to the chosen similarity measure. At the third level, for each of the K clusters obtained at the second level, K feature vectors are selected randomly from its pool of feature vectors as its next level cluster centers, resulting in K^2 cluster tree nodes at this level. This procedure continues until either all the feature vectors in a leaf node (the tree node that has no child nodes) belong to the same object class (a pure node) or the number of the feature vectors in a leaf node is below some limit, e.g., 50. Every feature vector has a class label associated with it.

Given a new feature vector, to search through the tree, its distances to K cluster centers at each level along a certain branch are calculated and the winner is the center among K ones that the new feature vector is nearest to. When a leaf node is reached, if it is pure, assign the associated label and then stop. Otherwise, do a nearest neighbor search over the feature vectors in the leaf node, assign the label associated with the feature vector which gives the minimum distance according to the chosen metric, and then stop.

9.3 An On-Line Novelty Detection Method

The visual words (i.e., the labeled feature vectors in the database) obtained by image segmentation correspond well to natural features in the image. It is generally agreed that data points which have natural association are contained in the same regions of the feature space. The distance to the closest region is actually an estimated confidence interval of an output given the input. The data points inside such regions significantly indicate the local fitting accuracy. Since the evaluation of data points only involves a relatively small portion of the tree, the detection of novelties becomes computationally efficient. As a result, the proposed on-line novelty detection technique will first find a threshold to detect potential outliers, then apply an eight-connected structure element to erode the obtained binary image in order to find the largest connected outlier groups, and finally, when those groups have sufficient number of novel feature vectors, insert them into the tree as an on-line augmentation. In the following, the methods for threshold selection, erosion operation, and tree insertion operation are detailed.

9.3.1 A Threshold Selection Method

Thresholding is one of the most popular methods for novelty detection. Manually tuning threshold values can be tedious. Because it is often impossible to anticipate possible abnormal situations, the main challenges of novelty detection aim to develop fully automatic systems that do not require task-specific thresholds and tuning. Being the most prominent probability distribution in statistics to describe real-valued random variables that cluster around a single mean value, a normal distribution is often used as a first approximation to complex phenomena. To detect those feature vectors with a potential possibility of coming from a novel object, a precise criterion should be obtained from empirically estimated statistical parameters (e.g., mean/median and standard deviation). Drawn from a normal distribution, about 68% of values are within one standard deviation σ away from the mean; about 95% of the values lie within two standard deviations, and about 99.7% are within three standard deviations. Therefore, for novelty detection based on a distance measure with respect to the database, the sum of the unbiased mean and one standard deviation can be a good choice for the threshold value.

To find such a suitable threshold, in the proposed approach, a sufficient number of normal pictures (i.e., with no novel objects in) are collected and the distances between feature vectors extracted from each picture and those in the database are calculated using the approximate nearest neighbor search tree. If the size of the pictures taken is width-by-height and a moving window of size $N \times N$ (pixels) is shifted by M pixels evenly in the row and column directions, for each picture, the number of feature vectors (FV) extracted, #FV, is

$$\begin{aligned} \text{width} &= \frac{(\text{Width} - N)}{M} + 1 \\ \text{height} &= \frac{(\text{Height} - N)}{M} + 1 \\ \#\text{FV} &= \text{width} \times \text{height} \end{aligned} \tag{9.5}$$

where *width* and *height* denote the size of the resulting segmented image. Since the approximate nearest neighbor search through the database tree is used, in order to reduce the sensitivity to noisy outliers, the median value, instead of the mean value, of these #FV distances is calculated as d_{median}. Then an L_1-norm-based statistic, std_{L1}, that loosely resembles the standard deviation is calculated as

$$\text{std}_{L1} = \frac{\sum_{i=1}^{\#\text{FV}} \left| d_{\text{FV},i} - d_{\text{median}} \right|}{\#\text{FV}} \tag{9.6}$$

where $d_{\text{FV},i}$ denotes the distance between the i-th feature vector (FV) and its approximate nearest neighbor in the database tree, $d_{\text{median}} = \text{median}(d_{\text{FV},1}, \ldots d_{\text{FV},\#\text{FV}})$, and #FV is the total number of feature vectors extracted per image. This results in multiple medians and multiple std_{L1}'s, each for a single training image. If l frames of normal images are used, the sum of the median of these l medians and the median of these l std_{L1}'s is used as the threshold T_{median}. The formal definition of the threshold T_{median} is the following sum

$$T_{\text{median}} = \text{median}\big(d_{\text{median},1}, \ldots, d_{\text{median},l}\big) + \text{median}\big(\text{std}_{L1,1}, \ldots, \text{std}_{L1,l}\big) \quad (9.7)$$

where the l in the last subscript in each term is the total number of normal images used to obtain the threshold. For comparison purpose, we use T_{mean} in the following to denote the threshold using the sum of the traditional mean and one standard deviation. That is

$$T_{\text{mean}} = \frac{\sum_{j=1}^{\#\text{FV}\times l} d_{\text{FV},j}}{\#\text{FV} \times l} + \sqrt{\frac{\sum_{j=1}^{\#\text{FV}\times l}\left(d_{\text{FV},j} - d_{\text{mean}}\right)^2}{\#\text{FV} \times l}} \quad (9.8)$$

9.3.2 Eight-Connected Structure Element Filter

After thresholding each of the width-by-height grayscale distance images, a binary image is obtained. To find those outliers that have high potentiality of coming from the novel objects, we are more interested in those outliers that are retained in a few largest connected groups. To locate such regions, an eight-connected structure element is applied to find such regions. In other words, in the binary image obtained through thresholding, if an element in the image has a value of 1 and all its 8 neighbors are 1, this element retains a value of 1. Otherwise, it retains a value of 0. Only when the largest connected groups contain more than a certain number of feature vectors, say 50, are these feature vectors collected and stored temporarily for the next step, insertion.

9.3.3 Tree Update Method

With the potential novel feature vectors being identified and collected, the next step is to insert them into the current database search tree for fast process. First, each novel feature vector (together with its new label) is appended to the end of the array of feature vectors holding the database. Its approximate nearest neighbor in a leaf node of the tree is next located. Finally, the index of the novel feature vector in the database list is added to the leaf node's member list and the flag for that leaf node is changed to impure if it is pure before this operation. Using this augmented approximate nearest neighbor search tree, the next image containing the novel object can be segmented.

The whole proposed on-line novelty detection method can be separated into the training phase, which models the normal data, and the testing phase, which identifies the novel data and adds them to the database tree.

9.4 Experiments and Results

In this section, we empirically investigate the performance of the proposed novel object detection mechanism by conducting a series of experiments in two different environments (an indoor one and an outdoor one). Each experiment consists of two consecutive phases: an exploration phase followed by an inspection phase. In the exploration phase, the robot collects images around the environment and builds a visual model of normality for it. In the inspection phase, the robot uses the learnt model to highlight any abnormal visual object that may appear in the environment.

9.4.1 Experiment I: An Indoor Environment

9.4.1.1 Setup

As shown in the left of Fig. 9.3, the first working environment is an indoor hallway. Typical percepts along the hallway include wood panel, white floor, yellow floor, blue wall, blue railing, black stripe tile, and light window. As shown in the right of Fig. 9.3, a Pioneer 2 AT four-wheel drive mobile robot, called Skeeter, is used in the experiment. The robot vision system is a video camera (SONY Digital Handycam DCR VX2000 digital video camera recorder) that sends the data to an on-board laptop computer through IEEE 1394 interface. The intelligent information processing system is an on-board laptop that runs the Fodera core 4 operating system.

Fig. 9.3 (left) Hallway; (right) mobile robot Skeeter with a camera and a laptop, reprinted from Ref. [65], copyright 2012, with permission from Elsevier

9.4.1.2 Exploration Phase

In this phase, a model of the environment is generated so that the robot vision system can identify the major percepts existing in this indoor environment. To build a percept database, 20 pictures of size 720-by-480 along the hallway are taken by the robot, from which 66,740 unlabelled feature vectors are extracted (with $N = 15$ and $M = 10$). According to a minimum spanning tree (MST)-based clustering algorithm given in [63], 11 perceptual groups are obtained. A detailed list of the major percepts present in the video sequences (not all objects appear on the same frames simultaneously but with some objects replacing others as the training sequence progresses) is provided in Table 9.1 together with their semantic meanings and a denoting color.

From Table 9.1, it can be seen that several perceptual groups obtained by the unsupervised learning algorithm actually correspond to the same object. To make the processed image visually pleasant, several finely differentiated colors are used to refer these different objects in the machine vision perspective. To use the obtained percept database to subsequently segment any image (taken in the environment) reasonably well and fast, ten vocabulary trees and ten proposed random trees are built. The average running times in seconds over ten runs and the corresponding standard deviations are summarized in Table 9.2. It can be seen that the proposed random trees take much less time to construct.

Table 9.1 Major percepts and their semantic meanings

Percept name	Percept symbol	Denoting color
Black stripe	Object 1	Black
Wood panel	Object 2	Light red
Light window	Object 3	Light gray
Wood panel	Object 4	Light red
White floor	Object 5	Dark white
Wood panel	Object 6	Dark red
Wood panel	Object 7	Red
Wood panel	Object 8	Light red
White floor	Object 9	White
Yellow floor	Object 10	Yellow
Yellow floor	Object 11	Blue

Reprinted from Ref. [65], copyright 2012, with permission from Elsevier

Table 9.2 Running time for trees' construction

	Vocabulary tree (s)	Random tree (s)
Mean	24,228	49
Std	41	2

Reprinted from Ref. [65], copyright 2012, with permission from Elsevier

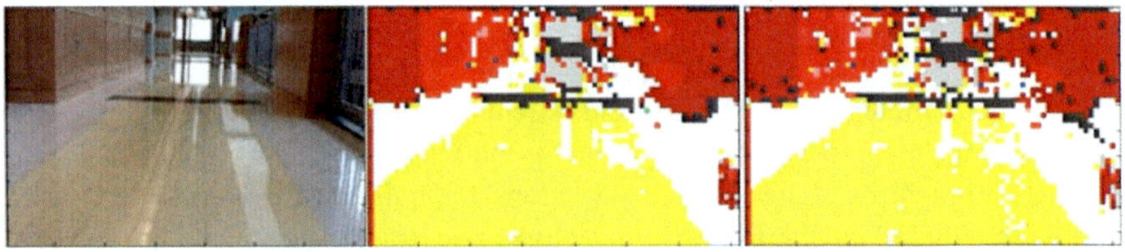

Fig. 9.4 (left) Original image; (middle) VT tree search result; (right) random tree search result, reprinted from Ref. [65], copyright 2012, with permission from Elsevier

To show the quality of these trees, segmentation results using vocabulary tree and the proposed random tree for a sample image are presented in Fig. 9.4, where the left one is the original image, the middle one is the segmented image using the vocabulary tree, and the right one is the segmented image using the proposed random tree.

From the figure, it can be seen both search trees work very well. In a qualitative visual evaluation, it can be seen that the metric used does a reasonably good job on the whole and clearly identifies the yellow floor, the white floor, and the wood panel wall, although it misclassifies the railing as the wood panel, which is not too surprising since the railing is a hard percept to model. To further quantify the search quality and the technical robustness of both trees, ten more images along the hallway are taken. Then for each of them, 3337 feature vectors are extracted which are classified first by an exact nearest neighbor search with respect to the whole database and then by an approximate nearest neighbor search through ten vocabulary trees (denoted by Vtree) and ten fast random search trees (denoted by Otree). The average percentages of correct classification (with respect to the exact search) and the corresponding standard deviations are plotted in Fig. 9.5. The horizontal axis denotes the image

Fig. 9.5 Average percentages of correct classification over ten runs and the corresponding standard deviations, reprinted from Ref. [65], copyright 2012, with permission from Elsevier

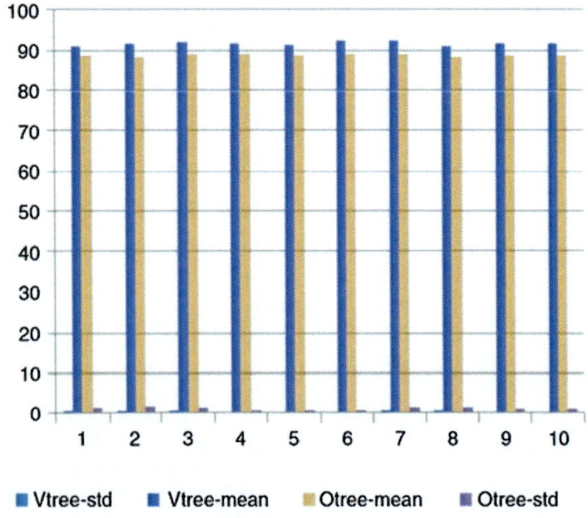

number. The vertical axis denotes the average percentages of correct classification over ten runs and the corresponding standard deviation. From the figure, it can be seen both vocabulary tree and the proposed random tree do a good job and are close in performance.

9.4.1.3 Parameter Extraction

The next step in this experiment is to extract parameters suitable for novelty detection. Threshold calculation is a crucial step. Using different thresholds may result in different detection performance. For this purpose, 80 normal images along the hallway are collected.

Two different sets of threshold values are extracted from them and shown in Table 9.3. In the table, OT-I, OT-II, VT-I, VT-II listed in the first column are used to denote the cases for which threshold values are calculated from Eqs. (9.7) and (9.8) using distances obtained by the proposed tree and vocabulary tree, respectively. The values in the second and third columns are the estimated values for unbiased mean and standard deviation, respectively. Those in the last column are the final thresholds to use.

Additionally, the performance of a neural network, Grow-When-Required (GWR), is also tested. As shown in Fig. 9.6, the GWR network consists of a clustering layer of nodes and a single output node. The connecting synapses are subject to habituation, which reduces in response to repeatedly presented stimuli. The inputs to GWR include some pre-defined parameters and the input feature vectors from the database.

Table 9.3 Threshold values for Novelty detection

Items	Mean	std	Threshold
OT-I	0.4950	0.2039	0.6989
OT-II	0.5810	0.4914	1.2820
VT-I	0.4475	0.1893	0.6368
VT-II	0.5177	0.3867	0.9044

Reprinted from Ref. [65], copyright 2012, with permission from Elsevier

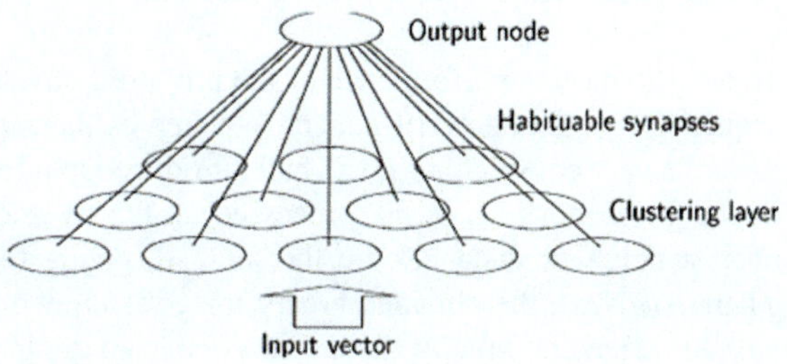

Fig. 9.6 GWR neural network architecture, reprinted from Ref. [65], copyright 2012, with permission from Elsevier

Table 9.4 Parameters used for GWR in this work

Parameter name	Parameter values
a_{T}	$\{0.3\ 0.4\ 0.5\ 0.6\ 0.7\ 0.8\ 0.9\}$
ε	$\{0.1\ 0.01\ 0.001\}$
h_{T}	0.1
α	1.05
τ_{n}	0.1
τ_{b}	0.3
$age_{\max}$	100
$neighbor_{\max}$	10

The training of GWR starts with two topologically unconnected nodes (e.g., any two input vectors), whose weights are randomly initialized. During training, following an unsupervised winner-take-all approach, the Euclidean distance from the input to each node currently in the clustering layer is first computed to find the best matching node, s, (called the winner) and the second best matching node, t. The weights and the output synapses of the winner node and its topological neighbors are updated. If a connection exists between the two nodes, its age is set to 0; otherwise a new connection is created. When the activation and habituation values of the winner node are below pre-defined thresholds a_{T} and h_{T}, the input is decided to be novel and a node can be added to the clustering layer. After inserting a new node, r, to the clustering layer, topological connections are updated by removing the link between the nodes s and t and inserting links between r and s and between r and t. The GWR training iteration is accomplished by removing nodes that no longer have any neighbors and edges whose age is greater than a pre-defined threshold $age_{\max}$. The set of parameters used in this work are given in Table 9.4. From empirical evidence, it can be observed that the larger the activity threshold, the larger the number of nodes, and the less the number of false negatives, however, the more time the novelty detection consumed.

9.4.1.4 On-Line Novel Percept Detection

To test the on-line performance of the proposed novelty detection method, a video sequence, consisting of 19 images, is taken as the robot moves along the hallway toward a very conspicuous green ball introduced in it for the robot to approach. Each of the grabbed 19 images is processed on the fly and the calculated approximate nearest neighbor distances are thresholded, giving rise to a binary image of size 71-by-47. Next, the obtained binary image is eroded twice by an eight-connected structure element. Finally, the largest connected group is identified and grows as the robot moves closer and closer to the green ball. If the number of feature vectors it has contained is larger than 50, they are retained and accumulated through temporary storage.

The novelty detection results are presented by five of the original 19 images in Fig. 9.7. The ones in the middle column show the results obtained when the robot uses the first 18 images as the training images to collect novel feature vectors, inserts them into the tree at the end of the 18th image, and then uses the updated database to segment the last (i.e., the 19th) image. The ones in the right column show that the robot uses the total set of 19 images as the training images to collect novel feature vectors, inserts them into the tree at the end of the 19th image, and then uses the

Fig. 9.7 (left) Original images; (middle) processed images; (right) processed images after learning, reprinted from Ref. [65], copyright 2012, with permission from Elsevier

updated database to segment the whole set (i.e., 19 images) all over again. As found in the experiments, about 600 feature vectors coming from the 19 images with the novel object in are added to the search tree after 2 s. This demonstrates that the proposed system is able to continue learning with only a modest processing delay.

9.4.1.5 Off-Line Novelty Detection Evaluation

To make further comparisons between both methods, 135 testing images (25 of which contains the novel object) are collected for off-line testing. Every round of the inspection phase consists of ten consecutive loops of these images. The segmented images based on the obtained percept database should yield a set of connected regions in the image. A connected-component labeling algorithm is then applied to find these regions, which can be regarded as the corresponding objects. Since the number of images used in the experiments is not very large, the target and nontarget images are manually labeled by human through visual examination.

For this round of experiments, the performance of each method is evaluated by the accumulated results which are given in the form of a contingency table. A contingency table is a two-dimensional table formed by cross-classifying subjects or events on two categorical variables used to relate the actual presence of novelty and the response of the system. One variable's categories define the rows while the other variable's categories define the columns. The intersection (cross-tabulation) of each row and column forms a cell, which displays the count (frequency) of cases classified as being in the applicable category of both variables. Next, the χ^2 analysis is performed, which is then followed by the computation of Cramer's V. The chi-square statistic (χ^2) is computed to determine whether or not the row and column categories for the table as a whole are independent of each other. If the computed χ^2 statistic exceeds a critical value, then we reject the null hypothesis and conclude that the variable categories are indeed related. Cramer's V is computed to test the strength of such relationships and is the nominal association measure of choice. Large values of Cramer's V's indicate better performance.

The performance evaluation of the proposed system vs. GWR network to detect the simple novel percept is shown in Table 9.5. The upper half of Table 9.5 lists the results when threshold OT-I is used. From it, it can be seen that among 250 test images that contain the novel object, 227 are detected and 23 are missed by the proposed system, and, among 1100 test images that are normal, none is detected. For GWR network approach, seven values for activity threshold and three values for ε have been tried, and the best results (in terms of Cramer's V value) are presented in Table 9.5. From the table, it can be seen that among 250 test images that contain the novel object, all are detected by GWR network, and, among 1100 test images that are normal, 20 are mistakenly detected, yielding statistically significant results.

Table 9.5 Results of the proposed method for green ball

OT-I	Novelty detected	Novelty not detected
Novelty present	227	23
Novelty not present	0	1100
Cramer's V = 0.9431		
GWR	**Novelty detected**	**Novelty not detected**
Novelty present	250	0
Novelty not present	20	1080
Cramer's V = 0.9535		
$a_T = 0.8$		

Reprinted from Ref. [65], copyright 2012, with permission from Elsevier

9.4.1.6 Other Threshold Values

The calculated values for Cramer's V obtained when other threshold values are considered are given in Fig. 9.8. There are four lines in the figure, each corresponding to one of four cases in Table 9.3. The vertical axis denotes the values of Cramer's V's. The horizontal axis denotes the number of standard deviations used in the threshold calculation (i.e., 1.0, 1.5, 2.0, 2.5, and 3.0). From the figure, it can be seen that the threshold values along those two lines starting with OT-I and VT-I achieve the best

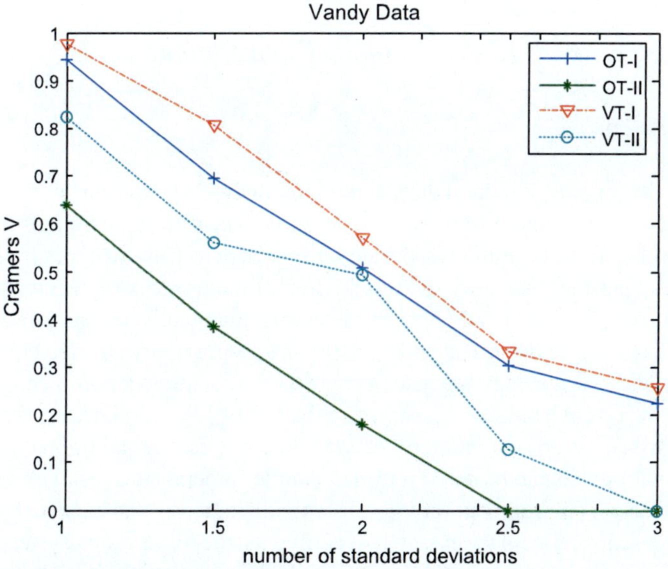

Fig. 9.8 Performance variations with different threshold values, reprinted from Ref. [65], copyright 2012, with permission from Elsevier

Table 9.6 Number of nodes

a_T	Training (# of nodes)	Testing (# of nodes)
0.7	2619	2936
0.8	4431	5023

results, indicating the proposed threshold selection criterion works well for both trees.

9.4.1.7 Running Time Issue

To compare the relative time issue of two approaches, for the database of size 66,740, it is observed that the proposed tree has so far at most 20 levels. As a result, at most $3 \times 20 + 50 = 110$ distance calculations are necessary. This is a considerable reduction. It takes no more than 2 s to find approximate nearest neighbors for the 3337 feature vectors obtained for each newly grabbed image. For GWR network, the number of nodes generated during training and testing are shown in Table 9.6, respectively. It can tell that GWR method takes several thousands of distance computations for each feature vector extracted from a newly grabbed image. This is to say, GWR method is very computationally expensive in high-dimensional feature space.

9.4.2 Experiment II: An Outdoor Environment

9.4.2.1 Setup

Although the proposed method does a good job in the last experiment, the detection of a single brightly colored green ball, which is very different from all the known objects in the frames, simplifies the novel object detection task significantly and makes the experiment somewhat artificial. For a more convincing demonstration of how the proposed model would perform detecting more realistic objects in a different and less structured environment and to justify the proposed novelty detection method is realistic and applicable in practice, an outdoor traffic intersection is chosen to be a second working environment, as shown in the left of Fig. 9.9. Further, the previous experiment is focused on mobile robots and does not use any additional constraints and information that can be derived from a mobile camera. As a result, the extension to a more general environment setting substantially increases the technical soundness of the proposed approach. In this set of experiments, the robot vision system is Canon IXUS 50HS digital video camera recorder, which is somewhat fixed and captures a video with three different vehicles coming in and leaving. The intelligent information processing system is a workstation computer that runs the Fodera core 4 operating

Fig. 9.9 (left) Original image; (middle) VT tree search result; (right) random tree search result, reprinted from Ref. [65], copyright 2012, with permission from Elsevier

system. From the video, 450 images of size 320-by-240 are extracted. The first 20 images are used as the training images to obtain the database. Then 80 images are used for threshold extraction. The last 350 images are the test images for novelty detection.

9.4.2.2 Exploration Phase

To build the percept database, the first 20 images in the video sequence are used, from which 93,220 feature vectors are extracted (with $N = 6$ and $M = 4$). The same MST-based clustering algorithm [63] is applied and 25 perceptual groups are obtained. Typical perceptual groups include those for gray road, white crossing, green grass, dark green trees, and blue sky. A list of major percepts present in the video sequences is provided in Table 9.7 together with their semantic meanings and displaying color.

Based on the obtained database, ten vocabulary trees and ten proposed random trees are constructed, respectively. Their average running times in seconds and the corresponding standard deviations are summarized in Table 9.8. From the table, it can be seen that the construction of the proposed random trees takes much less time.

Table 9.7 Major percepts and their semantic meanings

Percept name	Percept symbol	Denoting color
Blue Sky	Object 1	Blue
Blue Sky	Object 4	Light blue
Blue Sky	Object 5	Pale blue
Blue Sky	Object 7	Palest blue
Tree	Object 8	Green
Grass	Object 12	Dark green
Pole	Object 16	Dark gray
Crossing	Object 17	White
Ground	Object 22	Gray

Reprinted from Ref. [65], copyright 2012, with permission from Elsevier

Table 9.8 Running time for trees' construction

	Vocabulary tree (s)	Random tree (s)
Mean	34,961	64
Std	12	4

Reprinted from Ref. [65], copyright 2012, with permission from Elsevier

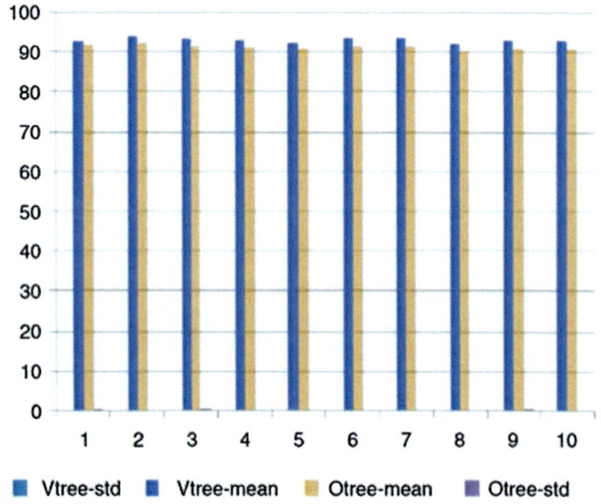

Fig. 9.10 Average percentages of correct classification and the corresponding standard deviations, reprinted from Ref. [65], copyright 2012, with permission from Elsevier

The segmentation results using vocabulary tree and the proposed random tree are shown in Fig. 9.9. Like the case in Experiment I, both search trees work very well.

To further quantify the search quality and the technical robustness of both trees, ten more normal images are chosen. Then for each of the ten images, 4661 feature vectors are extracted which are then classified first by an exact nearest neighbor search with respect to the whole database and then by an approximate nearest neighbor search through ten vocabulary trees and ten fast random search trees. To demonstrate the performance of these fast search trees, the average percentages of correct classification (with respect to the exact search) and the corresponding standard deviations are plotted in Fig. 9.10. The horizontal axis denotes the image number. The vertical axis denotes the average percentage of correct classification over ten runs and its standard deviation. From the figure, it can be seen both vocabulary tree and the proposed random tree do a good job. The performances of vocabulary tree and the proposed random tree are very close.

9.4.2.3 Parameter Extraction

To extract parameters suitable for novelty detection, similar to Table 9.3, the threshold values for the proposed method are extracted from the collected 80 normal images and shown in Table 9.9. In the table, OT-I, OT-II, VT-I, VT-II listed in the first

Table 9.9 Threshold values for novelty detection

Items	Mean	Std.	Threshold
OT-I	0.5150	0.5454	1.0604
OT-II	0.7176	0.5887	1.3063
VT-I	0.4560	0.4823	0.9383
VT-II	0.7940	0.8170	1.611

column are used to denote the cases for which threshold values are calculated from Eqs. (9.7) and (9.8) using distances obtained by the proposed tree and vocabulary tree, respectively. The values in the second and third columns are the estimated values for unbiased mean and standard deviation, respectively. Those in the last column are the final thresholds to use.

It can be clearly seen from the previous experiment that GWR method does not scale well with the size and the dimensionality of large databases. In this experiment, instead of GWR, we evaluate the performance of the proposed method with the more recently proposed E-SVDD. As a one-class classifier, support vector data description method is a very attractive kernel method and has become a standard novelty detection technique. Its run time complexity is linear in the number of support vectors, which is not very convenient for modern large databases. To improve, fast SVDD (F-SVDD) uses only one kernel term in its decision function. However, given the Gaussian kernel, its decision boundary becomes spherical in the input space, which is not suitable for many real cases. As a result, E-SVDD method first partitions the dataset into small clusters. As a result, the decision function of E-SVDD contains some crucial kernel terms, one for each cluster, and its time complexity is linear in the number of the clusters. The clustering algorithm used originally by E-SVDD is partitioning entropy-based kernel fuzzy c-means algorithm. For the high-dimensional database, K-means algorithm is used instead. To more thoroughly cover the data, the number of clusters is set to 100. Then for each cluster, the parameters used in the classifier such as the centroid are calculated. Finally, the decision function is computed according to F-SVDD.

9.4.2.4 On-Line Novel Percept Detection

To show how the proposed system would perform in the presence of multiple novel objects, this round of experiments works on a video sequence of 350 images, in which a traffic intersection is inspected when several vehicles come in. In this video sequence, the first novel object, a black car, begins to come in at about the 20th image. Then the second novel object, a purple truck, appears following the black car at about the 120th frame. Finally, the last novel object, a white jeep, shows up at about the 180th frame and goes away at about the 300th frame.

Each of the grabbed 350 images is processed on the fly. For the proposed method, the calculated approximate nearest neighbor distances are thresholded, giving rise to a binary image of size 79-by-59. For E-SVDD, the decision function is computed for each extracted feature vector. If the decision value is smaller than 0, the feature vector is normal. Otherwise, it is novel. Next, the obtained binary image is eroded twice by an eight-connected structure element as explained in Experiment I . Finally, the largest connected groups are identified. If the number of detected novel feature vectors in the largest connected groups is larger than 100, they are retained and accumulated through temporary storage. To detect multiple objects, spatial information about the position of the largest connected groups in the image space is utilized to separate novel percepts that are apart but appear in the same frame.

The novelty detection results are presented by six of the original 350 pictures in Fig. 9.11. The first column shows the original images of some typical novelty detection situations in the video sequence. The second column shows the segmented images using the database just before the novel feature vectors coming from the newly entered vehicles are learned. The third column shows the segmented images after all the novel objects are learned. That is, all 350 images are processed twice. In the first pass, they are used to collect novel feature vectors from three vehicles, which are labeled unsupervisedly. Then in the second pass, the images shown there are segmented ones using the updated database (with the novel feature vectors being added to the database).

More specifically, the images on the first row describe the normal situation when no vehicle comes in. The first image in the second column is a segmented image using the original database (i.e., with no novel feature vectors being added to it). And the first image in the third column is a segmented one using the database augmented with novel feature vectors detected from all three novel objects. The images on the second row describe the situation when the first novel object (i.e., the black car) comes in. The second image in the second column is a segmented one using the original database with no novel feature vectors augmented to it. The small region circled by an ellipse in this image denotes the place where the novel black car should otherwise appear.

When using the updated database with all three novel objects being learned, it can be seen the learned novel black car percept (denoted by black color) shows up clearly in the second image in the third column. The images on the third row manifest the situation when a purple truck enters the scene. The third image in the middle column is a segmented one using the database after the novel feature vectors from the black car have been identified and inserted into the tree. From the segmented one in the third column, the learned novel purple car (denoted by purple color) is partially seen as it shows up into the scene. The images on the fourth row further explore the situation when the purple truck comes deeper into the scene. Therefore, the fourth image in the second column is a segmented one using the database augmented only with the novel feature vectors detected from the black car. From the fourth image in the third column, the purple truck is clearly identified. The images on the fifth row show that a white jeep comes in. The fifth image in the second column is a segmented one using the database augmented with the novel feature vectors detected from the black

Fig. 9.11 (left) Original images; (middle) processed images; (right) processed images after learning, reprinted from Ref. [65], copyright 2012, with permission from Elsevier

car and the purple truck. From the fifth image in the third column, the white jeep can be clearly identified. The images on the last row show that the learning process is completed and all vehicles gradually disappear. More than 10,000 novel feature vectors from the black car are added to the search tree after 16 s. Then, more than 10,000 novel feature vectors from the purple truck are added to the search tree after 26 s. Finally, more than 10,000 novel feature vectors from the white jeep are added to the search tree after 31 s. This demonstrates that the proposed system is able to continue learning with only a modest processing delay.

To compare the detecting performances, the number of detected novel feature vectors over 350 images is shown in Fig. 9.12. We show in the upper-left plot of Fig. 9.12 the number of novel feature vectors, obtained by the proposed method (using the threshold referred to as OT-I in Table 9.9) and the E-SVDD method, respectively, for each of the 350 images. It can be seen that there are three separated peaks whose number of novel objects is larger than 100. Therefore, both methods

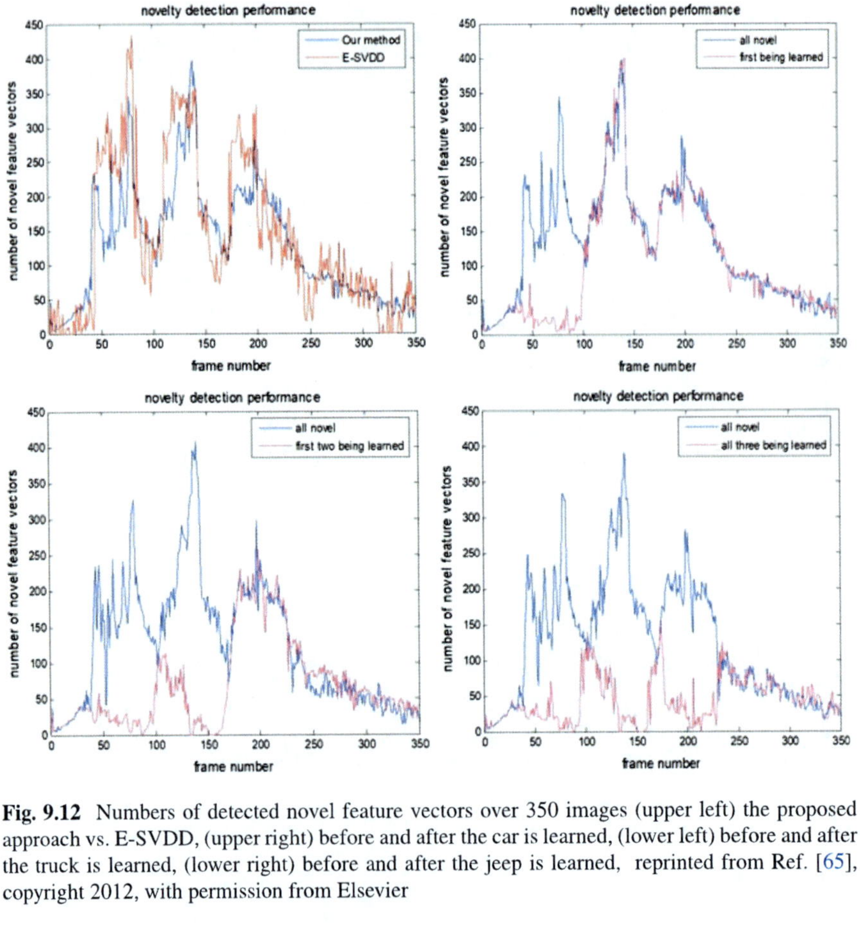

Fig. 9.12 Numbers of detected novel feature vectors over 350 images (upper left) the proposed approach vs. E-SVDD, (upper right) before and after the car is learned, (lower left) before and after the truck is learned, (lower right) before and after the jeep is learned, reprinted from Ref. [65], copyright 2012, with permission from Elsevier

successfully detect the three novel objects, consuming similar amount of time, i.e., two seconds for each frame. Shown in the upper-right plot is the number of novel feature vectors detected for each image before and after the collected novel feature vectors from the black car are inserted into the tree. Shown in the lower-left plot is the number of novel feature vectors detected for each image before and after the collected novel feature vectors from the purple truck are inserted into the tree. Shown in the lower-right plot is the number of novel feature vectors detected for each image before and after the collected novel feature vectors from the white jeep are inserted into the tree. As found in the experimental results, the total number of the novel feature vectors added to the database is more than 30,000, 31% of the original database. For this amount of feature vectors, it can be hard for many clustering algorithms to partition them into three clusters in real time.

9.4.2.5 Off-Line Novelty Detection Evaluation

As in Experiment I, to make fair comparisons by using the same dataset for off-line testing, we manually select a set of 254 testing images (154 of which contains the novel objects, i.e., the three vehicles) that we believe could be discriminated relatively well based on their visual feature distribution. Each round of the experiments consists of three consecutive test loops of these images. To evaluate the performance of each method, a contingency table is first used to relate the actual presence of novelty and the response of the system. Next, the χ^2 analysis is performed, which is then followed by the computation of Cramer's V.

The accumulated results obtained for this round of experiments are given in Table 9.10. It can be observed from both tables that both systems can detect the novelty percept, and statistically significant results are yielded.

Table 9.10 Results of the proposed method versus E-SVDD

OT-I	Novelty detected	Novelty Not detected
Novelty present	456	6
Novelty not present	0	300
Cramer's V = 0.9837		
GWR	**Novelty detected**	**Novelty not detected**
Novelty present	402	60
Novelty not present	3	297
Cramer's V = 0.8421		

Reprinted from Ref. [65], copyright 2012, with permission from Elsevier

9.4.2.6 Off-Line Multi-Novelty Detection Evaluation

Unlike the E-SVDD method which can detect the existence of novel object in the environment but requires an additional clustering algorithm to partition the obtained novel feature vectors into subgroups, the proposed method has the advantage of distinguishing between different objects on-line. To evaluate the recognition ability of the updated database (i.e., the augmented database), two more experiments are performed on the same test data set as in the previous subsection.

The accumulated results obtained for this round of experiments are given in Table 9.11 using two sets of threshold values referred to as OT-I and VT-I in Table 9.9. The rows in the table list all the possible combinations of the novel objects in the video sequence. The columns list all the possible responses of the proposed system. For example, on the first row, among all the 159 images (containing only the black car) presented to the system, 153 images are detected and six images are missed. For another example, on the second row, among all the 63 images containing both the black car and the purple truck presented to the system, 48 images are detected to have one novel object, and 15 images are detected to have both novel objects. It can be observed from the table that the proposed system can identify all the learned vehicles. However, when some of them appear at the same time in a test image, it happens that one is not detected in some images.

Table 9.11 Results of the proposed method

OT-I	1 Novelty detected	2 Novelty detected	Novelty not detected
Only car present	153	0	6
Car and truck present	48	15	0
Only truck present	96	0	0
Truck and jeep present	39	30	0
Only jeep present	75	0	0
No novelty present	0	0	300
Cramer's V = 0.7958			
VT-I	**1 Novelty detected**	**2 Novelty detected**	**Novelty not detected**
Only car present	159	0	0
Car and truck present	63	0	0
Only truck present	96	0	0
Truck and jeep present	39	30	0
Only jeep present	75	0	0
No novelty present	0	0	300
Cramer's V = 0.8526			

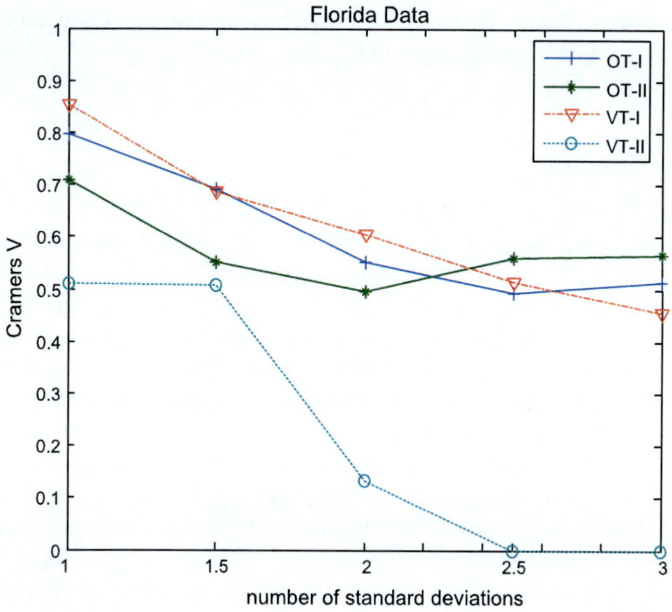

Fig. 9.13 Performance variations with different threshold values, reprinted from Ref. [65], copyright 2012, with permission from Elsevier

9.4.2.7 Other Threshold Values

The calculated values for Cramer's V obtained when other threshold values are considered are given in Fig. 9.13. There are four lines in the figure, each corresponding to one of four cases in Table 9.9. The vertical axis denotes the values of Cramer's V's. The horizontal axis denotes the number of standard deviations used in the threshold calculation (i.e., 1.0, 1.5, 2.0, 2.5, and 3.0). From the figure, it can be seen that the threshold values along those two lines starting with OT-I and VT-I achieve the best results, indicating the proposed threshold selection criterion works well for both trees.

9.4.2.8 Running Time Issue

To compare the relative time issue of the proposed approach and E-SVDD, the proposed tree has so far at most 18 levels for the database consisting of 93,220 labeled feature vectors from the experiments. As a result, we have had at most $3 \times 18 + 50 = 104$ distance calculations. This is a considerable reduction. It takes no more than 2 s to find approximate nearest neighbors for the 4661 feature vectors obtained for each newly grabbed image. For E-SVDD, the number of centroids used in the decision calculation is 100. It takes similar amount of computation time for

4661 feature vectors extracted from a newly grabbed image. Therefore, both methods have similar running time performances.

9.5 Conclusions

In this chapter, an efficient on-line novelty detection method is presented. Its performance is compared with two best known methods: GWR neural network and E-SVDD method. The proposed approach provides a better performance, while offering advantages of near real-time ability, which is extremely useful to assess which percepts of the environment are actually learnt by the system. The key innovations of the proposed approach are the use of a fast approximate nearest neighbor search facility (as opposed to the widely used vocabulary tree), a threshold extraction method for detecting novel feature vectors and the rather standard tree insertion procedure to cluster and unsupervisedly learn the novel feature vectors. In contrast to GWR neural network, the proposed approach is particularly suitable for real-time autonomous applications and has built-in robustness to noise and clutter. In contrast to E-SVDD, the proposed approach is able to use both the spatial information and feature information of novel objects to autonomously identify multiple precepts in a frame in real time. Overall, the proposed approach is autonomous and suitable for video-analytical tasks in different environments. Future investigations of the proposed method aim at how the system would perform in the presence of multiple objects of similar features but with different shapes in multiple video sequences.

Acknowledgements This chapter was modified from the paper published by our group in Robotics and Autonomous Systems [65]. The related contents are reused with permission.

References

1. Cover, T., & Hart, P. (1967). Nearest neighbor pattern classification. *IEEE Transactions on Information Theory, 13*(1), 21–27.
2. Liu, Z. G., Pan, Q., & Dezert, J. (2013). A new belief-based *k*-nearest neighbor classification method. *Pattern Recognition, 46,* 834–844.
3. Mitani, Y., & Hamamoto, Y. (2006). A local mean-based nonparametric classifier. *Pattern Recognition Letters, 27*(10), 1151–1159.
4. Gou, J., Yi, Z., Du, L., & Xiong, T. (2011). A local mean-based k-nearest centroid neighbor classifier. *Computer Journal, 55*(9), 1058–1071.
5. Sánchez, J. S., Pla, F., & Ferri, F. J. (1997). On the use of neighbourhood-based non-parametric classifiers. *Pattern Recognition Letters, 18*(11–13), 1179–1186.
6. Samsudin, N. A., & Bradley, A. P. (2010). Nearest neighbour group-based classification. *Pattern Recognition, 43*(10), 3458–3467.
7. Shanableh, T., Assaleh, K., & Al-Rousan, M. (2007). Spatio-temporal feature-extraction techniques for isolated gesture recognition in Arabic sign language. *IEEE Transactions on Systems, Man, and Cybernetics. Part B, Cybernetics, 37*(3), 641–650.

8. Xu, J., Yang, J., & Lai, Z. (2013). K-local hyperplane distance nearest neighbor classifier oriented local discriminant analysis. *Information Sciences, 232*, 11–26.
9. Maji, P. (2011). Fuzzy–rough supervised attribute clustering algorithm and classification of microarray data. *IEEE Transactions on Systems, Man, and Cybernetics. Part B, Cybernetics, 41*(1), 222–233.
10. Raymer, M. L., Doom, T. E., Kuhn, L. A., & Punch, W. F. (2003). Knowledge discovery in medical and biological datasets using a hybrid Bayes classifier/evolutionary algorithm. *IEEE Transactions on Systems, Man, and Cybernetics. Part B, Cybernetics, 33*(5), 802–813.
11. Dudczyk, J., Kawalec, A., & Owczarek, R. (2008). An application of iterated function system attractor for specific radar source identification. In *Proceedings of the 2008 17th International Conference on Microwaves, Radar and Wireless Communications* (MIKON'08), Wroclaw, Poland.
12. Dudczyk, J., Kawalec, A., & Cyrek, J. (2008). Applying the distance and similarity functions to radar signals identification. In *Proceedings of the 2008 International Radar Symposium* (IRS'08), Wroclaw, Poland.
13. Dudczyk, J., & Wnuk, M. (2004). The utilization of unintentional radiation for identification of the radiation sources. In *Proceedings of the 34th European Microwave Conference*, Amsterdam, The Netherlands (Vol. 2, Issue 2, pp. 777–780).
14. Dudczyk, J. (2017). A method of feature selection in the aspect of specific identification of radar signals. *Bulletin of the Polish Academy of Sciences: Technical Sciences, 65*(1), 113–119.
15. Mensink, T., Verbeek, J., Perronnin, F., & Csurka, G. (2013). Distance-based image classification: Generalizing to new classes at near-zero cost. *IEEE Transactions on Pattern Analysis and Machine Intelligence, 35*(11), 2624–2637.
16. Frigui, H., & Gader, P. (2009). Detection and discrimination of land mines in ground-penetrating radar based on edge histogram descriptors and a possibilistic k-nearest neighbor classifier. *IEEE Transactions on Fuzzy Systems, 17*(1), 185–199.
17. Ma, L., Crawford, M. M., & Tian, J. (2010). Local manifold learning-based k-nearest-neighbor for hyperspectral image classification. *IEEE Transactions on Geoscience and Remote Sensing, 48*(11), 4099–4109.
18. Manavalan, B., Shin, T. H., & Lee, G. (2018). PVP-SVM: Sequence-based prediction of phage virion proteins using a support vector machine. *Frontiers in Microbiology, 9*, 1–10.
19. Manavalan, B., Subramaniyam, S., Shin, T. H., & Kim, M. O. (2018). Machine-learning-based prediction of cell-penetrating peptides and their uptake efficiency with improved accuracy. *Journal of Proteome Research, 17*(8), 2715–2726.
20. Manavalan, B., Shin, T. H., & Lee, G. (2018). DHSpred: Support-vector-machine-based human DNase I hypersensitive sites prediction using the optimal features selected by random forest. *Oncotarget, 9*(2), 1944–1956.
21. Manavalan, B., Shin, T. H., Kim, M. O., & Lee, G. (2018). AIPpred: Sequence-based prediction of anti-inflammatory peptides using random forest. *Frontiers in Pharmacology, 9*, 1–12.
22. L. Tarassenko, P. Hayton, N. Cerneaz, M. Brady. Novelty detection for the identification of masses in mammograms. In *Proceedings of the 4th International Conference on Artificial Neural Networks*, Cambridge, UK (pp. 442–447).
23. Quinn, J., & Williams, C. (2007). Known unknowns: Novelty detection in condition monitoring. In *Proceedings of the 3rd Iberian Conference on Pattern Recognition and Image Analysis (IbPRIA'07)*, Girona, Spain, LNCS (Vol. 4477, Isusue 1, pp. 1–6).
24. Clifton, L., Clifton, D., Watkinson, P., & Tarassenko, L. (2011). Identification of patient deterioration in vital-sign data using one-class support vector machines. In *Proceedings of the 2001 Federated Conference on Computer Science and Information Systems (FedCSIS'11)*, Szczecin, Poland (pp. 125–131).
25. Tarassenko, L., Clifton, D., Bannister, P., King, S., & King, D. (2009). *Novelty detection*. Wiley.
26. Surace, C., & Worden, K. (2010). Novelty detection in a changing environment: A negative selection approach. *Mechanical Systems and Signal Processing, 24*(4), 1114–1128.
27. Patcha, A., & Park, J. (2007). An overview of anomaly detection techniques: Existing solutions and latest technological trends. *Computer Networks, 51*(12), 3448–3470.

28. Jyothsna, V., Prasad, V. V. R., & Prasad, K. M. (2011). A review of anomaly based intrusion detection systems. *International Journal of Computer Applications, 28*(7), 26–35.
29. Diehl, C., & Hampshire, J. (2002). Real-time object classification and novelty detection for collaborative video surveillance. In *Proceedings of the International Joint Conference on Neural Networks* (*IJCNN'02*) (Vol. 3, Issue 3, pp. 2620–2625).
30. Markou, M., & Singh, S. (2006). A neural network-based novelty detector for image sequence analysis. *IEEE Transactions on Pattern Analysis and Machine Intelligence, 28*(10), 1664–1677.
31. Vieira Neto, H., & Nehmzow, U. (2007). Real-time automated visual inspection using mobile robots. *Journal of Intelligent and Robotic Systems, 49*(3), 293–307.
32. Sofman, B., Neuman, B., Stentz, A., & Bagnell, J. (2011). Anytime online novelty and change detection for mobile robots. *Journal of Field Robotics, 28*(4), 589–618.
33. Zhang, Y., Meratnia, N., & Havinga, P. (2010). Outlier detection techniques for wireless sensor networks: A survey. *IEEE Communications Surveys & Tutorials, 12*(2), 159–170.
34. Dutta, H., Giannella, C., Borne, K., & Kargupta, H. (2007). Distributed top-k outlier detection from astronomy catalogs using the DEMAC system. In *Proceedings of the 7th SIAM International Conference on Data Mining*, Minneapolis, Minnesota, USA (pp. 473–478).
35. Escalante, H. (2005). A comparison of outlier detection algorithms for machine learning. In *Proceedings of the International Conference on Communications in Computing* (pp. 10–15).
36. Basu, S., Bilenko, M., & Mooney, R. (2004). A probabilistic framework for semi- supervised clustering. In *Proceedings of the 10th ACM International Conference on Knowledge Discovery and Data Mining* (*SIGKDD'04*), Seattle, WA, USA (pp. 59–68).
37. Markou, M., & Singh, S. (2003). Novelty detection: A review-part 1: Statistical approaches. *Signal Processing, 83*(12), 2481–2497.
38. Markou, M., & Singh, S. (2003). Novelty detection: A review-part 2: Neural network based approaches. *Signal Processing, 83*(12), 2499–2521.
39. Duda, R. O., Hart, P. E., & Stork, D. G. (2001). *Pattern classification*. NY, USA: Wiley.
40. Bishop, C. (2006). *Pattern recognition and machine learning*. New York: Springer.
41. Marsland, S., Nehmzow, U., & Shapiro, J. (2001). Vision-based environmental novelty detection on a mobile robot. In *Proceedings of the 8th International Conference on Neural Information Processing* (*ICONIP'01*), Shanghai, China (Vol. 1–3, pp. 69–74).
42. Marsland, S., Nehmzow, U., & Shapiro, J. (2002). Environment-specific novelty detection. In *From Animals to Animates: Proceedings of the 7th International Conference on the Simulation of Adaptive Behavior* (*SAB'02*), MIT, Edinburgh, UK (pp. 36–45).
43. Neto, H. V., & Nehmzow, U. (2004). Visual novelty detection for inspection tasks using mobile robots. In *Proceedings of the 8th Brazilian Symposium on Neural Networks* (*SBRN'04*), Sao Luis, Brazil.
44. Nehmzow, U., & Neto, H. V. (2004). Novelty-based visual inspection using mobile robots. In *Towards Autonomous Robotic Systems: Proceedings of the 5th British Conference on Mobile Robotics* (*TAROS'04*), Colchester, UK.
45. Neto, H. V., & Nehmzow, U. (2007). Visual novelty detection with automatic scale selection. *Robotics and Autonomous Systems, 55*(9), 693–701.
46. Tax, D. M. J., & Duin, R. P. W. (2004). Support vector data description. *Machine Learning, 54*(1), 45–66.
47. Liu, Y.-H., Liu, Y.-C., & Chen, Y.-J. (2010). Fast support vector data descriptions for novelty detection. *IEEE Transactions on Neural Networks, 21*(8), 1296–1313.
48. Peng, X., & Xu, D. (2010). Efficient support vector data descriptions for novelty detection. *Neural Computing and Applications, 21*(8), 2023–2032.
49. Nister, D., & Stewenius, H. (2006). Scalable recognition with a vocabulary tree. In *Proceedings of the IEEE International Conference on Computer Vision and Pattern Recognition* (*CVPR'06*), New York, NY, United states (Vol. 2, pp. 2161–2168).
50. Meeds, E. (2005). Novelty detection model selection using volume estimation. UTML-TR-2005–004, Technical Report, University of Toronto.
51. Olshausen, B. A., & Field, D. J. (1996). Emergence of simple-cell receptive field properties by learning a sparse code for natural images. *Nature, 381*(6583), 607–609.

52. Goldstone, R. (1998). Perceptual learning. *Annual Review of Psychology, 49,* 585–612.
53. Buhmann, J. M., Lange, T., & Ramacher, U. (2005). Image segmentation by networks of spiking neurons. *Neural Computation, 17*(5), 1010–1031.
54. Tugcu, M., Wang, X., Hunter, J. E., Phillips, J., Noelle, D., & Wilkes, D. M. (2007). A computational neuroscience model of working memory with application to robot perceptual learning. In *Proceedings of the 3rd IASTED International Conference on Computational Intelligence* (*CI'07*), Banff, AB, Canada (pp. 120–127).
55. Wang, X., Tugcu, M., Hunter, J. E., & Wilkes, D. M. (2009). Exploration of configural representation in landmark learning using working memory toolkit. *Pattern Recognition Letters, 30*(1), 66–79.
56. Swain, M. J., & Ballard, D. (1991). Color indexing. *International Journal of Computer Vision, 7*(1), 11–32.
57. Hunter, J. E. (2005). Human motion segmentation and object recognition using Fuzzy rules. In *Proceedings of 14th Annual IEEE International Workshop on Robot and Human Interactive Communication* (*RO-MAN'05*), Nashville, TN (pp. 210–216).
58. Gabor, D. (1946). Theory of communications. *Journal of Institute of Electrical Engineering, 93*(3), 429–457.
59. Hunter, J. E., Tugcu, M., Wang, X., Costello, C., & Wilkes, D. M. (2011). Exploiting sparse representations in very high-dimensional feature spaces obtained from patch-based processing. *Machine Vision and Applications, 22*(3), 449–460.
60. Shepard, R. N. (1964). Attention and the metric structure of the stimulus space. *Journal of Mathematical Psychology, 1*(1), 54–87.
61. Garner, W. R. (1974). *The processing of information and structure.* New York, USA: Wiley.
62. Shepard, R. N. (1987). Toward a universal law of generalization for psychological science. *Science, 237*(4820), 1317–1323.
63. Wang, X., Wang, X. L., & Wilkes, D. M. (2009). A divide-and-conquer approach for minimum spanning tree-based clustering. *IEEE Transactions on Knowledge and Data Engineering, 21*(7), 945–958.
64. Lowe, D. (2004). Distinctive image features from scale invariant keypoints. *International Journal of Computer Vision, 60*(2), 91–110.
65. Wang, X., Wang, X. L., & Wilkes, D. M. (2012). An automated vision based on-line novel percept detection method for mobile robot. *Robotics and Autonomous Systems, 60*(10), 1279–1294.

Chapter 10
Unsupervised Fraud Detection in Environmental Time Series Data

Abstract Time series often contains outliers and structural changes. These unexpected events are of the utmost importance in fraud detection as they may pinpoint suspicious activities. The presence of such unusual activities can easily mislead conventional time series analysis and yield erroneous conclusions. Traditionally, time series data are first divided into small chunks. *k*NN-based outlier detection approaches are then applied for monitoring behaviors over time in data mining. However, time series data are very large in size and cannot be scanned multiple times, and as they are produced continuously, new data are arriving. To cope with the speed they are coming, in this chapter, we propose a simple statistical parameter-based anomaly method for fraud detection in environmental time series data. The results of the experiments performed show that the proposed algorithm is effective and efficient .

Keywords Time series data · Fraud detection · Statistical parametric outlier detection method · *k* -nearest neighbor-based outlier detection approach

10.1 Introduction

Nowadays, more data are generated from different applications at a fast rate due to technology and hardware improvements. With the continuous development in environmental protection and the improvement of environmental monitoring level, a large number of environmental monitoring time series data have been produced. In recent years, the Chinese Ministry of Environmental Protection has paid more attention to the monitoring data and the inspection intensity is gradually increasing. This brings new opportunities to the environmental monitoring and testing industry. According to the prediction of Chinese Consulting Industry Research Center (CIRC), the market share of environmental monitoring industry may exceed ¥90 billion in 2020, with a compound growth rate of about 20% in five years. However, the environment protection has become more vulnerable to fraud activities, and environmental data fraud detection has gradually become a new challenge. Environmental fraud usually takes place when the companies try to manipulate their environmental data without regard

X. Wang et al., *New Developments in Unsupervised Outlier Detection*,
https://doi.org/10.1007/978-981-15-9519-6_10

for the environmental monitoring's own real interests. Detecting the frauds means identifying suspicious fraudulent and illegal activities against the law.

The generation of fraud data in environmental monitoring can generally be divided into three categories, namely fraud in sampling (such as watering around the sampling point), fraud in data processing (such as modifying and deleting data), and fraud in analytical instruments (such as modifying the upper and lower limits of analytical instruments). In the past, environmental monitoring equipment was maintained by the local government, so most of the falsifications in data processing were made. Nowadays, because the data are uploaded in real time and the instrument is also entrusted to a third party for operation and maintenance, it can only be fake in the sampling process.

The key principle underlying the generation of the environmental fraud data is that the environment's interests are secondary to the financial gain the individual company can make. To improve the ability of environmental protection personnel to identify fraud in the environmental time series data, the quality of environmental monitoring data is the lifeline of environmental protection work, which is related to the scientific decision-making of environmental governance. In order to cope with the inspection, some enterprises will cut off the sewage in advance and add pipes to inject qualified water into the inspection water tank so as to make the data up to the standard. In these areas, fog gun trucks are used to spray water near air monitoring points. According to the staff of the environmental protection department, the air quality monitoring substation mainly monitors the data of sulfur dioxide, PM10, PM2.5, etc., and the data will be distorted by sprinkling. Fake data will eventually be sent to the monitoring data platform. As shown in Fig. 10.1, common suspicious data forms include constant or sudden change of emission data, which requires analysis of data records in a long period of time.

Time series data are records accumulated over time and have a chronology that allows us to analyze their evolution (trend) as time goes by. Such data can be viewed as objects with an attribute time. The objects are the snapshots of entities with values that change over time. Time series analysis is to analyze data collected along time sequences, periodicity, and so on. It may also involve attribute relevance analysis, model construction, classification, and prediction. When analyzing time series, one often encounters unusual events such as outliers and structural changes. An object that does not obey the behavior of normal data object is called outliers. Outlier

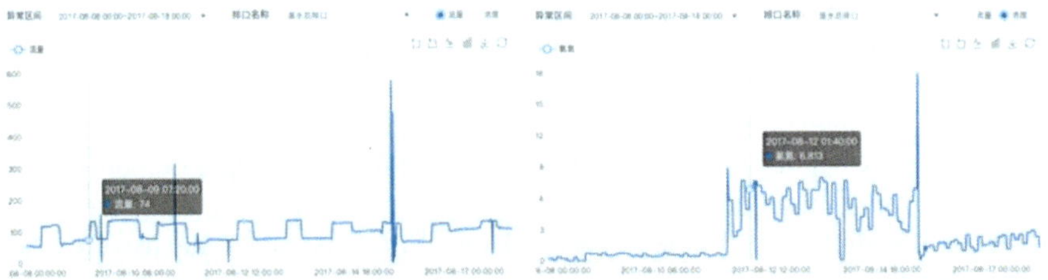

Fig. 10.1 Sample wastewater discharge concentration data

detection is used in different applications like fraud detection, intrusion detection, track environmental changes, and medical diagnosis so there is need to detect outliers from data streams. Various approaches are used for outlier detection. Fraud detection is of great importance to many time series applications. Thus, time series anomaly analysis explores most of the techniques developed in the data mining functions plus its own techniques. An approach for time series anomaly analysis is to convert the time series into sequences of events, then apply existing unsupervised outlier detection techniques to discover interesting patterns in the sequences. For example, data are first divided into small chunks and *k*NN-based outlier detection approaches are applied as an unsupervised method for monitoring behaviors over time in time series anomaly and fraud detection. Unfortunately, there is no definite science for identifying the hidden pattern in data. Accordingly, the methods should be chosen depending on the problem to be solved. Further, time series data are very large in size and cannot be scanned multiple times, and as they are produced continuously, new data arrive. For fraud detection, it is hard to collect labeled data, and also data are coming increasingly and data points may be asynchronous. As a result, it is better to use unsupervised anomaly detection where there is no need of class labels of time series data. To cope with the speed the time series are coming, we propose in this chapter a new outlier score for each time series developed upon statistical parameters.

The proposed method has been applied to the time series data which has been collected by JointSky Corporation to assess its performance on fraud detection in environmental monitoring time series data. We have observed the proposed technique can detect those abnormal behaviors of suddenly rising and dropping which happern in a different way with respect to other parts of the series to which they were supposed to be similar previously. We have also applied *k*NN-based outlier detection techniques to compare their effectiveness and efficiency. Experimental results show that the proposed algorithm is effective and reasonable.

In the following, we review some related work on time series anomaly detection in Sect. 10.2. The proposed statistical parameter-based anomaly detection method is presented in Sect. 10.3. Then in Sect. 10.4, the performance of the proposed fraud detection method is compared with those of *k*-nearest neighbor-based approach for an environmental monitoring time series data fraud detection task. Finally, conclusions are made in Sect. 10.5.

10.2 Related Work

With time series data being continuously produced from different applications, imposing different challenges like massive, infinite, high dimension, fast update speed, concept drift to processing, and being widely used in economy, meteorology, and medical and other fields, time series mining has gained attraction from many research areas and a lot of work in the mining of time series has been done in recent years. In time series mining, the purpose of most tasks is to find those frequent patterns and the corresponding rules. The current research in this chapter focuses on time series anomaly detection. Although the abnormal data are an infrequent event compared with the normal data and usually ignored as noise, some important information may be hidden behind them . The discovery of abnormal data often can bring people more valuable knowledge. The purpose of this section is to review some different approaches for outlier detection in time series data. Outliers can take several forms in time series. The problem of outlier detection in time series can be defined in the following two different ways.

10.2.1 Point Outliers

A point outlier is a sudden change in a sequence of time series values at a given timestamp. This problem is closely related to forecasting, because an outlier is defined as a significant deviation from the expected (or *forecasted*) values. Such outliers are referred to as contextual outliers because they are outliers in the context of their immediate history. Dependencies can be used to define expected values of data items. Deviations from these expected values can be regarded as outliers. For example, a sudden jump in the values of a time series may result in a point outlier at the specific spot at which the jump occurs. The idea in these methods is to use prediction-based techniques to forecast the value at that position. Significant deviation from the predicted value can be reported as a point outlier. Such outliers can be defined in the context of time series, spatial, and sequential data, where significant deviations from the corresponding neighborhoods can be detected using autoregressive [1], Markovian [2–6], or other models.

The formal definitions and a classification of outliers in a time series context were first proposed by Fox in 1972 [7]. Before this article, an assumption of independent, identically distributed observations was commonly made in outlier analysis. Autoregressive moving average (ARMA) models are a common framework for outlier detection and modeling in statistical time series analysis. Given a time series as $\{z_t$: $t = 1, 2, \ldots\}$, where each $z(t)$ is a d-dimensional column vector, an autoregressive (AR) model of the k-th order is given by

$$z_t = \sum_{i=1}^{k} A_i z_{t-i} + \varepsilon \tag{10.1}$$

where each A_i is the k-th order square matrix, ε is a normal random variable generated according to a Gaussian distribution with mean 0 and covariance matrix Σ, which we denote as $N(0, \Sigma)$.

In the standard outlier literature [8–10], a time series is modeled as ARMA plus outliers. Let us denote a time series which we observe as $\{x_t$: $t = 1, 2, \ldots\}$, the general model for an observed series x_t is

$$x_t = z_t + \mathrm{e}(t) \tag{10.2}$$

where z_t is a regular ARMA model. Outliers are incorporated through $e(t)$. Setting $\theta = (A_1, \ldots, A_k, e, \Sigma)$, for an estimation of θ, a variant of the maximum likelihood method is employed such that the weighted likelihood function is maximized where the weight decays exponentially in time t. Then the difference between x_t and the forecasted value of z_t can be used as an outlier score.

10.2.2 Shape Outliers

In this case, a consecutive pattern of data points in a contiguous window may be defined as an anomaly when considered together, although no individual point in the series may be considered an anomaly. Such outliers are referred to as collective outliers because they are defined by combining the patterns from multiple data items. One of the earliest methods for finding shape-based outliers is to define outliers on windows of the time series. A k-nearest neighbor based outlier detection method is then used to determine the outlier scores.

In other words, the outlier analysis is performed over windows of length W. Therefore, the approach reports windows of unusual shapes from a time series of data points. The first step is to extract all windows of length W from the time series by using a sliding-window approach. The analysis is then performed over these newly created data objects. For each extracted window, its Euclidean distance to the other nonoverlapping windows is computed. For example, based on distance-based outlier detection [11–13], the Euclidean distance of a data point to its k-th nearest neighbors is used to define the outlier score. Density-based outlier detection methods can also be applied to the extracted windows, including the LOF model [14], the COF method [15], the LOCI method [16], the SLOM method [17], the INFLO method [18], the ROF method [19], the LDOF method [20], the RBDA method [21], the KDEOS method [22], the NROF [23], and the RDOS method [24]. For high-dimensional data, an angle-based outlier detection method has been proposed [25]. Usually, nonoverlapping windows are used to minimize the impact of trivial matches to overlapping windows. The windows with the highest outlier scores are reported as outliers. However, the time complexity will scale with the square of the number of data points, i.e., the number of windows.

10.3 Method

10.3.1 A Simple Idea

At present, people generally use Hawkin's definition of time series anomaly [26]: Anomaly is the data deviating from most of the data in the data set, which makes people suspect that these data are generated by different mechanisms, rather than random deviation. Time series contains a large amount of data with high dimensions and fast data updates. Although most of the current research to detect the outliers in time series face challenges, in recent years, there have been some achievements in the research of time series outlier detections and many algorithms have been proposed to detect outliers in time series data. But there is a trade-off between processing time and accuracy. This section introduces a simple and fast algorithm based on window (or, block) mean and standard deviation to detect a time series outlier in relatively less time with appreciable accuracy.

It is very difficult to detect abnormal patterns directly on the original time series. To solve this problem, a long time series is usually divided into several relatively short but nonoverlapping subsequences, and each subsequence is transformed into a high-level multi-dimensional data representation.

Definition 1 (*Exception mode*). An exception mode in a time series is a pattern with abnormal behavior which has significant difference from other patterns in a time series. By calculating the local abnormal coefficient of each mode, the value of the local abnormal coefficient associated with the exception mode is significantly larger than that of the normal modes.

Time series data are the sequence data measured or recorded in chronological order, which contains the operation mode according to internal law of the development. For example, sensors installed at the pollution monitoring stations to check the pollution level due to sulfur dioxide, carbon dioxide, particulate matter, etc. generally produce an output which follows nearly a uniform distribution. Therefore, the method proposed in this chapter to detect an outlier in time series assumes that the outlier sequence occurs in a uniform signal or a nearly uniform signal. On the other hand, time series data can usually be affected by internal variability and external interferences in the measurement process. For example, when fraud happens, a series of sudden jumps in the value of a time series may result in an outlier sequence at the specific spot time at which the jump occurs, as manifested in Fig. 10.1.

In order to facilitate the description, the following symbols are used to describe the data involved in the study. The symbol t is used to represent any time point in the time series. In this chapter, t is transformed from a standard time such as coordinated universal time (UTC) to a relative time based on a certain time point. The measured data value is represented by y. Before building the detection model, we are given the time series sequence data $D = \{(t_1, y_1), (t_2, y_2), ..., (t_N, y_N)\}$. In other words, there are N groups of data in the set for analysis. Two data items y and t in each group correspond to each other.

Definition 2 Given a time series sequence data, D, the mean and the standard deviation of the sequence can be computed and their ratio can be defined as the time series sequence shaper outlier scores as

$$\text{Mean} = \frac{1}{N} \sum_{i=1}^{N} y_i \tag{10.3}$$

$$\mathrm{Std} = \sqrt{\frac{1}{N}\sum_{i=1}^{N}(y_i - \mathrm{Mean})^2} \tag{10.4}$$

$$\mathrm{SOM} = \frac{\mathrm{Std}}{\mathrm{Mean}} \tag{10.5}$$

With the outlier scores being defined, therefore, in order to detect an outlier in a time series data, the following steps are involved:

(1) For a large time series data to be checked for the presence of anomaly, the data are divided into equal sized blocks in a sliding-window fashion.
(2) The mean, standard deviation, and the corresponding window-based time series sequence outlier score are calculated for each and every block in the time domain.
(3) Then the blocks are sorted according to the calculated scores, and the time series sequences with top scores are returned as anomalies.

10.3.2 Selecting an Appropriate Threshold for Fraud Detection

Given a set of time series sequences and the corresponding outlier scores, to return the set of outliers as a result of fraud detection, sort the outlier scores obtained in Sect. 10.3.1 in a descending order and select those set of time series sequences whose scores are above the point at which the outlier factors drop sharply as the separation point of outliers and normal modes. The mechanism for the determination of outlier time series sequences to return is illustrated in Fig. 10.2. In the figure, the horizontal axis denotes the number of data objects, the vertical axis denotes the outlier scores, and the two data objects with top outlier scores significantly larger than those of the normal mode are returned as outliers.

Fig. 10.2 Determination of the number of outliers

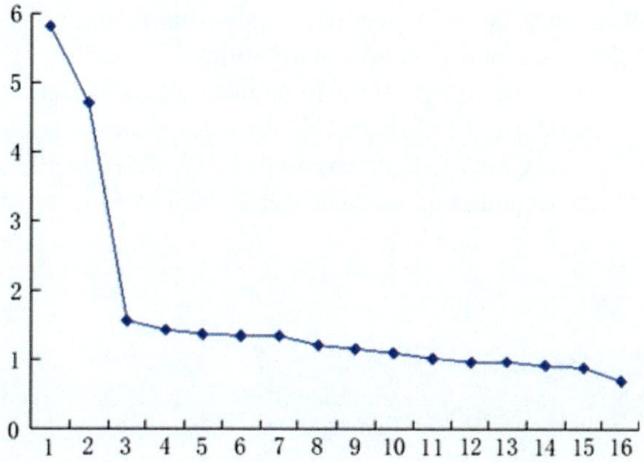

10.3.3 The Complexity Analysis

To compare the relative time and space issues of the proposed approach, from the description in the previous subsections, it can be seen that *k*NN based algorithms for time series outlier detection mainly consist of three steps, the division of the time series data into N smaller-sized blocks, a *k*-nearest neighbors search for each of the blocks, and the calculation of the outlier scores and finally top outliers' mining. Therefore, most of the calculation time is spent on the second step, *k*-nearest neighbors' search. As a result, the total time complexity for the average case can be $O(N^2)$ or O(NlogN) if some index structure can be used. Physically, the resource consumed by *k*NN based algorithms includes the space for hold the N data blocks in memory, space to store their k nearest neighbors and the scores, and some space for the index structure if used. To summarize, traditional *k*NN based approaches are both time and space consuming. Fortunately, the proposed approach can not only detect the time series outliers in real time by dividing the time series data into small-sized blocks in a sliding-window fashion and calculating the block-wise mean and standard deviation and then the outlier scores as the data arrive in an incremental manner, that is, in O(1) time, but also do so without the need to store the data in the previous blocks, which can be a significant advantage over the *k*NN based approaches.

10.4 Experiments and Results

In this section, we empirically investigate the performance of the proposed time series fraud detection mechanism by conducting two sets of experiments. In order to verify the effect of the proposed method on fraud detection in environmental time series data, the hourly data of wastewater discharge concentration and the hourly data of gas emission concentration collected by JointSky corporation are used to study the anomaly detection in the flow time series data of environmental monitoring, including the flow statistics within each hour starting from the whole point 24 h a day and the manual annotation of whether each flow data is abnormal. Each experiment consists of testings by two categories of algorithms, the proposed algorithm and the set of *k*NN-based outlier detection algorithms [12–25].

In the experiments, we implement the proposed algorithm in MATLAB of the version R2013a and all the *k*NN-based algorithms in Java which are run on a computer with Intel Core 2 Duo Processor E6550 2.33 GHz CPU and 2 GB RAM. The operating system running on this computer is Windows 7 professional.

10.4.1 Fraud Detection on Wastewater Discharge Concentration Data

In this set of experiments, the 79 segments of data used for fraud detection come from the flow statistics of the sewage outlet nodes collected by the company in the real scene. The time series data include the traffic statistics within each hour starting from the whole time 24 h a day and also contains the manual annotation of whether each flow data are abnormal, that is, whether fraud exists in any of all the 79 segments of time series data.

The outlier analysis is performed over windows of length $W = 10$, 12, 15, 20, and 24. Therefore, the approach reports windows of unusual shapes from a time series of data points. The first step is to extract all windows of length W from the time series by using a sliding-window approach. The analysis is then performed over these newly created data subsequences . For each extracted window, its kNN-based outlier factors to the other nonoverlapping windows are computed. The windows with the highest k-nearest neighbor-based outlier values are reported as outliers. The reason for using nonoverlapping windows is to minimize the impact of trivial matches to overlapping windows. However, in comparison with the proposed algorithm, the time complexity of kNN-based algorithms will scale with the square of the number of data points. The relative effectiveness of our proposed outlier detection algorithm is compared with those of the DB-Max [12], DB [13], LOF [14], COF [15], INFLO [18], LDOF [20], RBDA [21], and ABOD [25] methods. The final outlier scores are casted into the range [0, 100]. Shown in Fig. 10.3 are the example outlier scores for

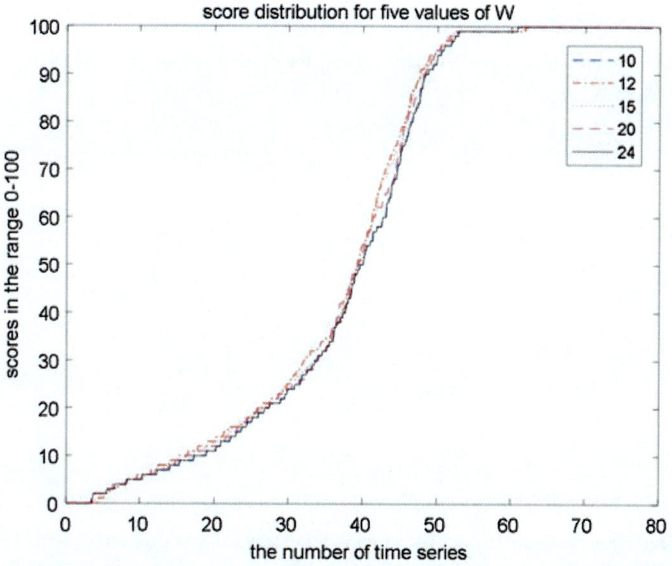

Fig. 10.3 Distribution of the scores for five values of window Ws

five values of window size calculated according to each outlier definition and sorted in a nondecreasing order (instead of a nonincreasing order). From the figure, we can see a larger number of high outlier scores.

Shown in left plot of Fig. 10.4 are the sorted (in a nondecreasing order) DB outlier scores for the given 79 time series segments for which the detected number of anomalies is the highest with a window size of $W = 20$. Shown in the right plot of the figure is an illustration of the highest 42 (higher than the threshold value plotted in the red line) out of 79 anomalies being detected for $W = 24$ among five window sizes.

Shown in left plot of Fig. 10.5 are the sorted (in a nondecreasing order) DB-Max outlier scores for the given 79 time series segments when the detected number of anomalies is the highest with a window size of $W = 24$. Shown in the right plot of the figure is an illustration of the highest 42 (higher than the threshold value plotted

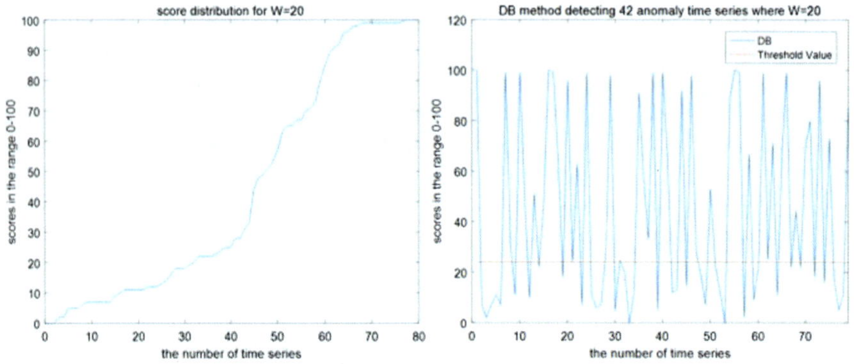

Fig. 10.4 Detection results using DB method (left) the score distribution (right) 42 anomalies are detected for $W = 20$

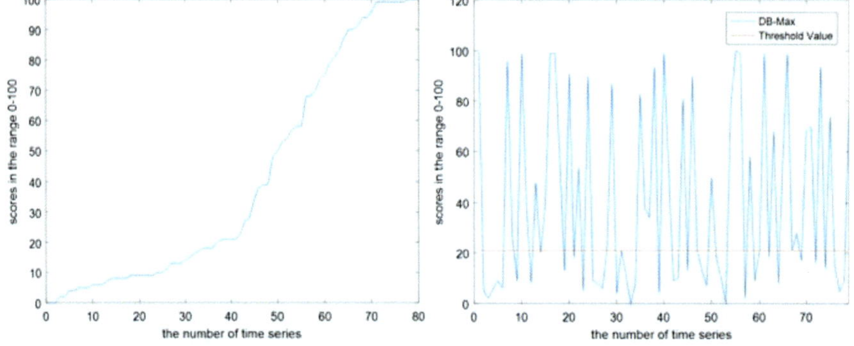

Fig. 10.5 Detection results using DB-Max method (left) the score distribution (right) 42 anomalies are detected for $W = 24$

in the red line) out of 79 anomalies being detected for $W = 24$ among five window sizes.

Shown in left plot of Fig. 10.6 are the sorted (in a nondecreasing order) INFLO outlier scores for the given 79 time series segments when the detected number of anomalies is the highest with a window size of $W = 24$. Shown in the right plot of the figure is an illustration of the highest 50 (higher than the threshold value plotted in the red line) out of 79 anomalies being detected for $W = 24$ among five window sizes.

Shown in left plot of Fig. 10.7 are the sorted (in a nondecreasing order) LDOF outlier scores for the given 79 time series segments when the detected number of anomalies is the highest with all window sizes. Shown in the right plot of the figure is an illustration of all 79 anomalies being detected among all five window sizes.

Shown in left plot of Fig. 10.8 are the sorted (in a nondecreasing order) LOF outlier scores for the given 79 time series segments when the detected number of

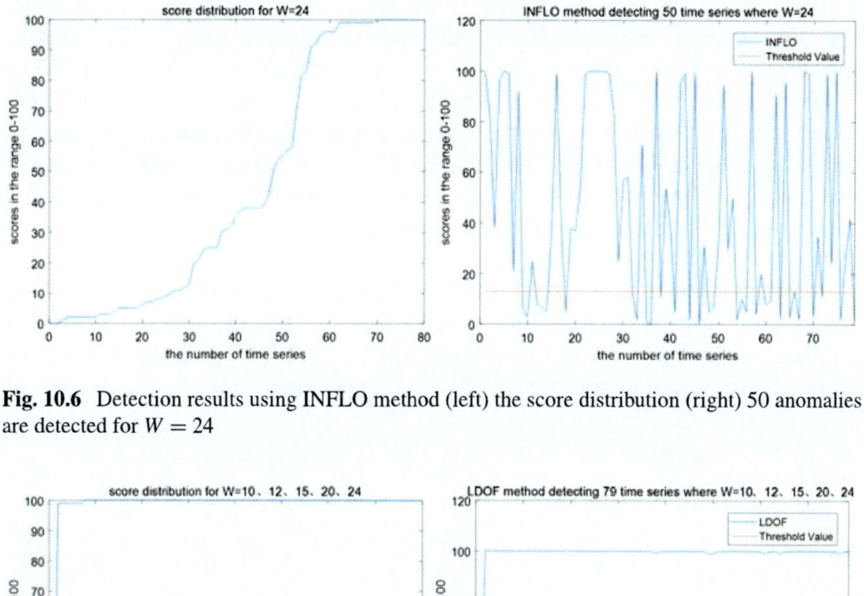

Fig. 10.6 Detection results using INFLO method (left) the score distribution (right) 50 anomalies are detected for $W = 24$

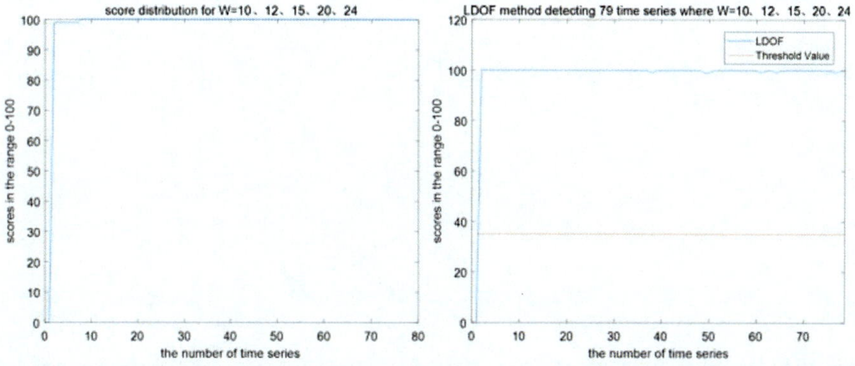

Fig. 10.7 Detection results using LDOF method (left) the score distribution (right) 79 anomalies are detected for all window sizes

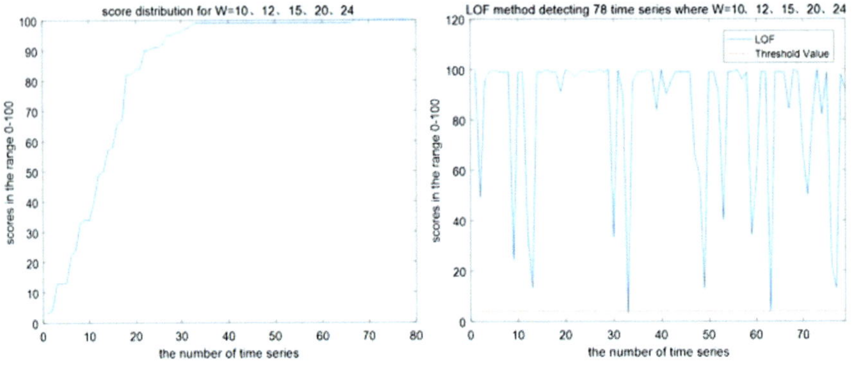

Fig. 10.8 Detection results using LOF method (left) the score distribution (right) 78 anomalies are detected for all window sizes

anomalies is the highest with all window sizes. Shown in the right plot of the figure is an illustration of the highest 78 anomalies being detected by using all five window sizes.

Shown in left plot of Fig. 10.9 are the sorted (in a nondecreasing order) COF outlier scores for the given 79 time series segments when the detected number of anomalies is the highest with a window size of $W = 10$. Shown in the right plot of the figure is an illustration of the highest 40 (higher than the threshold value plotted in the red line) out of 79 anomalies being detected for $W = 10$ among five window sizes.

Shown in left plot of Fig. 10.10 are the sorted (in a nondecreasing order) RBDA outlier scores for the given 79 time series segments when the detected number of anomalies is the highest with a window size of $W = 12$. Shown in the right plot of the figure is an illustration of the highest 13 (higher than the threshold value plotted

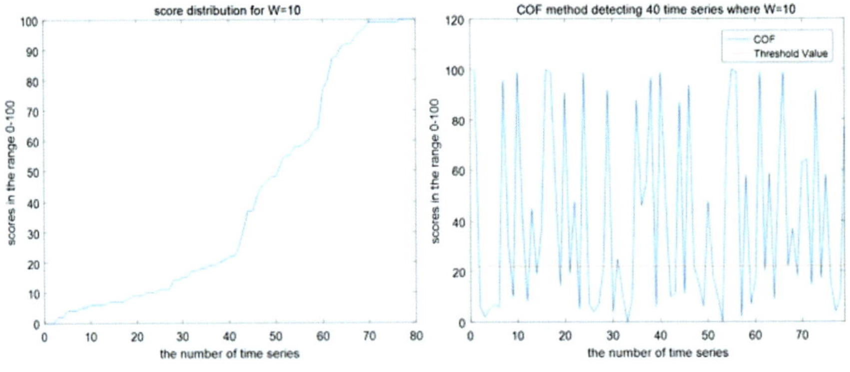

Fig. 10.9 Detection results using COF method (left) the score distribution (right) 40 anomalies are detected for $W = 10$

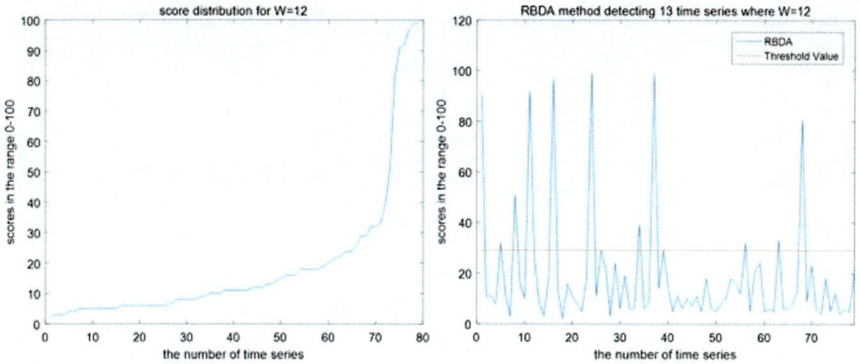

Fig. 10.10 Detection results using RBDA method (left) the score distribution (right) 13 anomalies are detected for $W = 12$

in the red line) out of 79 anomalies being detected for $W = 12$ among five window sizes.

Shown in left plot of Fig. 10.11 are the sorted (in a nondecreasing order) ABOD outlier scores for the given 79 time series segments when the detected number of anomalies is the highest with all window sizes. Shown in the right plot of the figure is an illustration that all 79 anomalies are detected by using all five window sizes.

Shown in left plot of Fig. 10.12 are the sorted (in a nondecreasing order) proposed outlier scores, SOM, for the given 79 time series segments when the detected number of anomalies is the highest with all window sizes. Shown in the right plot of the figure is an illustration that 44 (higher than the threshold value plotted in the red line) out of 79 anomalies are detected for $W = 24$ among five window sizes.

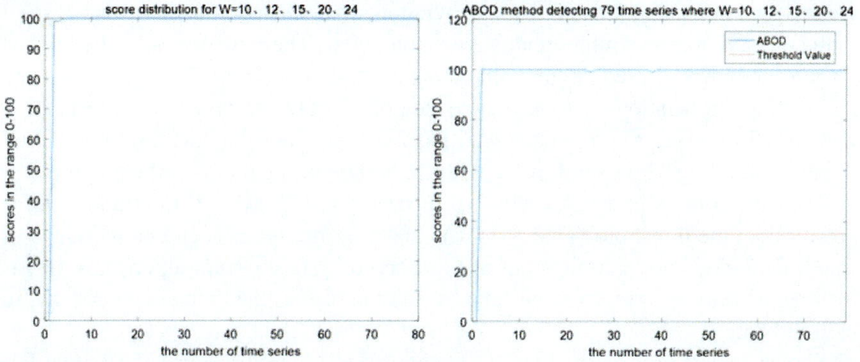

Fig. 10.11 Detection results using ABOD method (left) the score distribution (right) all 79 anomalies are detected for all window sizes

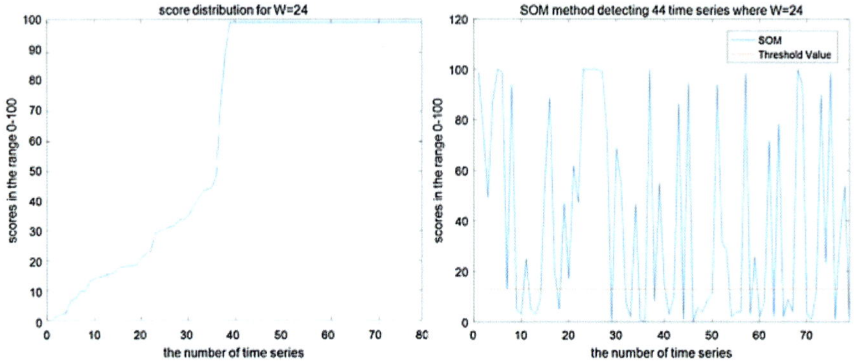

Fig. 10.12 Detection results using the proposed method (left) the score distribution (right) 44 anomalies are detected for $W = 24$

10.4.2 Fraud Detection on Gas Emission Concentration Data

In this set of experiments, the 306 groups of data used for fraud detection come from the flow statistics of the gas emission concentration hourly data collected by the company in the real scene. The time series data include the traffic statistics within each hour starting from the whole time 24 h a day and also contains the manual annotation of whether each flow data is abnormal, that is, there are 87 true fraud time series segments.

The outlier analysis is performed over windows of length $W = 10, 12, 15, 20$, and 24. Therefore, the approach reports windows of unusual shapes from a time series of data points. The first step is to extract all windows of length W from the time series by using a sliding-window approach. The analysis is then performed over these newly created data objects. For each extracted window, its kNN-based outliers with respect to the other nonoverlapping windows are computed. The windows with the highest k-nearest neighbor-based outlier values are reported as outliers. The reason for using nonoverlapping windows is to minimize the impact of trivial matches to overlapping windows. However, in comparison with the proposed algorithm, the time complexity of kNN-based algorithms will scale with the square of the number of data points.

The threshold value is calculated as illustrated in Fig. 10.2. If the fraud score is greater than the threshold value, it means that the time series segment is judged as fraud. Each data may get different results according to different algorithms, but as long as the data are judged to be fake by any algorithm, the data are considered to be fraud data.

Shown in left plot of Fig. 10.14 are the sorted (in a nondecreasing order) DB outlier scores for the given 306 time series segments when the detected number of anomalies is the highest with a window size of $W = 15$. Shown in the right plot of the figure is an illustration of the highest 37 (higher than the threshold value plotted in the red line) out of 79 anomalies being detected for W=15 among five window sizes.

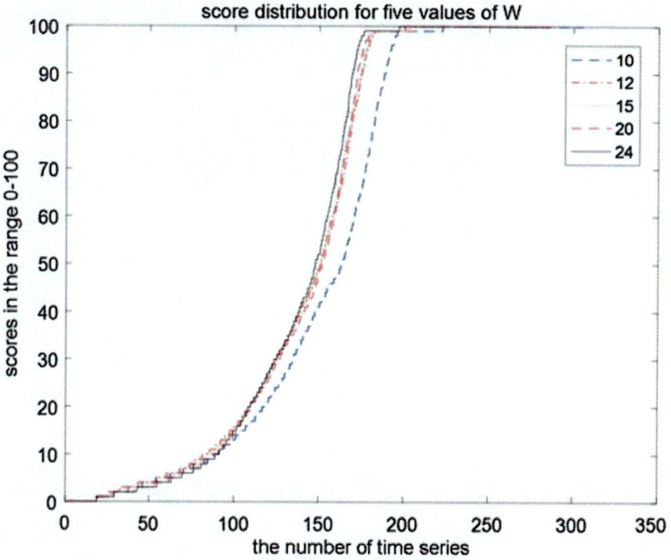

Fig. 10.13 Distribution of outlier scores for five values of window Ws

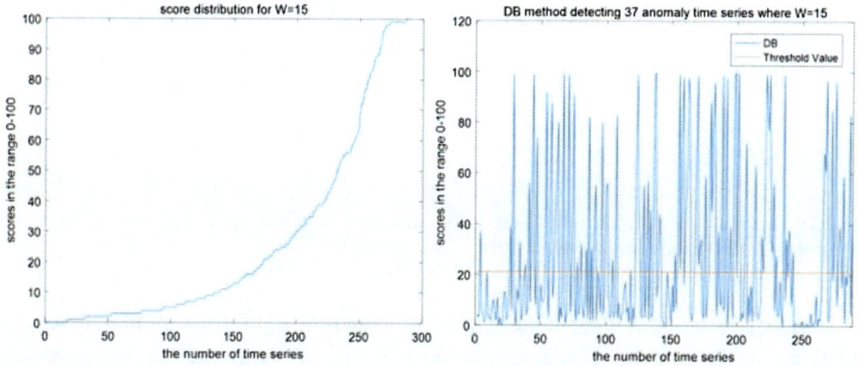

Fig. 10.14 Fraud detection results using DB method (left) the score distribution (right) 37 anomalies are detected for $W = 15$

Shown in left plot of Fig. 10.15 are the sorted (in a nondecreasing order) DB-Max outlier scores for the given 306 time series segments when the detected number of anomalies is the highest with a window size of $W = 15$. Shown in the right plot of the figure is an illustration of the highest 30 (higher than the threshold value plotted in the red line) out of 87 anomalies being detected for W=15 among five window sizes.

Shown in left plot of Fig. 10.16 are the sorted (in a nondecreasing order) INFLO outlier scores for the given 306 time series segments when the detected number of anomalies is the highest for all window sizes. Shown in the right plot of the figure

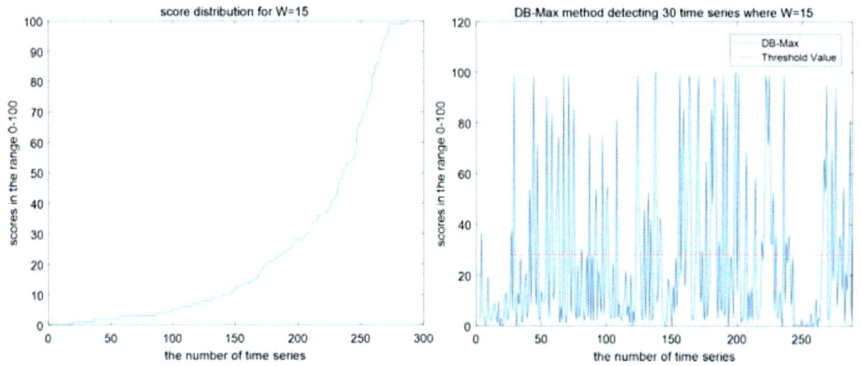

Fig. 10.15 Fraud detection results using DB-Max method (left) the score distribution (right) 30 anomalies are detected for $W = 15$

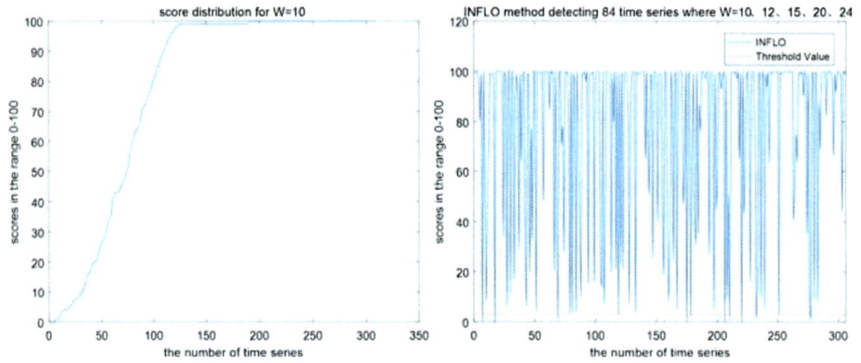

Fig. 10.16 Fraud detection results using INFLO method (left) the score distribution (right) 84 anomalies are detected for all window sizes

is an illustration of the highest 84 (higher than the threshold value plotted in the red line) out of 87 anomalies being detected by using all five window sizes even though a lot of false negatives happen as well.

Shown in left plot of Fig. 10.17 are the sorted (in a nondecreasing order) LDOF outlier scores for the given 306 time series segments when the detected number of anomalies is the highest with all window sizes of $W = 10$, 12, 15, 20, and 24. Shown in the right plot of the figure is an illustration of all 87 anomalies being detected by using all five window sizes even though a lot of false negatives happen as well.

Shown in left plot of Fig. 10.18 are the sorted (in a nondecreasing order) LOF outlier scores for the given 306 time series segments when the detected number of anomalies is the highest with all window sizes. Shown in the right plot of the figure is an illustration of the highest 86 (higher than the threshold value plotted in the red line) out of 87 anomalies being detected by using all five window sizes even though a lot of false negatives happen as well.

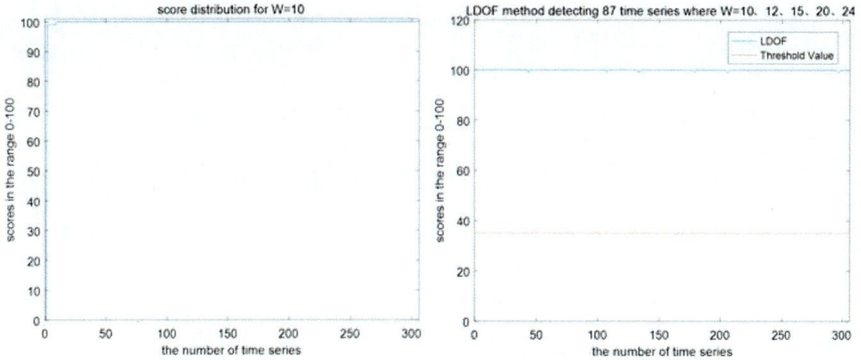

Fig. 10.17 Fraud detection results using LDOF method (left) the score distribution (right) 87 anomalies are detected for all window sizes

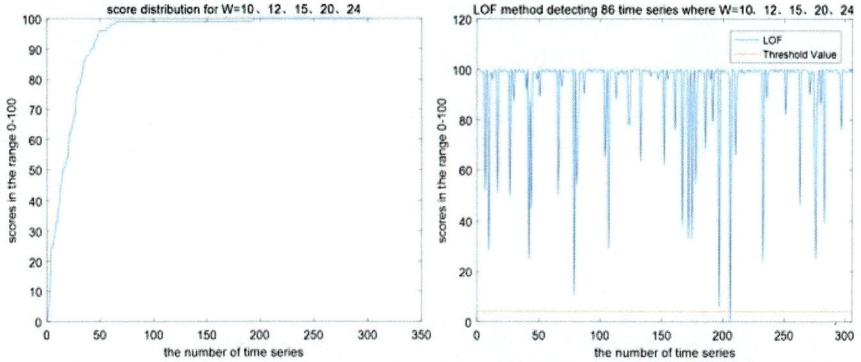

Fig. 10.18 Fraud detection results using LOF method for $W = 10, 12, 15, 20, 24$ (left) the score distribution (right) 86 anomalies are detected for all window sizes

Shown in left plot of Fig. 10.19 are the sorted (in a nondecreasing order) COF outlier scores for the given 306 time series segments when the detected number of anomalies is the highest with a window size of $W = 10$. Shown in the right plot of the figure is an illustration of the highest 36 (higher than the threshold value plotted in the red line) out of 87 anomalies being detected for $W = 10$ among five window sizes.

Shown in left plot of Fig. 10.20 are the sorted (in a nondecreasing order) RBDA outlier scores for the given 306 time series segments when the detected number of anomalies is the highest with a window size of $W = 15, 20, 24$. Shown in the right plot of the figure is an illustration of the highest 22 (higher than the threshold value plotted in the red line) out of 87 anomalies being detected for $W = 15, 20, 24$ among five window sizes.

Shown in left plot of Fig. 10.21 are the sorted (in a nondecreasing order) ABOD outlier scores for the given 306 time series segments when the detected number of

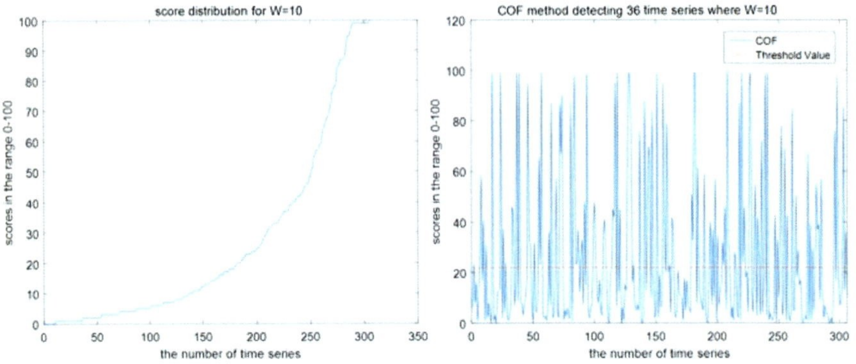

Fig. 10.19 Fraud detection results using COF method for $W = 10$ (left) the score distribution (right) 36 anomalies are detected for $W = 10$

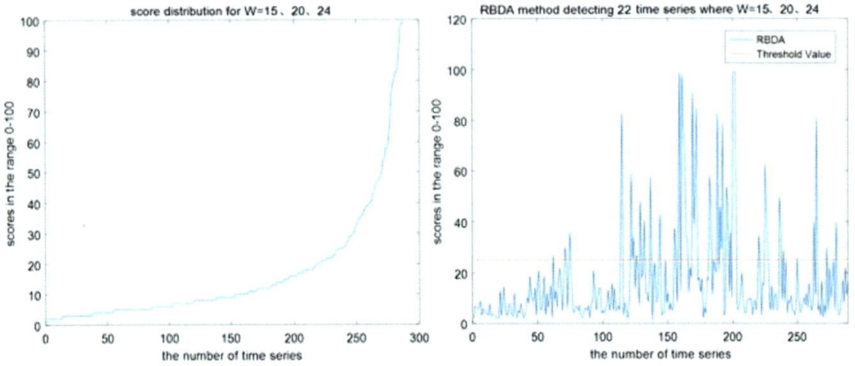

Fig. 10.20 Fraud detection results using RBDA method for $W = 15, 20, 24$ (left) the score distribution (right) 22 anomalies are detected for $W = 15, 20, 24$

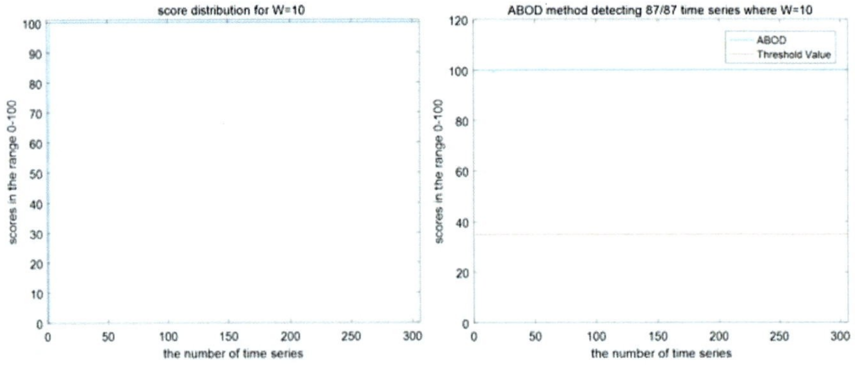

Fig. 10.21 Fraud detection results using ABOD method for $W = 10$ (left) the score distribution (right) all 87 anomalies are detected for $W = 10$

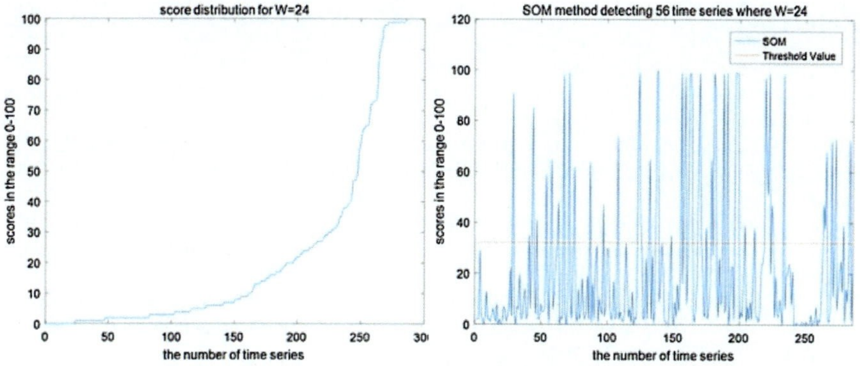

Fig. 10.22 Detection results using the proposed method (left) the score distribution (right) 56 anomalies are detected for $W = 24$

anomalies is the highest with all window sizes. Shown in the right plot of the figure is an illustration that all 87 anomalies are detected for $W = 10$ among five window sizes even though a lot of false negatives happen as well.

Shown in left plot of Fig. 10.22 are the sorted (in a nondecreasing order) proposed outlier scores, SOM, for the given 306 time series segments when the detected number of anomalies is the highest with a window size of $W = 24$. Shown in the right plot of the figure is an illustration that 56 (higher than the threshold value plotted in the red line) out of 87 anomalies are detected for $W = 24$ among five window sizes.

10.5 Conclusions

Fraud detection is of great importance to many applications. This chapter is concerned with the problem of fraud detection in environmental time series data. We present an unsupervised anomaly detection technique based on statistical parameters extracted from the subsequences of the data. We begin with an introduction to the concept of environmental time series data anomaly detection in general and shape-based time series anomaly detection in specific. The proposed statistical parameter-based anomaly detection method is next presented. Finally, the performance of the proposed anomaly detection method is compared with that of k-nearest neighbor-based approaches for a fraud detection task in an environmental time series data. The results of the experiments performed show that the proposed algorithm can fulfill the task effectively and in real time when the environmental time series data come in.

References

1. Akaike, H. (1969). Fitting autoregressive models for prediction. *Annals of the Institute of Statistical Mathematics, 21*(1), 243–247.
2. Ye, N. (2000). A Markov chain model of temporal behavior for anomaly detection. In *Proceedings of the 2000 IEEE SMC Information Assurance and Security Workshop* (Vol. 166, pp. 171–174).
3. Yang, J., & Wang, W. (2003). CLUSEQ: Efficient and effective sequence clustering. In *Proceedings of the 19th IEEE International Conference on Data Engineering* (ICDE'03), Bangalore, India (pp. 101–112).
4. Sun, P., Chawla, S., & Arunasalam, B. (2006). Mining for outliers in sequential databases. In *Proceedings of the 6th SIAM International Conference on Data Mining* (SDM'06), Bethesda, MD, United states (pp. 94–105).
5. Eskin, E., Lee, W., & Stolfo, S. (2001). Modeling system calls for intrusion detection with dynamic window sizes. In *Proceedings of DARPA Information Survivability Conference and Exposition II* (DISCEX'01), Anaheim, CA, United states (Vol. 1, pp. 165–175).
6. Lee, W., Stolfo, S. J., & Chan, P. K. (1997). Learning patterns from unix process execution traces for intrusion detection. In *Proceedings of the AAAI Workshop on AI Approaches Fraud Detection and Risk Management* (pp. 50–56).
7. Fox, A. J. (1972). Outliers in time series. *Journal of the Royal Statistical Society: Series B, 34,* 350–363.
8. Tsay, R. S. (1986). Time series model specification in the presence of outliers. *Journal of the American Statistical Association, 81,* 132–141.
9. Tsay, R. S. (1988). Outliers, level shifts, and variance changes in time series. *Journal of Forecasting, 7,* 1–20.
10. Chen, C., & Liu, L.-M. (1993). Joint estimation of model parameters and outlier effects in time series. *Journal of the American Statistical Association, 88,* 284–297.
11. Knorr, E. M. and Ng, R. T. (1998). Algorithms for mining distance-based outliers in large datasets. In *Proceedings of the International Conference on Very Large Data Bases* (VLDB'98), New York, pp. 392–403.
12. Ramaswamy, S., Rastogi, R., & Shim, K. (2000). Efficient algorithms for mining outliers from large data sets. In *Proceedings of the 2000 ACM International Conference on Management of Data* (SIGMOD'00), Dallas (pp. 427–438).
13. Angiulli, F., & Pizzuti, C. (2002). Fast outlier detection in high dimensional spaces. In *Proceedings of the 6th European Conference on Principles of Data Mining and Knowledge Discovery* (PKDD'02), Helsinki (pp. 15–26).
14. Breuning, M. M., Kriegel, H. P., Ng, R. T., & Sander, J. (2000). LOF: Identifying density-based local outliers. In *Proceedings of the 2000 ACM SIGMOD International Conference on Management of Data* (SIGMOD'00), Dallas, TX, United states (pp. 93–104).
15. Tang, J., Chen, Z., Fu, A. W. C., & Cheung, D. W. (2002). Enhancing effectiveness of outlier detections for low density patterns. In *Proceedings of the 6th Pacific-Asia Conference on Knowledge Discovery and Data Mining* (PAKDD'02), Taipei, Taiwan (pp. 535–548).
16. Gibbons, P. B., Papadimitriou, S., Kitagawa, H., & Faloutsos, C. (2003). LOCI: Fast outlier detection using the local correlation integral. In *Proceedings of the IEEE 19th International Conference on Data Engineering* (ICDE'03), Bangalore, India (pp. 315–326).
17. Sun, P., & Chawla, S. (2004). On local spatial outliers. In *Proceedings of the 4th IEEE International Conference on Data Mining* (ICDM'04), Brighton, UK (pp. 209–216).
18. Jin, W., Tung, A. K. H., Han, J., & Wang, W. (2006). Ranking outliers using symmetric neighborhood relationship. In *Proceedings of the 10th Pacific-Asia Conference on Knowledge Discovery and Data Mining* (PAKDD'06), Singapore (pp. 577–593).
19. Fan, H., Zaiane, O. R., Foss, A., & Wu, J. (2006). A nonparametric outlier detection for efficiently discovering top-N outliers from engineering data. In *Proceedings of the 10th Pacific-Asia Conference on Advances in Knowledge Discovery and Data Mining* (PAKDD'06), Singapore, Singapore (pp. 557–566).

20. Zhang, K., Hutter, M., & Jin, H. (2009). A new local distance-based outlier detection approach for scattered real-world data. In *Proceedings of the 13th Pacific-Asia Conference on Knowledge Discovery and Data Mining* (PAKDD'09) (pp. 813–822).
21. Huang, H., Mehrotra, K., & Mohan, C. K. (2013). Rank-based outlier detection. *Journal of Statistical Computation and Simulation, 83*(3), 518–531.
22. Schubert, E., Zimek, A., & Kriegel H. P. (2014). Generalized outlier detection with flexible kernel density estimates. In *Proceedings of the 14th Siam International Conference on Data Mining* (SDM'14), Philadelphia (pp. 542–550).
23. Ru, X., Liu, Z., Huang, Z., et al. (2016). Normalized residual-based constant false-alarm rate outlier detection. *Pattern Recognition Letters, 69,* 1–7.
24. Tang, B., & He, H. (2017). A local density-based approach for outlier detection. *Neurocomputing, 241,* 171–180.
25. Kriegel, H.-P., Schubert, M., & Zimek, A. (2008). Angle-based outlier detection in high-dimensional data. In *Proceedings of the 14th ACM SIGKDD International Conference on Knowledge Discovery and Data Mining* (KDD'08), Las Vegas, Nevada, USA (pp. 444–452).
26. Hawkins, D. M. (1980). *Identification of outliers.* London: Chapman and Hall.